DATE DUE

Insect
Neurohormones

Insect Neurohormones

Marie Raabe
National Center for Scientific Research
Pierre and Marie Curie University
Paris, France

Translated from French by Nissim Marshall
Illustrated by Daisy Chervin

PLENUM PRESS • NEW YORK AND LONDON

Library of Congress Cataloging in Publication Data

Raabe, Marie, Date-
Insect neurohormones.

Bibliography: p.
Includes indexes.
1. Insect hormones. 2. Neurosecretion. I. Title. II. Title: Neurohormones.
QL495.R2813   595.7'7'01'88   82-7535
ISBN 0-306-40782-5   AACR2

©1982 Plenum Press, New York
A Division of Plenum Publishing Corporation
233 Spring Street, New York, N.Y. 10013

Printed in the United States of America

Preface

The discovery of insect neurohormones dates from the earliest experimental investigations in insect endocrines, and the matter cannot be discussed without evoking the names of its pioneers—Kopec, Wigglesworth, Fraenkel. Whereas the experiments demonstrated the existence of the first known neurohormones, the formulation of the concept of neurosecretion was of fundamental importance to further progress, and tribute must be paid to Ernst and Berta Scharrer.

The recent proliferation of investigations into insect neurohormones has created the need for an overall review of the data. Our knowledge of the subject is voluminous, and the evidence clearly demonstrates that neurohormones play a part in most insect regulatory processes.

This book analyzes and synthesizes the data, starting from neurosecretion (i.e., source sites and release modes of neurohormones) and continuing through the various functions in which neurohormones have been shown to be involved: endocrine gland activity; diapause; reproduction; visceral muscle functioning; color change; behavior; water and ion balance; protein, sugar, and lipid metabolism; and tanning and other processes occurring at the cuticle level.

In each chapter, besides the experimental information, technical procedures as well as recent information concerning purification of the particular neurohormones and their mode of action are reported. Numerous exhaustive tables allow the reader to get an overview of the matter while the major findings of the moment are presented in the conclusion of each chapter.

The fundamental role played by neurohormones in insect physiology makes this volume particularly useful for teachers, researchers, and students, engaged both in this field and in others, such as zoology, invertebrate endocrinology, and neurobiology. They will appreciate having at their disposal an easy-to-use source of precise information.

v

Acknowledgments

I am most indebted to all authors who contributed to the illustrations of this book by providing me with prints or allowing me to reproduce their published and unpublished micrographs: Drs. N. Baudry-Partiaoglou, J.-P. Grillot, M. Lafon-Cazal, M. Mesnier, E. N'Kouka, Ch. Noirot, A. Provansal-Baudez, H. Schooneveld, and A. Thomas.

I wish to thank those co-workers, colleagues, and friends from whom I received support during the writing of this book and Nadine Claquin, Marcha Schlee, Jocelyne Guilleminot, and Daisy Chervin for their important participation in preparing the manuscript.

The translation is by Nissim Marshall and the figures are by Daisy Chervin.

I am indebted to Kirk Jensen and the staff at Plenum Press for their valuable work on all aspects of the production of the volume.

Last, I am grateful to the Centre National de la Recherche Scientifique for sponsoring my work.

Contents

Chapter 3
Diapause   89

Chapter 4
Reproduction   101

Insect
Neurohormones

Synthesis, Storage, and Release of Neurohormones

Neurohormones are synthesized in the neurosecretory cells of the nervous system and in certain peripheral neurons. They are released in different neurohemal organs and sometimes in the effector itself. The ensuing discussion will provide basic information about the location and structure of the neurosecretory cells and neurohemal organs, as well as a review of our knowledge concerning their mode of operation and their regulation.

1.1. THE NEUROSECRETORY CELLS

1.1.1. General Features of Neurosecretory Cells

Neurosecretory cells are characterized by the presence of secretory material, detectable histologically as chromophilic granules. Electron microscopy (Table I) reveals the presence of characteristic dense elementary granules in the neurosecretory cells, probably representing the neurohormones associated with their carrier protein.

The morphology of the neurosecretory cells displays few particular characteristics, at least at the level of the nerve centers. The perikaryon is not necessarily large, and the nucleocytoplasmic ratio is variable. The dendrites, like those of all insect neurons, originate in the proximal part of the axon and are difficult to distinguish from the collaterals. They penetrate the neuropil, where they enter into synaptic contact with interneurons running from the sensory centers or association centers. The neurosecretory axons, which are often very long, branch repeatedly at their distal end, where they exhibit preterminal swellings (Fig. 1).

The perikaryon, dendrites, collaterals, and axon are surrounded by a nar-

1

TABLE I. Some Major Electron Microscopical Studies

Order/insect	Brain			gg	cc	po	Reference
	pi	pl	tr				
Dictyoptera							
Leucophaea					+		Scharrer (1963, 1968)
Blaberus					+	+	Scharrer (1968)
Periplaneta					+		Scharrer (1968)
					+		Scharrer and Kater (1969)
						+	Brady and Maddrell (1967)
Phasmoptera							
Carausius					+		Smith and Smith (1966)
						+	Raabe and Ramade (1967)
						+	Brady and Maddrell (1967)
Orthoptera							
Locusta	+						Girardie and Girardie (1967)
		+	+				Girardie (1973, 1975)
					+		Cassier and Fain-Maurel (1970a,b)
					+		Cazal *et al.* (1971)
				+		+	Chalaye (1974a,b, 1975)
Schistocerca					+		Brady and Maddrell (1967)
Acheta	+						Geldiay and Edwards (1973)
Coleoptera							
Sitophilus		+					Sandifer and Tombes (1972)
Leptinotarsa	+						Schooneveld (1974a,b)
Lepidoptera							
Platysamia	+						Nishiitsutsuji-Uwo (1961)
Bombyx	+						Nishiitsutsuji-Uwo (1961)
	+				+		Bassurmanova and Panov (1967)
Diptera							
Calliphora	+						Bloch *et al.* (1966)
					+		Normann (1965)
Musca	+						Ramade (1966, 1968, 1969b)
Glossina						+	Baudry-Partiaoglou and Grillot (1975)
Hymenoptera							
Vespula						+	Provansal *et al.* (1970)
Vespa						+	Raabe *et al.* (1970)
Polistes	+						Strambi and Strambi (1973)
Homoptera							
Myzus					+		Bowers and Johnson (1966)
Heteroptera							
Oncopeltus	+				+		Unnithan *et al.* (1971)
Rhodnius	+						Morris and Steel (1975)

[a] cc, Corpora cardiaca; gg, ganglia; pi, pars intercerebralis neurosecretory cells; pl, protocerebral lateral neurosecretory cells; po, perisympathetic organs; tr, tritocerebral neurosecretory cells.

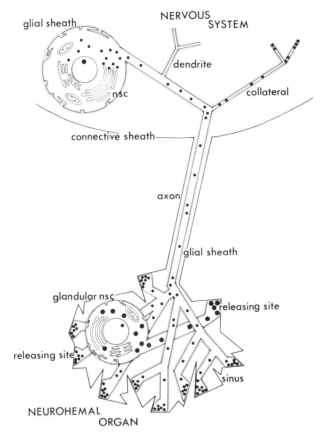

FIGURE 1. Schematic view of the complex formed by a neurosecretory cell and its neurohemal organ. nsc, Neurosecretory cells.

row glial sheath, which disappears in the terminal areas where neurohormones are released. Neurohormones are synthesized in the perikaryon in the rough endoplasmic reticulum, and then in the Golgi bodies. The granules are transported along the axons up to their endings. Their speed of delivery has been estimated in adult females of the locust *Schistocerca gregaria* by the injection of [^{35}S] cysteine. During vitellogenesis, which is accompanied by rapid release, the rate is 3.2 mm/hr, while in mature females, it is only 1 mm/hr (Highnam, 1976). Lysosomes, which are autophagous organelles, resorb the material synthesized in excess in the perikaryon and in the axons.

The neurosecretory cell axons usually terminate in neurohemal organs, which store the neurohormones and release them into the bloodstream, in response to appropriate stimuli. The release points are characterized by the presence of synaptoid sites formed by 30- to 50-nm electron-lucent vesicle clusters lying against a densified portion of the axon membrane. The synaptoid sites are often found in the bare peripheral endings isolated in the connective tissue, thus implying the direct passage of the neurohormones into the hemolymph. Sometimes, however, they are located in contact with glial cells or other axons.

The mechanisms implicated in the release of neurosecretory products at the axon endings have been a subject of considerable controversy. According to some investigators, who noted a dedensification of the granules at the endings, release occurs by the diffusion of active factors through the granule membrane, and then through the axon membrane. Other investigators suggest a fragmentation of the elementary granules into microvesicles. However, most authors support the view that the release process is one of exocytosis, occurring by fusion of the elementary granule membrane with that of the axon ending. The different phases of this process—binding of the elementary granules to the axon membrane, modification of the axon profile with the sporadic appearance of omega profiles—have actually been observed in periods of neurohormone release. It is generally considered that the synaptoid vesicles result from a micropinocytotic mechanism and ensure the maintenance of a constant surface at the axon endings. In fact, exocytosis brings about a significant increase in the membrane surface, and requires the existence of regulatory mechanisms. By ferritin injections into a stick insect, Smith (1971) revealed the passage of this substance from the outer periphery of the axons into the vesicle membrane. Rapidly after their formation the vesicles are destroyed, and their recovered material probably follows an antidromic flux and return to the perikaryon, where it is reutilized.

The neurosecretory cells display electrical activity and transport the nerve impulses. They exhibit a low resting potential of $20-50$ mV and an action potential similar to that of nonneurosecretory neurons (about 60 mV) but much slower ($3-6$ msec as opposed to $0.6-2.5$ msec) (Gosbee et al., 1968; Normann, 1973). Neurohormone release is elicited by the arrival of the nerve impulse at the axon endings. In fact, their artificial depolarization by potassium-rich solutions causes neurohormone release that can be detected by histology, electron microscopy, and the occurrence of physiologically active substances in the incubation medium (Gersch et al., 1970; Maddrell and Gee, 1974). The latter may also be obtained by electrical stimulations of the brain (Hodgson and Geldiay, 1959; Highnam, 1961b; Girardie et al., 1975, 1976; Girardie and Girardie, 1977).

The way in which depolarization acts on release has been thoroughly investigated in vertebrates, where it is believed to involve the penetration of Ca^{2+} ions. This has been confirmed in insects, where the presence of Ca^{2+} is essential for hormone release by electrical or potassium stimulation (Gersch et al., 1970; Maddrell and Gee, 1974; Normann, 1974).

1.1.2. Activity of Neurosecretory Cells and Its Regulation

Histological examination of neurosecretory cells in various physiological states reveals the existence of significant changes in their size, their nucleocytoplasmic ratio, and their neurosecretory material load (Fig. 2.) These observations are important, because they provide information about the mechanisms to which neurosecretory cell activity is related. However, it is difficult to interpret the quantitative variations in neurosecretory products stored in the somata—a heavily loaded cell can be in a period of intense secretory activity, accompanied by hormone release into the internal environment, but it may also be in a storage phase. The simultaneous study of variations in nuclear size and neurosecretory material or product load of the perikarya supplies evidence to support this view (Table II). However, one interpretation or the other may be correct depending on the specific case, for the presence of secretory material in the cell body depends on the difference between the rates of synthesis and release, and also on the location of the storage sites.

Some investigators, attempting to better understand the cell activity, have

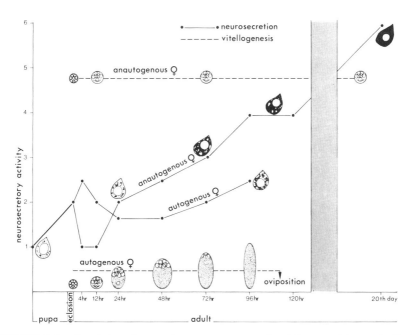

FIGURE 2. Load variations in type A neurosecretory cells of the pars intercerebralis, in autogenous and anautogenous *Aedes detritus* females. (Modified from Guilvard *et al.,* 1976.)

TABLE II. Changes Related to Secretory Activity in Type A Neurosecretory Cells of Pars Intercerebralis

Insect	Reference	Nuclear size	Cell size	nsm stored[a,b]	Nutritional or sexual state	Vitellogenesis[c]
Locusta migratoria	McCaffery and Highnam (1975a)	Small		+ + + +	Fasting	0
		Large		+	Eating	+ + +
Calliphora erythrocephala	Thomsen (1965)	Small		+ + + +	Sugar-fed	0
		Large		+	Meat-fed	+ + +
Aedes detritus	Guilvard *et al.* (1976)		Large	+ + + +	Fasting	0
			Small	+	Blood meal-fed	+ + +
Stilbocoris natalensis	Furtado (1971b)		Large	+ + + +	Virgin females	0
			Small	+	Mated females	+ + +

[a]nsm, Neurosecretory material stored in the perikarya.
[b]+, Slight load; + + +, important load.
[c]0, No vitellogenesis; + + +, normal vitellogenesis.

simultaneously observed the cell bodies and the proximal and distal parts of the axons. This method sometimes helps to distinguish three successive stages: the first in which only the perikaryon is loaded, the second in which it is depleted while the origin of the axon is filled, and the third in which the neurosecretory material is only present at the axon endings in the neurohemal organs. In general, however, images as simple as this are not obtained because the storage processes may take place at different points. Thus, in the stick insects, storage occurs in the cell bodies in *Clitumnus extradentatus* and in the proximal and distal parts of the axons in *Carausius morosus* (Dupont-Raabe, 1958). In locusts (Highnam and West, 1971), the axons originating in the type A cells of the pars intercerebralis form a series of loops in the neuropil at the origin of the nccI, thus constituting a neurosecretory reservoir. This reservoir is far more developed in *Locusta* than in *Schistocerca,* which means that in a fasting period, the perikarya are heavily loaded in *Schistocerca,* but not in *Locusta,* where storage takes place in the reservoir.

The electron microscope yields data about variations in the synthesis and the release activity of the neurosecretory cells. The abundance of elementary granules is estimated more precisely by this method than by histology, in which the lysosomes may be confused with neurosecretory products. The intensity of syntheses is ascertained by the state of development of the rough endoplasmic reticulum and the Golgi apparatus. Neurosecretory product re-sorption processes are detectable by the presence of autophagous organelles. The numerical variations in omega profiles and synaptic sites allow an esti-mate of the extent of release processes, as shown by Scharrer and Kater (1969) and Normann (1969), who simultaneously carried out ultrastructural studies and investigations into the release of hypertrehalosemic and cardioac-celerator neurohormones.

Because the neurosecretory products or carrier proteins of certain neu-rosecretory cells contain cystine or cysteine, [^{35}S]cysteine can be used in injections to investigate their functioning. The rate of incorporation in the perikaryon reflects the level of synthetic activity, while the speed of transit toward the corpora cardiaca supplies information about release mechanisms. Such investigations have been carried out on locusts and grasshoppers: *Schistocerca* and *Locusta* (Highnam, 1962, 1976; Highnam and Mordue, 1970, 1974; Girardie and Girardie, 1972, 1977; Mordue and Highnam, 1973; Girar-die *et al.,* 1975, 1976), *Anacridium aegyptium* (Geldiay, 1970; Geldiay and Edwards, 1973), and *Melanoplus sanguinipes* (Gillott *et al.,* 1970; Dogra and Gillott, 1971; Dogra, 1975).

The activity of neurosecretory cells includes the synthesis of neurohor-mones, their transport along the axon, their release at the endings, and sometimes autophagous mechanisms. The regulation of these steps is com-plex, depends on different factors, and follows various procedures.

Neurosecretory processes are influenced by environmental conditions such as temperature and photoperiod, and by internal factors such as the repletion of the digestive tract, the metabolic composition of the hemolymph, and the level of circulating hormones. Although information is not always available concerning the transmission mode of these processes, a number of possible pathways are known.

The investigations of Brousse-Gaury (1968, 1971a,b) in cockroaches clearly demonstrated the existence of connections between the sensory organs and the brain neurosecretory centers: ocellar receptors and neurosecretory cells of the pars intercerebralis or the pars lateralis, antennal contact receptors and deutocerebral neurosecretory cells, tegumentary photoreceptors or labrum receptors, and deuto- and tritocerebral neurosecretory cells. Furthermore, by surgical procedures, Brousse-Gaury confirmed the action of sensory receptors on the synthesis and the release activity of brain neurosecretory cells.

The anterior sympathetic nervous system also exerts control over neurosecretory activity. Removal of the frontal ganglion or its disconnection from the brain seriously disturbs neurosecretory material release. This frontal ganglion transmits to the brain sensory impulses from the osmotic and tension receptors of the anterior part of the digestive tract. The hypocerebral ganglion, closely linked to the corpora cardiaca, also performs a function in the regulation of neurohormone release. Finally, the sensory peripheral neurons may also play a part in release mechanisms (Finlayson and Osborne, 1968).

The neurosecretory cell has many axo-dendritic and axo-axonal contacts in the neuropil of the brain and ventral ganglia, along the nerves and in the neurohemal organs. Most of the sensory data are transmitted in the neuropil by means of interconnecting neurons in true synapses, but the neurosecretory cells also receive and transmit data from synaptoid contacts with other peptidergic or aminergic nerve fibers. These contacts, which have often been described since their earliest observation by B. Scharrer in 1963, appear to modulate neurohormone release, and also their synthesis and destruction.

Coordination between the different neurosecretory centers, the pars intercerebralis, lateral protocerebral and tritocerebral cells, and cells of the ventral nerve cord, is probably brought about by the collaterals of the neurosecretory cells which run from one to the other. Furthermore, a neuropil area located on each side of the axon bundles issuing from the pars intercerebralis was discovered in the cockroach *Periplaneta americana*. It is formed by the collaterals of the neurosecretory cells of the pars intercerebralis (Adiyodi and Bern, 1968) and the collaterals of the neurosecretory cells of the pars lateralis (Fraser and Pipa, 1977; Pipa, 1978). Hence it is possible that synaptoid contacts exist between the two categories of neurosecretory cells. In *Locusta* (Rademakers, 1977b), such structures have been described as in many other insects.

Independent of the regulatory pathways described above, the neurosecretory cells also exhibit cyclic activity governed by circadian rhythms, as indicated by investigations carried out on the cricket *Acheta domesticus* (Cymborowski and Dutkowski, 1969, 1970a), the beetle *Carabus nemoralis* (Klug, 1958), the fruit fly *Drosophila melanogaster* (Rensing, 1966), and the bee *Apis mellifica* (Heinzeller, 1976) (See Chapter 7).

A completely different regulatory mechanism is represented by the feedback effects exerted by the hormones. Hence, ecdysone stimulates brain hormone secretion, *in vivo* and *in vitro*, while juvenile hormone inhibits the secretion of brain hormone in the pupal instar, but stimulates it in the adult and larva (see Chapters 2 and 3).

Despite the existence of all these regulatory mechanisms, experiments utilizing electrical stimulation applied at various sites in insects injected with [^{35}S]cysteine show that the neurosecretory cells are also self-regulating, depending on the load state of their axon endings, which modifies the scope of the processes of synthesis, transport, and destruction of the substances already formed. The neurosecretory cell appears to secrete continuously according to a circadian rhythm, but, in the absence of release, resorption processes are linked to synthesis activity. The release of neurosecretory materials at the axon endings appears to regulate synthesis and transport mechanisms by accelerating them (Girardie *et al.*, 1975, 1976; Girardie and Girardie, 1977). This may be achieved by the passage of antidromic impulses from the axon ending to the perikaryon, as suggested by Finlayson and Osborne (1970) and Anwyl and Finlayson (1974).

1.1.3. Neurosecretory Products, Their Staining Methods, and Classification

Neurosecretory products are often revealed by standard histological techniques. Nevertheless, various staining methods are available to demonstrate their existence very clearly. Chrome hematoxylin phloxine was the first method employed intensively to investigate neurosecretory processes. Other techniques yielding more or less similar results were subsequently developed: paraldehyde fuchsin, alcian blue, paraldehyde thionine–phloxine, paraldehyde thionine–paraldehyde fuchsin, pseudoisocyanine, and for *in toto* staining, Victoria blue and fuchsin resorcin. These methods successively involve a basic dye, which is fixed on the acidic structures, and an acid dye, which is fixed on the basic structures. Neurosecretory products are generally basic and therefore stained by acidic dyes. After strong oxidation, however, always carried out in these techniques, some of them are modified. Initially basic, they become acidic and retain basic dyes such as chrome hematoxylin, paraldehyde fuchsin, and alcian blue; these are called type A materials. However, other neurosecretory products remain basic and hence acidophilic, and are stained by the acidic dyes used after the basic dyes: phloxine after chrome hematoxylin, or alcian blue, chromotrope 2R, orange G, light green, or picro-indigo-carmine after paraldehyde fuchsin; these are type B products (Table III).

Not all neurosecretory products are stained after the use of these methods, particularly when fixation is carried out with Bouin's fluid. One such product is clearly demonstrated by fixation with Helly's fluid followed by azan staining. The cells belonging to this category have been called type C cells. (Raabe, 1965b). We designate them here by the denomination Cr. Their secretions strongly retain azocarmine and sometimes show slight affinity for phloxine and paraldehyde fuchsin.

The dyes mentioned above display varying sensitivities and are not entirely retained when the quantities of neurosecretory material are small, which is often the case in the axons. Because it is important to determine their pathway, special methods have been developed that are more efficient than the standard technique of silver impregnation or vital staining with methylene blue: acridine orange used in vital staining, procion yellow, and a recently developed technique for filling sectioned axons with a cobalt chloride solution and then precipitating the cobalt as a black sulfide that invades all parts of the neuron, revealing its numerous processes.

A very accurate technique is currently employed in vertebrates to identify the source of the different neurohormones: antibodies are combined with fluorescent substances and fixed on those cells that synthesize their antigens, which can thus be recognized. We are just beginning to isolate insect neurohormones, and investigations of this type are thus at their inception (Eckert et al., 1971; Gersch et al., 1977). However, other investigations have been conducted using antibodies of vertebrates. They have shown that certain

TABLE III. Staining Properties of Some Neurosecretory Cells[a]

Staining method	Stains used	nsc responses[b]		
		A nsc	B nsc	C nsc
CHP	Chrome hematoxylin	+++	0	0
	Phloxine	0	+++	0 or +
PF	Paraldehyde fuchsin	+++	0	0 or +
	Picro-indigo-carmine			
	Chromotrope 2R	0	+++	0 or +
	Light green			
	Orange G			
Azan	Azocarmine		++	+++
	Orange G	+++		
	Aniline blue		+	

[a]Abbreviations used: CHP, chrome hematoxylin phloxine; nsc, neurosecretory cells; PF, paraldehyde fuchsin.
[b]0, No staining; +, ++, +++, weak, medium, strong staining.

neurosecretory cells located in the pars intercerebralis (Doerr-Schott *et al.*, 1978) and especially two medioventral neurosecretory cells of the subesophageal ganglion are vasopressinlike in the stick insect *Clitumnus extradentatus*, the locust *Locusta migratoria*, and the cricket *Acheta domesticus* (Rémy *et al.*, 1977, 1979; Strambi *et al.*, 1979; Rémy and Girardie, 1980; Girardie and Rémy, 1980). On the other hand, type Cr cells located in the subesophageal ganglion of the moth *Thaumetopoea pityocampa* are α-endorphinlike (Rémy *et al.*, 1978), and numerous cells in the brain and the subesophageal ganglion of the blowfly *Calliphora erythrocephala* are pancreatic polypeptide-like (Duve and Thorpe, 1980). Moreover, the occurrence of immunoreactive neurons to a mollusc cardioexcitatory tetrapeptide has been revealed in the brain and the corpora cardiaca in several insects (Boer *et al.*, 1980).

Many histochemical investigations have been carried out on insect neurosecretory cells in attempts to clarify the problem of types A and B, which some authors hold to be two distinct functional states of the same cell (Nayar, 1955; de Lerma, 1956; Herlant-Meewis and Paquet, 1956; Köpf, 1957; Thomsen, 1965; Girardie and Girardie, 1967), while according to others they correspond to independent categories (Johansson, 1958; Highnam, 1961a; Prabhu, 1966; Ramade, 1969b; Prentø, 1972). The presence of lipids or carbohydrates in neurosecretory products is hotly disputed (Arvy and Gabe, 1962), but the protein nature of neurosecretory products has been established by enzymatic destruction (pepsin, trypsin, pronase) and by histochemical reactions (see Table IV). The most solidly established data concern the type A cells of the pars intercerebralis, where investigators agree concerning the presence of cysteine and/or cystine. In the ventral nerve cord, the thoracoabdominal A cells react irregularly, but the two medioventral cells of the subesophageal ganglion and the abdominal cells react strongly. In the type B cells of the pars intercerebralis, the presence of tyrosine and tryptophan has been reported, and sometimes also small amounts of SS or SH groups. The reactions of some B cells of the ventral nerve cord are not different. Type Cr neurosecretory cells, which probably correspond to type B cells of the pars intercerebralis in Polyneoptera, sometimes contain tryptophan. Their essential feature is the presence of strongly basic amino acids, demonstrated by fast green staining (Raabe and Monjo, 1970). According to Smalley (1970), these cells selectively incorporate tritiated dopamine. As for the glandular cells of the corpora cardiaca, they contain tryptophan (*Locusta, Carausius*) and tyrosine (*Locusta*).

In many species, ultrastructural studies have revealed the existence of several types of elementary granules, distributed among distinct cells and axons. However, the results differ from one species to another (Table V). Most neurosecretory cells contain dense granules, among which more or less dedensified or even transparent elements may be encountered. They represent

TABLE IV. Histochemical Reactions of Neurosecretory Products[a]

Order/insect	Reference	Protein	Peptidic groups	SS and SH groups	Phenol groups	Indol groups	Basic amino acids	Carbohydrates	Lipids
PARS INTERCEREBRALIS AND CORPORA CARDIACA									
Pars Intercerebralis A Cells									
Ephemeroptera									
Ecdyonurus fluminum, Heptagenia flava, and Siphlonurus lacustris	Arvy and Gabe (1962)	++		+++	++	+		0/+	0
Odonata									
Aeshna cyanea	Arvy and Gabe (1962)	++		+++	+++	++		+	0
Orthetrum chrysis	Tembhare and Thakare (1977)		+++	+++			+	+	0
Plecoptera									
Perla maxima	Arvy and Gabe (1962)	++		+++	+	0/+		++	0
Orthoptera									
Locusta migratoria	Arvy and Gabe (1962)	++		+++	++	0/+		0/+	0
	Girardie and Girardie (1967)	++						0	0
	Prentø (1972)		+	+++					+
Acheta domesticus	Arvy and Gabe (1962)	+		+++	+	0/+		0/+	0
Gryllus bimaculatus	Anwar and Ismail (1979)				+		+		++
Phasmoptera									
Carausius morosus	Arvy and Gabe (1962)	+		0/+	0/+	0/+		0	0

Leptinotarsa decemlineata	Arvy and Gabe (1962)	++	++	+	+		+	0
Tenebrio molitor	Arvy and Gabe (1962)	++	++	+	+		0	0
Hydrous triangularis	Govardhan et al. (1978)	+++	+++	0	0	+++	++	0
Lepidoptera								
Bombyx mori	Arvy and Gabe (1962)	+++	+++	+++	+		++	0
Galleria mellonella	Arvy and Gabe (1962)	++	++	++	+		++	0
Ephestia kuehniella	Rehm (1955)	+						+
	Rehm (1955)	+						+
Noctua pronuba and Agrotis ipsilon	Busselet (1968)				++			
	Hinks (1967)				++			
Diptera								
Calliphora erythrocephala	Arvy and Gabe (1962)	+	+	+	0		0	0
Musca domestica	Ramade (1969b)	++	+++		++	++	0	+
	Ramade and Rivière (1970)	+++	+++	0	++	++	+	+
Heteroptera								
Notonecta glauca	Arvy and Gabe (1962)	+++	+++	+++	0/+		+	0
Naucoris cimicoides	Arvy and Gabe (1962)	++	++	++	+		++	0
Oncopeltus fasciatus	Schreiner (1966)	+++	+++				0	0
Rhodnius prolixus	Baudry and Baehr (1970)	+++	+++		0		0	0
Dysdercus fasciatus	Gupta (1971)	+			0			+
Sphaerodema rusticum, Ranatra filiformis, Enithares indica, and Spilostethus pandurus	Faruqui (1977a)	+++	0				++	++

[a]The neurosecretory cells are classified according to the criteria and nomenclature proposed on p. 21.

(Continued)

TABLE IV. (Continued)

Order/insect	Reference	Protein	Peptidic groups	SS and SH groups	Phenol groups	Indol groups	Basic amino acids	Carbo-hydrates	Lipids
		Pars Intercerebralis B Cells							
Orthoptera *Gryllus bimaculatus*	Anwar and Ismail (1979)				+	++			+
Coleoptera *Hydrous triangularis*	Govardhan et al. (1978)	++	++	++	0	0		+	0
Lepidoptera *Galleria mellonella* and *Ephestia kuehniella*	Rehm (1955)	+					+++		+
Diptera *Musca domestica*	Ramade (1969b)	++		++					++
	Ramade and Rivière (1970)	+++		++	0	+	++	0	+
Heteroptera *Sphaerodema rusticum, Ranatra filiformis, Enithares indica,* and *Spilostethus pandurus*	Faruqui (1977a)	++		0				++	++

Pars Intercerebralis "B" Cells (= Cr Cells)

Orthontera

Orthoptera							
Gryllus bimaculatus	Anwar and Ismail (1979)			++		++	+
Phasmoptera							
Carausius morosus and *Clitumnus extradentatus*	Raabe and Monjo (1970)	+++	0	++	+/++	+++	
Diptera							
Musca domestica	Ramade (1969b)	+++	+/0	0			
	Ramade and Rivière (1970)	+++	0	+		+	+

Corpora Cardiaca Glandular Cells

Orthoptera							
Locusta migratoria	Prentø (1972)		+	+++	++	+	0
Phasmoptera							
Carausius morosus	Mouton (1969)	+			+++	++	
Clitumnus extradentatus	Raabe and Monjo (1970)			++	+++	+	++
Heteroptera							
Sphaerodema rusticum, Ranatra filiformis, Enithares indica, and *Spilostethus pandurus*	Faruqui (1977b)	+++	0	+++		++	++

(Continued)

TABLE IV. (Continued)

Order/insect	Reference	Protein	Peptidic groups	SS and SH groups	Phenol groups	Indol groups	Basic amino acids	Carbo-hydrates	Lipids
			VENTRAL NERVE CORD						
		Subesophageal Median A Cells							
Odonata									
Orthetrum chrysis	Tembhare and Thakare (1977)			++				0	0
Orthoptera									
Locusta migratoria	Chalaye (1965, 1975)			0	++		+	0	0
Phasmoptera									
Carausius morosus	Raabe (1965b)			+++		0		0	0
	Mouton (1969)			+++					
Clitumnus extradentatus	Raabe (1965b)			+++		0			
Dictyoptera									
Periplaneta americana	de Bessé (1965)			++					
		Thoracoabdominal Lateral A Cells							
Odonata									
Orthetrum chrysis	Tembhare and Thakare (1977)			++				0	0
Orthoptera									0

	Reference					
Phasmoptera						
Carausius morosus	Raabe (1965b)	+/0	0		0	0
	Mouton (1969)	+	0	0		0
Clitumnus extradentatus	Raabe (1965b)	+/0	0		0	
Dictyoptera						
Periplaneta americana	de Bessé (1965)	++				
Heteroptera						
Rhodnius prolixus	Baudry and Baehr (1970)	++	0	0	0	

Abdominal Cells

	Reference		
Orthoptera			
Locusta migratoria	Chalaye (1965)	+++	0
Phasmoptera			
Carausius morosus	Raabe (1965b)	+++	0
Clitumnus extradentatus	Raabe (1965b)	+++	0
Dictyoptera			
Periplaneta americana	de Bessé (1965)	0	0
Heteroptera			
Rhodnius prolixus	Baudry and Baehr (1970)	0	+++	0	0

(Continued)

TABLE IV. (Continued)

Order/insect	Reference	Protein	Peptidic groups	SS and SH groups	Phenol groups	Indol groups	Basic amino acids	Carbo-hydrates	Lipids
			C Cells						
Odonata									
Orthetrum chrysis	Tembhare and Thakare (1977)			0/+					
Orthoptera									
Locusta migratoria	Chalaye (1965, 1975)			0	++		+	0	
Phasmoptera									
Carausius morosus	Raabe (1965b)			0		++	+++		
Clitumnus extradentatus	Raabe (1965b)			0		++	+++		
Heteroptera									
Rhodnius prolixus	Baudry and Baehr (1970)			0		0/++		0	0
	Faruqui (1977)			+++				++	+

TABLE V. Size of Elementary Granules Contained in Neurosecretory Cells and Corpora Cardiaca in Some Species

Order/insect	Reference	Organ studied[a]	Elementary granule size (nm)[b]
Dictyoptera			
Leucophaea maderae	Scharrer (1963, 1968)	cc	90–150, 220–360
Periplaneta americana	Scharrer and Kater (1969)	cc	150, 300
Phasmoptera			
Carausius morosus	Smith and Smith (1966)	cc ns ax	100–150, 100–300
		cc gl cells	100–300
Orthoptera			
Locusta migratoria	Girardie and Girardie (1967)	pi	100, 200 (Cj), 200, 300 (A, B)
	Cassier and Fain-Maurel (1970a,b)	cc ns ax	250, 250–300
			280, 200 × 650, 350–400
		cc gl lobe	600
	Cazal *et al.* (1971)	cc ns ax	110, 200, [200], 300
	Lafon-Cazal (1976)	cc gl lobe	400
Acheta domesticus	Geldiay and Edwards (1973)	pi	60–100, 70–160 (Cj), 150–300 (A)
Coleoptera			
Leptinotarsa decemlineata	Schooneveld (1974a)	pi	125 (A), 120 (B), 210 (A′), [170 (Cj)]
Diptera			
Musca domestica	Ramade (1966, 1968, 1969a,b)	pi	150 (Cr), 250 (A, B)
Calliphora erythrocephala	Normann (1965)	cc ns ax	150–200
		cc gl lobe	150–200

[a] cc, Corpora cardiaca; gl lobe, glandular lobe; ns ax, neurosecretory axons; pi, pars intercerebralis; [ ], transparent granules.
[b] A, A′, B, Cj, Cr, neurosecretory cell types.

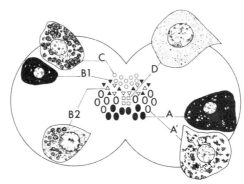

FIGURE 3. Schematic view of the neurosecretory cell types in the pars intercerebralis of the bug *Roscius*. Letters refer to neurosecretory cell types.

steps in the synthetic cycle (Schooneveld, 1974a), or, on the contrary, partial depletion by diffusion. Granules containing small vesicles or a crystalline structure are also found. They probably result from resorption of neurosecretory material. Neurosecretory cells with transparent vesicles also exist, at least in certain technical conditions. The various aspects of neurosecretory products may derive not only from the state of activity of the neurons, but also from fixation conditions, for the cell organelles are destroyed very rapidly, especially by osmotic stress (Normann, 1970). Apart from the elementary granules, the neurons often contain large electron-dense droplets, which probably result from their fusion. Like the dedensified granules, these droplets may be artifacts.

Three major types can be distinguished among neurosecretory cells, recognizable by their staining affinities (Fig. 3), and within these three types, several subtypes can be distinguished, usually characterized by their location and the general appearance of the cell and its secretory products. Hence, in the pars intercerebralis, in addition to types A, B, and Cr, at least three distinct subtypes of A may be distinguished, as reported by Johansson (1958), and noted subsequently by different investigators. These type A neurons were first designated A, C, and D, but subtype C was subsequently called A′ owing to its resemblance to subtype A, and to avoid confusion with type Cr azocarminophilic neurosecretory cells.

Recently, using thionine paraldehyde–phloxine, and especially paraldehyde thionine–paraldehyde fuchsin and alcian blue–alcian yellow, differences in affinity were observed between the A and the A′ cells of many insects (Raabe *et al.*, 1979). With these staining techniques, the two cell types are easily identified and seen to contain distinct neurosecretory products, rich in strong acids in the A cells, and weak acids in the A′ cells. In addition, electron microscopy has revealed the occurrence of three to five subtypes within the Cr type (Chalaye, 1974a).

Some neurosecretory products are stained by paraldehyde fuchsin but not by chrome hematoxylin (de Lerma, 1956; Raabe, 1965a; Mahon and Nair, 1975). Hence these two techniques cannot be completely superimposed, contrary to what is usually thought, and this is understandable because only paraldehyde fuchsin stains the aldehyde groups liberated by oxidation (Gabe, 1955). In certain technical conditions, paraldehyde fuchsin stains the neurosecretory products in the early synthetic phase (de Lerma, 1956; Fraser 1959; Raabe, 1965a; Charlet et al., 1974). In fact, if a comparison is made by double staining the same cell with chrome hematoxylin and paraldehyde fuchsin, it is observed that both dyes stain the large round granules, but that only paraldehyde fuchsin also stains the small grains, whose disposition recalls the shape of the Golgi bodies (Raabe, 1965a).

The terminology employed to designate the types and subtypes of neurosecretory cells lacks uniformity. The terminology given by Johansson (1958) for the milkweed bug Oncopeltus fasciatus (A, B, C, and D) has the drawback of designating cells belonging to type A by three different letters. This terminology, applied to locusts by Highnam (1961a), is often used and raises problems because the B cells of Heteroptera are quite different from those of primitive insects which appear to be rare, and because the designation B is generally given to A2 or Cr cells.

Finally, the "C" cells of Locusta (Girardie, 1970) represent a special category of A cells, which is here termed Cj.

It is difficult to propose a terminology that satisfies every available case, in view of the rich variety of the neurosecretory system of insects and the diversification of species. The best answer at the present time appears to be to use the designations A, B, and Cr for the three major types, and to distinguish numbered subtypes (A1, A2, B1, B2, etc.) within each type as done by certain authors.

1.1.4. Distribution of Neurosecretory Cells — Axon Pathways

Neurosecretory cells are present throughout the central nervous system, the sympathetic nervous system, and even outside the ganglia among the peripheral sensory cells.

The most abundant data concern the brain, and within it, the pars intercerebralis containing the richest group of neurosecretory cells, which give rise to the first pair of corpora cardiaca nerves (nccI). Other groups, the lateral protocerebral group and the tritocerebral group, are the source of the two other pairs of corpora cardiaca nerves (nccII and nccIII) (Fig. 4).

1.1.4.1. Brain Neurosecretory Cells

1.1.4.1a. *Pars Intercerebralis.* The pars intercerebralis is located mediodorsally at the junction of the two protocerebral lobes. We have seen that it

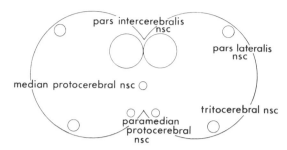

FIGURE 4. Main locations of brain neurosecretory cells. nsc, Neurosecretory cells.

contains many types of neurosecretory cells. In Lepidoptera, Diptera, Coleoptera, and Heteroptera, the neurosecretory cells, probably polyploid, are large and few in number; in Dictyoptera, Phasmoptera, and Orthoptera, they are small but very numerous. While there are 24 type A cells in Heteroptera and Lepidoptera (Panov and Kind, 1963), 1000 and more are found in Orthoptera [2400 in *Schistocerca* according to Highnam (1961a), 900 in *Acheta* according to Geldiay and Edwards (1973)].

The pars intercerebralis is not always homologous from one species to another and neither from one instar to another. In a number of insects, the neurosecretory cells of the pars intercerebralis, which form a single group in the adult, are divided into three distinct groups in the larva, and a migration occurs during development. This occurs in Lepidoptera (Panov and Kind, 1963; Panov, 1975, 1976; Panov and Davydova 1976) (Fig. 5), Diptera (Dogra and Tandan, 1965; Gieryng, 1976; Khan *et al.*, 1978b), Odonata, and Orthoptera. Furthermore, as shown by Nijhout (1975) the largest lateral group (L1 of Panov) is paramedian in the larva of *Manduca sexta* and lateral in the pupa.

The axon pathway from the pars intercerebralis is not as simple as initially thought. It has long been assumed that the two fiber bundles of the right and left lobes of the pars intercerebralis cross in front of the brain, just before emerging, and then form the two voluminous nccI each of which branches into the contralateral corpus cardiacum. While most fibers originating in the pars intercerebralis do take this pathway, some of them follow a different one and send one branch into the ipsilateral bundle (Schooneveld, 1974b, working on the Colorado beetle *Leptinotarsa decemlineata*). Moreover, the collaterals of the neurosecretory cells enter the protocerebral neuropil, where they branch repeatedly in many species: cockroaches (Adiyodi and Bern, 1968; Farley and Evans, 1972), crickets (Geldiay and Edwards, 1973), bees (Delye, 1972), beetles (Schooneveld, 1974b), bugs (Ewen, 1962; Unnithan *et al.*, 1971), and aphids (Johnson, 1963; Srivastava, 1969). In *Schistocerca, Periplaneta,* and *Carausius,* the axons form wide loops entering the neuropil before resuming the ipsilateral and contralateral axon pathways as mentioned above. The collaterals and axons also take longer paths up to the tritocerebrum and the subesophageal ganglion (Mason, 1973, *Schistocerca*) and by the circumesophageal connectives along the ventral nerve cord ganglia at least in mosquitoes and aphids (Füller, 1960, *Corethra plumicornis* larva; Bowers and Johnson, 1966, *Myzus persicae;* Burgess, 1971, 1973, *Culiseta inornata* and *Culex tarsalis*; Steel, 1977, *Megoura viciae*).

1.1.4.1b. *Protocerebral Median and Paramedian Areas.* Some neurosecretory cells located in the median plane outside the pars intercerebralis have been known for a long time in Diptera and Lepidoptera. They were described recently under different names in *Leucophaea* (de Bessé, 1978), *Gryllus bimaculatus* (Anwar and Ismail, 1979), *Melanoplus* (Dogra and Ewen, 1970), *Acheta* (Geldiay and Edwards, 1973), and *Locusta* (Girardie, 1970), where their location was shown to be at the source of the median ocellar nerve. Three groups can be identified in dragonfly larvae, one mediodorsal consisting of four type A cells, and the other two paramedian comprising type A and B cells (Schaller and Meunier, 1968; Charlet and Schaller, 1974). The axon pathways of these cells have been described in *Acheta* by Geldiay and

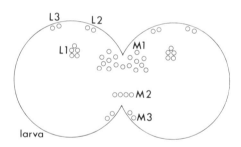

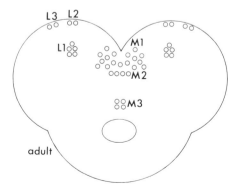

FIGURE 5. Migration of the neurosecretory cells during development in the moth *Spilosoma menthastri*. The median M2 and M3 clusters are further from the M1 clusters in the larvae, but move closer together during growth. The lateral neurosecretory clusters (L1, L2, and L3) also change their position during development. (Modified from Panov and Kind, 1963.)

Edwards (1973), who reported that they terminate in a neurohemal region of the brain (see Section 1.2.2).

1.1.4.1c. *Lateral Protocerebral Area (Pars Lateralis)*. Neurosecretory cells are present in the lateral parts of the protocerebrum, in the vicinity of the calyces of the pedunculate bodies. Few in number, they often form two distinct groups. Their staining properties have been described as corresponding to type Cr (Dupont-Raabe, 1956a), subtypes A, A' (=Cj), and D (Girardie, 1973).

The axons of the lateral protocerebral neurosecretory cells follow a sinuous path and emerge ventrally from the brain, constituting the nccII, which in evolved species fuse with the nccI. In *Periplaneta*, the lateral neurosecretory cells send collaterals toward the pars intercerebralis, in the vicinity of which they branch repeatedly and end in close contact with the collaterals of the neurosecretory cells of the pars intercerebralis (Pipa, 1978).

1.1.4.1d. *Optic Lobes, Deutocerebrum*. Deutocerebral neurosecretory cells have been described by Brousse-Gaury (1967) and Khan (1976b) in cockroaches. A distinct pair of corpora cardiaca nerves corresponds to them (Brousse-Gaury, 1967). Other neurosecretory cells were reported in the optic lobes of termites (Noirot, 1957), moths (Mitsuhashi, 1963), flies (Thomsen, 1965), and cockroaches (Beattie, 1971; Khan, 1976b).

1.1.4.1e. *Tritocerebrum*. The tritocerebrum, investigated in many species (Dupont-Raabe, 1957; Raabe, 1963a, 1964), always contains a small number of type Cr neurosecretory cells. According to ultrastructural studies, five distinct types are present in the tritocerebrum of *Locusta* (Girardie, 1975). Some of the axons of tritocerebral neurosecretory cells enter the nccIII (Raabe, 1963b); other pass in the nccI (Mason, 1973, Girardie, 1975). Before entering the nccI, the axons branch intensely in the neuropil, and some of them run toward the pars intercerebralis.

1.1.4.2. Neurosecretory Cells of the Ventral Nerve Cord

The presence of neurosecretory cells in the different ventral nerve cord ganglia has been noted in all the species examined, i.e., in all insect orders. The most thorough investigations have been conducted on dragonflies (Charlet, 1969; Tembhare and Thakare, 1977), stick insects (Raabe, 1965a,b, 1967), cockroaches (de Bessé 1965), locusts (Fréon, 1964; Delphin, 1965; Chalaye, 1965, 1974a, 1975), the beetle *Blaps mucronata* (Fletcher, 1969), the moth *Galleria mellonella* (Delépine, 1965), and the blood-sucking bug *Rhodnius prolixus* (Baudry, 1968).

The neurosecretory cells of the ventral nerve cord are less numerous and more widely dispersed than the brain cells, but they belong to many types and form rich and diversified systems. A comparative examination of the most thoroughly investigated species reveals that four major categories of neurosecretory cells are found in almost all of the insect groups (see Table VI) (Figs. 6 and 7).

The first category consists of a single pair of cells located medioventrally in the subesophageal ganglion, and is present in almost all of the groups. The axons of both cells run into the circumesophageal connectives and terminate within the brain (*Calliphora*, Vijverberg, 1970; *Teleogryllus commodus*, Weinbörmair *et al.*, 1975; and *Acheta*, Strambi *et al.*, 1979). In *Acheta*, they terminate in a definitive neurohemal area (Geldiay and Karaçali, 1980). Immunocytological studies indicate that in *Locusta* these cells also send

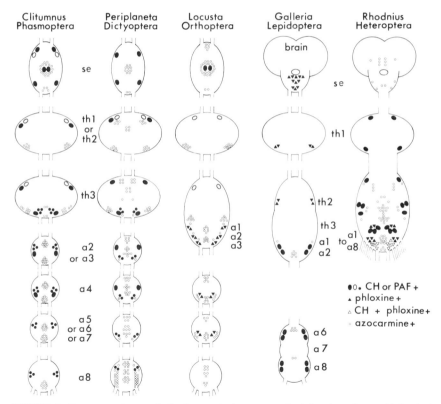

FIGURE 6. Neurosecretory cells in the ventral nerve cord of five insects. a, Abdominal ganglion; se, subesophageal ganglion; th, thoracic ganglion; CH, chrome hematoxylin; PAF, paraldehyde fuchsin. (From data of Charlet, 1969; Raabe, 1965b; de Bessé, 1965; Fréon, 1964; Chalaye, 1965; Delépine, 1965; Baehr, 1968; Baudry, 1968.)

TABLE VI. Distribution of Neurosecretory Cells in the Ventral Nerve Cord Ganglia[a,b]

Order/species	Reference	Subesophageal ganglion	Thoracic ganglia 1, 2 or 3	Abdominal ganglia 1, 2, 3 or 4, 5, 6, 7 or 8
Ephemeroptera				
Various species	Arvy and Gabe (1953)	1 pair‡	Not studied	Not studied
Odonata				
Various species	Arvy and Gabe (1953)	1 pair‡ / 1–2 pairs‡		
Aeshna cyanea larva	Charlet (1969); Charlet et al. (1974) (Tembhare and Thakare, 1977)	4 pairs	2 pairs	1 pair / 2 pairs / 5–7 pairs*
		3 lateral pairs†		1 central area†
Dictyoptera				
Blaberus craniifer	Geldiay (1959)		3–5 pairs / 1 pair*	
Periplaneta americana	Füller (1960)	1 pair‡ / 3 pairs / 0–2 pairs‡	2 pairs	1 pair
	de Bessé (1965) (Brady, 1967b)	4 pairs	2 pairs	1 pair / 2 pairs
		Central area†	3–5 lateral pairs†	Central area†
Leucophaea maderae	de Bessé (1965)	2 pairs	2 pairs	1 pair
		Central area†	3–5 lateral pairs†	1 pair†

———————————— More or less numerous according to the caste ————————————

Order and species	Reference	Neurosecretory cell distribution[b]
Isoptera		
Calotermes flavicollis	Noirot (1957)	
Reticulitermes lucifugus	Bernardini-Mosconi and Vecchi (1961)	
Phasmoptera		
Carausius morosus and *Clitumnus extradentatus*	Raabe (1965b) (Naisse and Mouton, 1965)	1 pair‡ ←——→ 1 pair; 3 pairs ←— 2 pairs —→ 3 pairs; Central area† ←— 5 lateral pairs† —→ Central area†
Orthoptera		
Numerous species	Panov (1962, 1964) (Huignard, 1964)	2 pairs ←— 1–2 pairs —→ 2–4 pairs; 1 pair‡ ←——→ 2–4 pairs
Locusta migratoria	Fréon (1964); Chalaye (1965)	1 pair‡ ←— 1 pair —→ 2 pairs*; Central area† ←— 6 lateral pairs† —→ 20 central area†; 1 pair‡
Schistocerca gregaria	Delphin (1965)	1 pair ←——→ 2 pairs; As in *Locusta*†
Acheta domesticus	Gaude (1975) (Khattar, 1972)	1 pair‡ ←— 1 pair —→ 1 pair; 2 pairs ←— 1–2 pairs —→ 2 pairs*
Coleoptera		
Tenebrio molitor larva and pupa	Arvy and Gabe (1953)	2 groups§ ←——→ Not studied

[a]Where necessary, the nomenclature of the neurosecretory cells has been changed according to that proposed on p. 21.

[b]Notation: unmarked, A cells; †, B cells; *, C cells; ‡, medioventral subesophageal cells; §, cells difficult to classify. Arrows indicate the distribution of neurosecretory cells throughout ventral nerve cord ganglia. In the Reference column, the second or incomplete descriptor is shown in parentheses.

TABLE VI. (Continued)

Order/species	Reference	Subesophageal ganglion	Thoracic ganglia 1, 2 or 3	Abdominal ganglia 1, 2, 3 or 4, 5, 6, 7 or 8
Coleoptera (continued)				
Galeruca tanaceti	Siew (1965)	1 pair‡ / 4 pairs*	Not studied	Not studied
Blaps mucronata adult	Fletcher (1969)	1 pair‡ / 4 pairs / Central area†	4 pairs / 10 laterals†	2 pairs / 2 pairs / 6–10†
Hymenoptera				
Synagris calida	Thomsen (1954)	2–3 cells§	Not studied	
Camponotus vagus adult	Delye (1972)	1 pair‡		
Lepidoptera				
Bombyx mori pupa and adult	Bounhiol et al. (1953) Fukuda and Takeuchi (1967)	Ventral area§ / 1 pair‡ / 1–2 pairs* / 3–4 pairs*	1 pair* / 1–2 pairs†	1 pair
Galleria mellonella Prepupa	Delépine (1965)	0–1pair†	2 pairs	2 pairs
Pupa		1 pair* / 3 pairs*	2 pairs*	2 pairs
Adult				Area†

Diptera

Corethra plumicornis larva	Füller (1960)		2–3 pairs → 1 pair →
Chironomus plumosus larva	Credland and Scales (1976)		2–3 pairs → Not studied
Lucilia caesar	Fraser (1959)	{	2 pairs →, 3 pairs* →, 1 pair →
Drosophila melanogaster larva	Köpf (1957)		5–6 cells* →
Phormia regina adult	Hsiao and Fraenkel (1966)	{ 2 pairs‡ / 1 pair	1–2 pairs → / 1–2 median cells → / 1–2 pairs* →

Homoptera

Megoura viciae	Steel (1977)	{ 1 pair‡ / 2 pairs	1 pair →

Heteroptera

Oncopeltus fasciatus	Johansson (1958)	3 pairs* →	2–4 pairs → 1 pair → / 5 pairs* →
Adelphocoris lineolatus	Ewen (1962)	2 groups§	6 groups →
Lethocerus indicum	Bhargava (1967)	{	
Iphita limbata	Seshan (1968) (Nayar, 1955)	1 pair‡ / Area§	2–4 pairs →
Rhodnius prolixus	Baehr (1968); Baudry (1968)	2–4 pairs →	Large areas† / 4 pairs → / 5 pairs →
Nezara viridula	Awasthi (1972)	1 pair‡	
Chrysocoris stolli	Singh et al. (1978)	1 pair‡	

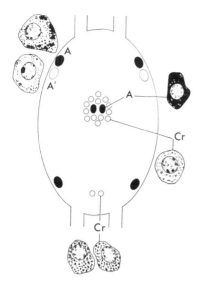

FIGURE 7. Schematic view of the neurosecretory cell types in the subesophageal ganglion of the stick insect *Carausius morosus*. Letters refer to neurosecretory cell types. (From data of Raabe, 1965b.)

processes back toward the thoracic and abdominal ganglia (Rémy and Girardie, 1980) and terminate in the somatic nerves (Girardie and Rémy, 1980).

A second category is represented by the "thoracic" cells that form two or three lateral pairs. These are large pyriform neurons whose secretion assumes the appearance of granules. In contact with a first anterior pair, the presence of a second pair, smaller and less loaded (dragonflies, cockroaches, stick insects), has often been noted, recalling the A and A' cells of the pars intercerebralis. These cells are limited to the thoracic region in certain insects, but they can extend to the subesophageal ganglion, where they form a supplementary pair. They are also encountered in the abdominal ganglia, sometimes only in the first four. Their number is often characteristic and constant: three subesophageal pairs (4 A, 2 A'), two anterior thoracic pairs (2 A, 2 A') and one abdominal pair (A). In Heteroptera, however, their distribution is less uniform.

A third category is exclusively abdominal. It includes two to five pairs of posterior lateral cells, morphologically distinct from the former and generally smaller. In Heteroptera, two types of abdominal cells are observed rather than one.

A fourth category is represented by the Cr cells present in all the ganglia of the species investigated. Far more numerous than the cells of the other categories, they form small lateroposterior clusters in the thoracic ganglia, and wide ventromedian areas in the subesophageal and abdominal ganglia.

Apart from the Cr cells, the neurosecretory cells of the ventral nerve cord are generally of type A. The type B affinities, rare in the subesophageal and thoracic categories, are more frequently encountered in the abdominal category.

1.1.4.3. Neurosecretory Cells of the Anterior Sympathetic Nervous System

The presence of neurosecretory cells has sometimes been demonstrated in the frontal and hypocerebral ganglia of cockroaches, locusts, and stick insects (Smith, 1968; Cazal *et al.*, 1971; Chanussot, 1972; Dorn, 1978; Khan, 1976a) and in the frontal ganglion of moth larvae (Borg *et al.*, 1973; Bell *et al.*, 1974; Yin and Chippendale, 1975). Dense granules of 105 nm have been observed in *Corethra* (Tombes and Malone, 1977), and dense granules from 100 to 300 nm and from 90 to 120 nm in the hypocerebral ganglion of the bug *Oncopeltus* (Unnithan *et al.*, 1971). Based on the size and density of the elementary granules described, one may conclude that two types of secretion exist, a peptidergic secretion and an aminergic secretion, the existence of which has been confirmed by use of the argentophilic method on the frontal ganglion of the stick insect (Dupont-Raabe, 1958) and by the Falck and Hillarp technique on the sympathetic ganglia of trichopterans (Klemm, 1968b), cockroaches, and locusts (Chanussot *et al.*, 1969; Klemm, 1971; Chanussot, 1972).

It is not known at present exactly where the neurosecretory products of the sympathetic ganglia are delivered. Those of the frontal ganglion pass either toward the brain (Tombes and Malone, 1977) or toward the recurrent nerve, from where they probably reach esophageal effectors. In fact, neurosecretory granules have been reported in the esophageal nerves of *Melanoplus* (Dogra and Ewen, 1970) and *Calliphora* (Thomsen, 1969). A detailed study of the neuronal connections of the frontal ganglion has been carried out in *Periplaneta* (Gundel and Penzlin, 1978).

1.1.4.4. Neurosecretory Peripheral Neurons

Insects possess many peripheral bipolar and multipolar neurons, some of which are neurosecretory. These neurons are located in the vicinity of the visceral muscles, midgut (Wright *et al.*, 1970), rectum (Nagy, 1978), heart (Miller and Thomson, 1968), and alary muscles (Hinks, 1975), where their secretory products undoubtedly play a part in the contractility of the visceral muscles. They are often also located along neurohemal nerves such as the lateral cardiac nerves of aphids (Bowers and Johnson, 1966) and cockroaches (Johnson, 1966b; Miller and Thomson, 1968), the proctodeal nerve of *Manduca* (Reinecke *et al.*, 1973), and also near or within neurohemal organs as in *Carausius* (Finlayson and Osborne, 1968), *Locusta* (Thomas and Raabe,

1974), and various dipterans (Grillot, 1977) and heteropterans (Baudry-Partiaoglou, 1978), where they probably release their secretions. The peripheral neurosecretory cells can be distinguished from the other peripheral neurons by their electrical activity, which is comparable to that of ganglionic neurosecretory cells (Orchard and Finlayson, 1976, 1977).

1.2. NEUROHORMONE RELEASE SITES

1.2.1. The Corpora Cardiaca

The corpora cardiaca represent the most important organs involved in the release of neurosecretory material from the brain. They are located behind the brain and connected to it by three pairs of nerves (the nccI, II, and III) (Fig. 8). They are often situated close to the corpora allata, which are innervated by nerves coming from the corpora cardiaca. They are also connected to the hypocerebral ganglion with which they sometimes fuse. An exchange of fibers running in both directions has been reported in cockroaches (Willey, 1961).

The corpora cardiaca are usually attached to the wall of the aorta and

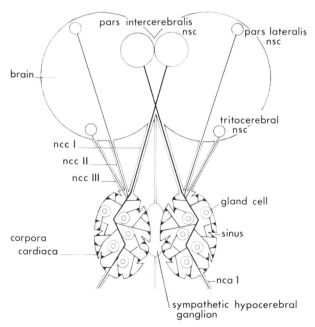

FIGURE 8. Schematic representation of insect corpora cardiaca, storage sites of brain neurosecretory products, and endocrine organ. nsc, Neurosecretory cells.

appear as a swelling of it. They are often present as a pair and sometimes fused. In higher Diptera, they are incorporated in Weismann's ring, which also contains the corpora allata and the molting gland. In some cases, the corpora cardiaca are just partly in contact with the aorta or entirely independent of it; their morphological variations have been thoroughly studied in all orders by Cazal (1948).

The cerebral neurosecretory products are transported along the neurosecretory axons contained within the corpora cardiaca nerves. In the corpora cardiaca, one may distinguish many axons forming a bundle in the center of the organ and branching repeatedly on either side of it in all directions (Fig. 9). Surrounded by glial cells forming numerous folds, the neurosecretory axons make contact with the basement membrane or its invaginations, in sites where the glial sheath is lacking and where the neurosecretory products are released by exocytosis. These sites, termed synaptoid sites, are characterized by a thickening of the axolemma and an accumulation of electron-transparent vesicles of 40–50 nm. They are considered to be the precise release sites of neurohormones.

The glial cells, the thin invaginations of the basement membrane, the branched neurosecretory axons loaded with neurosecretory material, the numerous endings, and the occurrence of release sites are characteristic of neurohemal organs.

The corpora cardiaca contain glandular cells also termed intrinsic cells, parenchymatous cells, or chromophilic cells, the processes of which, like the neurosecretory endings, make contact with the basement membrane or its invaginations. Krogh (1973) in *Schistocerca* succeeded in separating the glandular cells by incubating the corpora cardiaca in saline containing hydrolytic enzymes. Following this treatment she observed the cells in phase-contrast microscopy and dark-field illumination and found them to be unipolar neurons with thin elongated axons.

The glandular cells of the corpora cardiaca contain electron-dense neurosecretory granules that sometimes are of the same size as the brain neurosecretory granules, as for example in *Carausius* (Smith, 1968) and *Calliphora* (Normann, 1965); often, however, they are larger, as in *Locusta* (Cassier and Fain-Maurel, 1970a; Cazal *et al.*, 1971). In some species, such as *Schistocerca* (Krogh, 1973) and *Leptinotarsa* (Schooneveld, 1970), two distinct glandular cell types can be distinguished according to granule size; in bugs, these two types are recognized, even through histological techniques (Furtado, 1971a; Awasthi, 1972). In *Oncopeltus*, some cells contain 130- to 300-nm dense granules, while other more numerous and smaller cells contain 100- to 230-nm dense granules (Unnithan *et al.*, 1971).

The corpora cardiaca also contain some neurons, which were described in stick insects (Pflugfelder, 1937), in the beetle *Hydrous* (de Lerma, 1956),

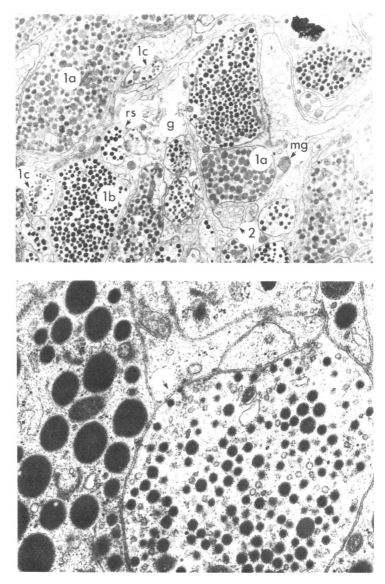

FIGURE 9. The corpora cardiaca of the locust *Locusta migratoria*. (Top) The neurohemal region containing several neurosecretory axon types with 300-nm dense granules (1a), 200-nm dense granules (1b), 100-nm dense granules (1c), and 200-nm transparent vesicles (2). g, Glia; mg, glial mitochondria; rs, release site. (Bottom) The glandular region. Note the glandular cell on the left containing conspicuous dense granules and on the right, a neurohemal ending. (Courtesy of Lafon-Cazal.)

and in other species (Cazal, 1948). More recently, they have been reported in *Locusta* (Cazal *et al.*, 1971) and the bug *Oncopeltus* (Unnithan *et al.*, 1971). The relationships between the different components of the corpora cardiaca pose interesting problems. It is generally considered that neurosecretory products are released in response to nerve impulses transported along the neurosecretory axon. Possibly, the release may in addition be modulated by axons that make contact with the neurosecretory axon, for axo-axonal and axo-somatic contacts between two different axons, or between axons and glandular cells, have been described in several species (Scharrer, 1963, 1968; Normann, 1965; Cazal *et al.*, 1971). Some of these contacts involve the presence of amines, namely octopamine (Evans, 1978, 1980).

In most species, the glandular cells of the corpora cardiaca are dispersed among the neurosecretory axons, but in some of them, the corpora cardiaca have two distinct parts as in *Locusta*, a neurohemal lobe and a glandular lobe; in other species, the corpora cardiaca contain only glandular cells, and the neurosecretory axons end in other sites. Such arrangements provide good tools for experimental studies in permitting the separation of material from neurosecretory endings and glandular cells; when this disposition is lacking, the technique of "regenerates" allows the same separation in sectioning the nccI or nccII and comparing after some time the activity of the corpora cardiaca to that of the regenerated neurohemal organ formed by the sectioned axons.

1.2.2. Other Release Sites of Brain Neurosecretory Products

Recent investigations have demonstrated the occurrence of brain neurosecretory product release areas located outside the corpora cardiaca (Fig. 10).

In the cricket *Acheta*, the ventral part of the brain lying against the aorta wall has a neurohemal structure (Geldiay and Edwards, 1973; Geldiay and Karaçali, 1980). The neurosecretory axons penetrate between the perilemma cells up to the basement membrane, and synaptoid sites characteristic of release areas can be observed.

In the blowfly *Calliphora* and the Colorado beetle *Leptinotarsa*, corpora cardiaca nerves are surrounded by an elongated or globular neurohemal area (Schooneveld, 1974b; Kaiser, 1979); thus in these species, a portion of the brain neurosecretory products is released before entering the corpora cardiaca.

Cardiac nerves issuing from the corpora cardiaca and the esophageal nerves are neurohemal in cockroaches and some other insects. Their axons originate both in the brain and in the ventral ganglia.

In crickets, the nerves joining the corpora allata to the subesophageal ganglion closely resemble the corpora cardiaca. Hence in these species the corpora cardiaca extend beyond the corpora allata (Thomsen, 1943; Huignard,

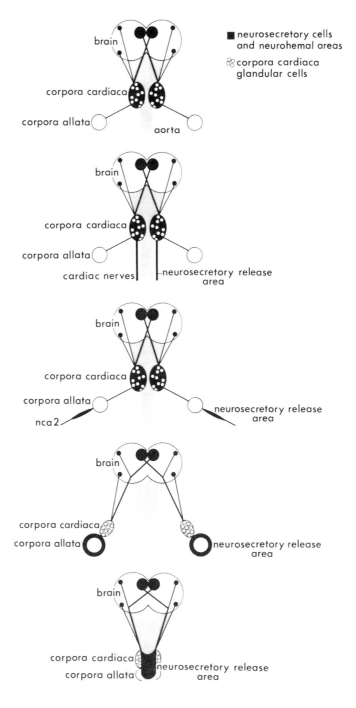

1964; Gaude and Weber, 1966; Awasthi, 1968; Weber and Gaude, 1971). In the cockroach *Periplaneta,* similar but reduced structures have been described (Pipa and Novak, 1979); moreover, a neurohemal plexus extends between the corpora cardiaca and the corpora allata (Adiyodi, 1974).

In some Lepidoptera and Coleoptera, the corpora cardiaca have no neurohemal part at all and the neurohemal areas are located around the corpora allata (Kind, 1965; Panov and Bassurmanova, 1967; Tombes and Smith, 1970; Srivastava *et al.,* 1975). This unusual feature is important for an understanding of the experiments involving corpora allata (see Chapter 2).

In some insect species, the corpora cardiaca are composed mostly of glandular cells and the wall of the aorta serves a neurohemal function. This has been observed in Heteroptera (Johansson, 1958; Dogra, 1967a,b, 1969; Unnithan *et al.,* 1971; Faruqui, 1974; Ghosh and Faruqui, 1977; Singh *et al.,* 1978), Diptera (Normann, 1965; Burgess and Rempel, 1966; Burgess, 1971), and Dermaptera (Awasthi, 1975, 1976).

1.2.3. The Perisympathetic Organs

The corpora cardiaca and the other release centers that have been mentioned are mainly concerned with brain neurosecretory products. Ventral nerve cord neurosecretory products are delivered mostly in other neurohemal organs (Raabe, 1965b) (Fig. 11).

The perisympathetic organs typically display a metameric distribution. They are located on the transverse or median nerve of the posterior sympathetic nervous system, and thus their designation as perisympathetic organs.

The search for perisympathetic organs in the principal insect orders has revealed that they are always present (see Raabe *et al.,* 1971a; Raabe, 1975) but widely vary in disposition (Grillot *et al.,* 1971; Grillot, 1976b). In primitive orders, the perisympathetic organs show a relatively uniform distribution along the course of the nerves of the posterior sympathetic nervous system, the median nerve or the transverse nerves, or both. The most evolved groups exhibit considerable diversity in relation to the concentration of the central nervous system and to the lack of a distinct posterior sympathetic nervous system. The perisympathetic organs are either associated with somatic nerves or remain at the dorsal side of the ganglia, beneath the neurilemma. In the latter case, they may retain a metameric distribution as in the beetle

FIGURE 10. Various arrangements of brain neurosecretory product release sites. From top to bottom: in the corpora cardiaca; in the corpora cardiaca and cardiac nerves; in the corpora cardiaca and around the nca2 nerves; in a neurohemal area surrounding the corpora allata; in the heart wall.

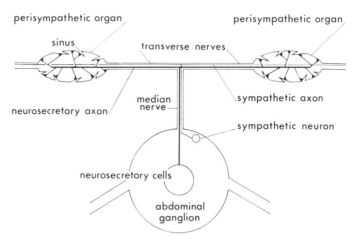

perisympathetic organ   perisympathetic organ

sinus   transverse nerves

median   sympathetic axon
neurosecretory axon   nerve

sympathetic neuron

neurosecretory cells

abdominal
ganglion

FIGURE 11. Schematic representation of insect perisympathetic organs linked to ventral nerve cord neurosecretory cells.

Oryctes rhinoceros or fuse in a single organ as in the tsetse fly *Glossina fuscipes*.

Moreover, the perisympathetic organs may divide into several parts, a median part usually located around the median nerve or within the ganglia, and two or more lateral ones surrounding the transverse nerves or the segmental nerves. In the case of the primitive hymenopteran *Diprion pini*, it has been demonstrated that the discontinuous structure of the adult results from a continuous larval structure (Provansal, 1971). Sometimes the lateral organs are located far from the nerve ganglia as in the bug *Roscius* sp. and the tsetse fly, in which they are called distal perisympathetic organs (Fig. 12).

The general structure of the perisympathetic organs strongly resembles the structure of the corpora cardiaca. They contain many neurosecretory axons surrounded by glial cells, and their connective sheath is very thin and may invaginate deeply, creating sinuses.

According to their anatomical position, the perisympathetic organs present a number of distinct structural features. In primitive orders like Ephemera and Odonata (Raabe and Provansal, 1972), the neurohemal structure may form a fine sheath over a great distance along the sympathetic nerves. The same is found in some evolved insects such as *Rhodnius* (Maddrell, 1966) and *Tenebrio* (Grillot, 1971a) along somatic nerves (Fig. 13) and also around thin distal plexuses as in *Roscius* (Baudry-Partiaoglou, 1978) (Figs. 12 and 15). In this case, the release of neurosecretory products takes place mostly at the surface, which presents a relatively large area of contact with the internal environment. This also occurs in Lepidoptera and Dermaptera; in these insects

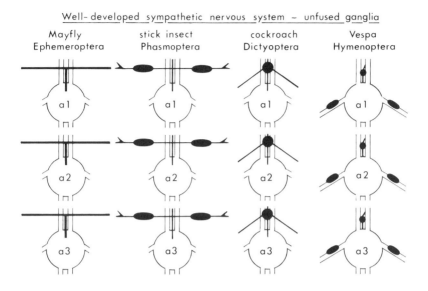

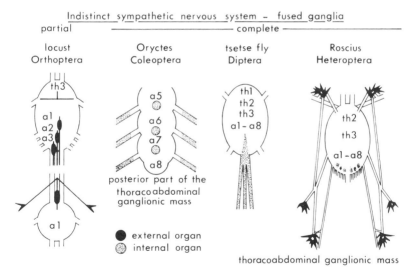

FIGURE 12. Figure showing the diversity displayed in the location and arrangement of the perisympathetic organs. a, Abdominal ganglion; th, thoracic ganglion. (From data of Raabe and Provansal, 1972; Raabe, 1966b; de Bessé, 1966; Provansal, 1968; Chalaye, 1966; Grillot, 1968, 1971b; Grillot and Raabe, 1973; Baudry-Partiaoglou, 1978).

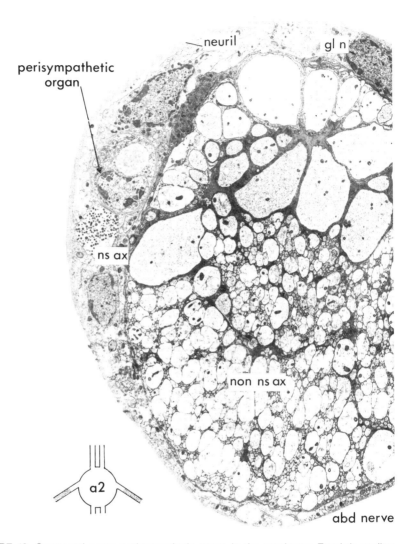

FIGURE 13. Segmental nerve perisympathetic organ in the mealworm *Tenebrio molitor* (cross section). a2, Second abdominal ganglion; abd nerve, abdominal nerve; gl n, glial nucleus; neuril, neurilemma; non ns ax, nonneurosecretory axons; ns ax, neurosecretory axons. (Courtesy of Grillot.)

the perisympathetic organs form a very thin lamina intermingling with the ventral diaphragm (Provansal, 1972; Raabe, 1972).

In many species, in which the perisympathetic organs are short and thick, the passage of neurohormones into the bloodstream takes place at the surface but above all through the numerous internal sinuses (Raabe and Ramade, 1967; Raabe *et al.*, 1970) (Fig. 14).

Globular perisympathetic organs apparently less adapted to a neurohemal function are found among the median nerve organs. Ultrastructural studies show that they are generally formed of two zones: a central zone consisting

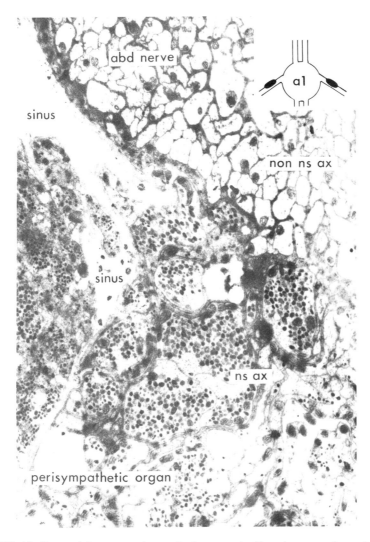

FIGURE 14. Segmental nerve perisympathetic organ in *Vespula germanica*. a1, First abdominal ganglion; abd nerve, abdominal nerve; non ns ax, nonneurosecretory axons; ns ax, neurosecretory axons. (Courtesy of Baudry-Partiaoglou and Provansal-Baudez.)

mainly of nerve fibers and a peripheral zone with few sinuses, where the neurosecretory terminals are most often surrounded by glial cells (Provansal *et al.*, 1970).

Perisympathetic organs may contain little type A and B neurosecretory products, but type Cr neurosecretory products are abundant. By applying the cobalt chloride retrograde diffusion technique to the cockroach *Periplaneta* (Ali and Pipa, 1978), it was observed that the axons projecting within the perisympathetic organs originate in four mediodorsal clusters where type Cr neurosecretory cells were identified (de Bessé, 1966).

Although perisympathetic organs are mostly loaded with type Cr neurosecretory products, they also contain, especially in some species, type A and B neurosecretory products. Ultrastructural studies performed on stick insects, cockroaches, locusts, wasps, beetles, and tsetse flies have revealed the presence of endings loaded with different-sized granules (Raabe and Ramade, 1967; Brady and Maddrell, 1967; Scharrer, 1968; Provansal *et al.*, 1970; Raabe *et al.*, 1970; Chalaye, 1974a; Baudry-Partiaoglou and Grillot, 1975; Grillot, 1976b, 1977) (Fig. 15). This indicates that they probably contain more than one neurosecretory product.

As for the corpora cardiaca, some perisympathetic organs contain neurosecretory cells (Provansal and Grillot, 1972; Baudry, 1972b; Grillot, 1976a,b, 1977; Baudry-Partiaoglou, 1978), which may be sensory peripheral neurons similar to those described by Finlayson and Osborne (1968) on the link nerve of stick insects. The relationship between these cells and the perisympathetic organs seems to vary. Sometimes they are located within the perisympathetic organs, sometimes nearby.

As in the corpora cardiaca, octopamine has been found in the perisympathetic organs (Evans, 1978, 1980), thus suggesting a possible role of amines in the release of neurosecretory material.

1.2.4. Distal Neurohemal Organs

We have seen above that the perisympathetic organs may divide, which results in the occurrence of a classical organ located close to the nervous system and in two remote distal organs. Such dispositions are encountered in Diptera and Heteroptera whose central nervous system is very concentrated. In Heteroptera, the distal perisympathetic organs usually appear as very thin plexuses (Baudry, 1972a,b; Baudry-Partiaoglou, 1978) (Fig. 15). The distal perisympathetic organs of Diptera are often located near chordotonal organs or sensory neurons (Grillot, 1977). They have also been observed in the vicinity of Malpighian tubules in Glossina (Maddrell and Gee, 1974).

Distal perisympathetic organs have also been found in primitive insects such as stick insects (Finlayson and Osborne, 1968) and locusts (Thomas and Raabe, 1974). In both cases, the neurohemal areas surround the transverse

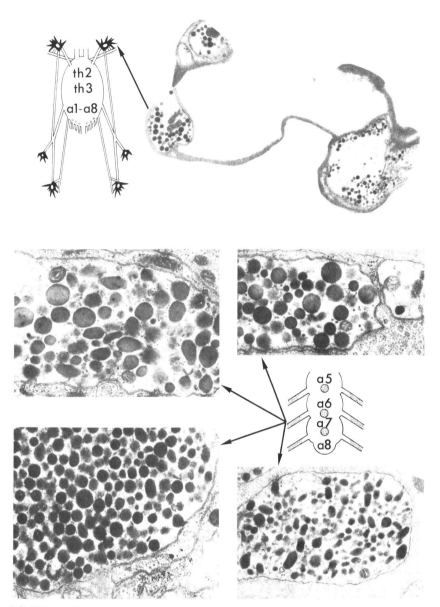

FIGURE 15. (Top) *Roscius* distal perisympathetic organ. (Bottom) Neurosecretory axon diversity in segmental nerve perisympathetic organ of the beetle *Oryctes nasicornis*. a, Abdominal ganglion; th, thoracic ganglion. (Top, courtesy of Baudry-Partiaoglou; bottom, courtesy of Baudry-Partiaoglou and Provansal-Baudez.)

nerves and their branches at some distance from the perisympathetic organs, and are associated with multipolar neurosecretory neurons.

1.2.5. Neuroeffector Junctions

Some neurosecretory products are not released in neurohemal organs but are directly delivered to their effector organs (Fig. 16, Table VII). The occurrence of neurosecretory axons has been reported in different organs, namely endocrine glands, visceral muscles of the heart, gut, salivary glands, and even in the epidermis and in skeletal muscles. The neurosecretory axons make synaptoid contact either with the connective sheath of the organs or with the cells themselves. These contacts, termed by Knowles and Bern (1966) neurosecretomotor junctions, seem to represent a mode of release of neurosecretory products in which the active factors are no longer transmitted by a circulatory pathway but are delivered directly to the right place, either to the effector cells or to the connective tissue where they diffuse in limited areas.

1.2.6. Transport of Neurosecretory Products along the Ventral Nerve Cord

Brain neurosecretory cells sometimes have very long axons that run throughout the length of the ventral nerve cord, send collaterals to each ganglion, and end up in the neuropil (Table VII, Fig. 17). This has been observed in mosquitoes (Füller, 1960; Burgess, 1971, 1973; Panov, 1979), in a bug (Seshan, 1968), and in aphids (Johnson, 1963; Steel, 1977) where two or three pairs of brain neurosecretory cells send their processes to the ventral nerve cord. These cells may at the same time branch in the brain neuropil and send a process to the corpora cardiaca. An opposite pathway was observed in *Acheta* for the medial pair of neurosecretory cells in the subesophageal ganglion, which terminate in a brain neurohemal area (Geldiay and Edwards, 1973; Strambi *et al.*, 1979) but which also proceed along the ventral nerve cord and send collaterals to the successive ganglia (Rémy and Girardie, 1980).

The significance of these neurosecretory collaterals located in the neuropil is not quite clear. Probably the neurosecretory products are released there and modulate nervous or neurosecretory activity.

1.3. NEUROHORMONES AND BIOGENIC AMINES

Electron microscopy, autoradiography, and the Falck and Hillarp formaldehyde method used for the demonstration of biogenic amines have led to considerable progress in the field, permitting the identification of aminergic

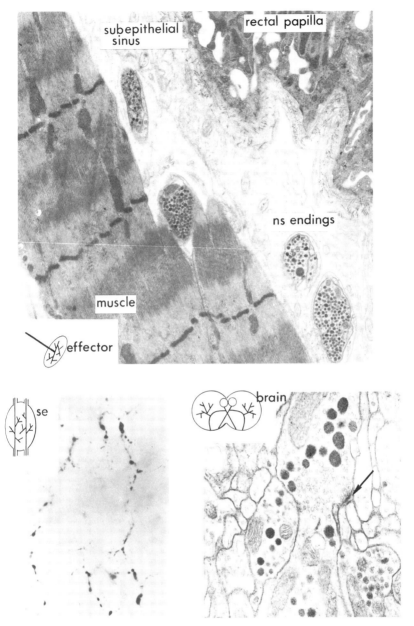

FIGURE 16. (Top) Neurosecretory endings in the hindgut of the termite *Calotermes flavicollis.* (Bottom) Collateral swelling containing neurosecretory granules. (Left) In the dorsal neuropil of the subesophageal ganglion of the stick insect *Carausius morosus;* (right) in the brain neuropil of the Colorado beetle *Leptinotarsa decemlineata.* Note site of contact between the neurosecretory collaterals and both a neurosecretory and a non-neurosecretory axon. ns endings, Neurosecretory endings; se, subesophageal ganglion. (Top, courtesy of Noirot and Noirot-Thimothée; left, from Raabe, 1965b; right, courtesy of Schooneveld, 1974b.)

TABLE VII. Demonstration of Neurosecretory Axon Endings in Various Organs

Order/insect	Reference	Brain	Ventral ganglia	Connectives	Hypo-cerebral ganglion	Corpora allata	Molting gland	Heart	Hindgut	Other organs
Ephemeroptera	Arvy and Gabe (1953)					+				
Odonata							+			
Dictyoptera										
Leucophaea maderae	Scharrer (1964a,b)					+	+			
Blaberus cranifera	Geldiay (1959)		+							
Blaberus craniifer	Chanussot (1972)			+						
Periplaneta americana	Johnson (1966b)				+					
	Brown (1967)							+		Cardiac nerves
	Adiyodi and Bern (1968)	+							+	
	Miller and Thomson (1968)							+		Cardiac nerves
	Gupta and Smith (1969)									Spermatheca
	Oschman and Wall (1969)								+	
	Wright et al. (1970)									Midgut
	Whitehead (1971)									Salivary gland
	Farley and Evans (1972)	+	+							
	Adams et al. (1973)									Alary muscles
	Miller and Adams (1974)									Hyperneural muscle
	Beattie (1976)									Pulsatile organ
	Raziuddin et al.									

and *Braberus*
giganteus (1975)

Isoptera
Kalotermes
flavicollis — Quennedey (1969), Noirot and Noirot-Thimothée (1976) — Sternal gland +

Phasmoptera
Carausius morosus — Dupont-Raabe (1956) +, Raabe (1965b), Pasteels (1965), Osborne et al. (1971), Fifield and Finlayson (1978) — Skeletal muscles

Orthoptera
Locusta migratoria — Panov (1962, 1964), Strong (1966), Cassier and Fain-Maurel (1970b), Chalaye (1966, 1975) — Foregut muscles
Schistocerca gregaria — Highnam (1961a), Delphin (1965), Osborne et al. (1971)
Poekilocerus pictus — Raziuddin et al. (1979) +
Acheta domesticus — Geldiay and Edwards (1973) + — Skeletal muscles

TABLE VII. (Continued)

Order/insect	Reference	Brain	Ventral ganglia	Connectives	Hypo-cerebral ganglion	Corpora allata	Molting gland	Heart	Hindgut	Other organs
Dermaptera										
Euborellia annulipes	Awasthi (1976, 1979)					+				
Hymenoptera										
Synagris calida	Thomsen (1954)	+								
Apis mellifica	Formigoni (1956)					+				
Camponotus vagus	Delye (1972)		+							
Coleoptera										
Hydrous piceus	de Lerma (1956)	+								
Blaps mucronata	Fletcher (1969)		+	+					+	Genital organs
Leptinotarsa decemlineata	Schooneveld (1974b)	+	+	+						
Tenebrio molitor	Arvy and Gabe (1953)					+	+			
	Mordue (1965a)					+				
Sitophilus granarius	Tombes and Smith (1970)					+				
Hypera postica	Tombes (1976)					+				Spermatheca
Photuris pennsylvanica	Smith (1963)									Light organs
Lepidoptera										
Celerio lineata	Schultz (1960)					+				
Hyalophora cecropia	Waku and Gilbert (1964)					+				
Silkworms	Bassurmanova and Panov (1967)					+				
Bombyx mori	Bounhiol et al.					+				

						Foregut			Skeletal muscles
Antheraea pernyi	Busselet (1969)				+				
Cerura vinula	Hintze-Podufal (1970)						+		
Galleria mellonella	Blazsek *et al.* (1975)						+		
Manduca sexta	Reinecke and Adams (1977)			+					
Diptera									
Corethra plumicornis	Füller (1960)	+							
Aedes aegypti	Burgess and Rempel (1966)			+	+				
Culesita inormata	Burgess (1971)	+		+	+				
Culex tarsalis	Burgess (1973)	+		+					
Chironomus riparius	Credland and Phillips (1976)		+						
Chironomus plumosus	Panov (1979)	+		+					
Chaetodacus cucurbitae	Nayar (1954)	+		+			+		
Phormia regina	Hsiao and Fraenkel (1966)	+		+					
Phormia terrae-novae	Osborne *et al.* (1971)						+		
Calliphora stygia	Johnson (1966a)			+					
Calliphora erythrocephala	Normann (1965)			+					
Gupta and Berridge (1966)	Gupta and Berridge (1966)							+	

(Continued)

TABLE VII. (Continued)

Order/Insect	Reference	Brain	Ventral ganglia	Connectives	Hypo-cerebral ganglion	Corpora allata	Molting gland	Heart	Hindgut	Other organs
Diptera (Continued)										
Calliphora erythrocephala	King et al. (1966)					+				
	Thomsen and Thomsen (1970)					+				
Homoptera										
Myzus persicae	Johnson (1963)	+	+	+		+				
	Bowers and Johnson (1966)							+		Pericardial cells, muscles
Aphis craccivora	Elliott (1976)	+	+			+		+		
Megoura viciae	Steel (1977)	+	+	+						
Pyrilla perpusilla	Srivastava (1969)	+								
Heteroptera										
Oncopeltus fasciatus	Johansson (1958)	+	+			0[a]				
	Unnithan et al. (1971)			+						
Adelphocoris lineolatus	Ewen (1962)	+	+							
Rhodnius prolixus	Maddrell (1965)									Epidermis
	Maddrell (1969)									Malpighian tubules
	Busselet (1969)					+				
	Anwyl and Finlayson (1973)									Skeletal muscles
Iphita limbata	Nayar (1956a)					+				
	Seshan (1968)		+	+						
Lethocerus indicum	Bhargava (1970)					+				
Sphaerodema rusticum	Faruqui (1974)					+				

[a] 0, No neurosecretory axons observed.

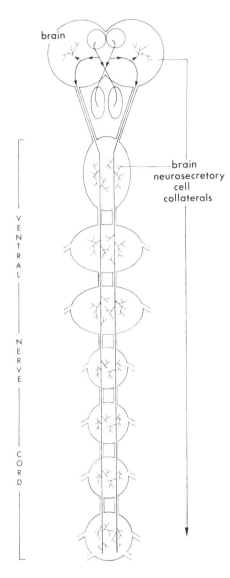

FIGURE 17. Pathway of certain cerebral neurosecretory axons sending collaterals to ventral nerve cord ganglia.

neurons and aminergic endings in nerve ganglia and neurohemal organs (see Klemm, 1976; Lafon-Cazal, 1978; Evans, 1980).

The insect brain has been thoroughly investigated by means of the Falck and Hillarp method (Plotnikova and Govyrin, 1966; Frontali, 1968; Klemm 1968a, 1974; Plotnikova, 1968; Mancini and Frontali, 1970; Klemm and Axelsson, 1973; Klemm and Schneider, 1975; Klemm and Falck, 1978). An aminergic fluorescence was found in the associative parts of the brain, for example the protocerebral or tritocerebral neuropil; aminergic neurons were discovered both in purely nervous parts and in the pars intercerebralis of several species, such as *Musca* (Ramade, 1966, 1969a; Ramade and L'Hermite, 1971), *Calliphora* (Bloch *et al.*, 1966), *Locusta* (Girardie and Girardie, 1966, 1967), *Acheta* (Geldiay and Edwards, 1973), and *Leptinotarsa* (Schooneveld, 1974a); in the pars lateralis of *Locusta* (Rademakers, 1977b); and in the ventral nerve cord of this insect (Chalaye, 1974 a,b).

Microgranular axons have also been observed in the sympathetic nervous system (Chanussot, 1972), in the corpora cardiaca (Normann, 1965; Johnson, 1966a; Scharrer, 1968; Cassier and Fain-Maurel, 1970a; Cazal *et al.*, 1971), and in the perisympathetic organs (Evans, 1978, 1980).

Within the corpora cardiaca, aminergic axons make axo-axonal and axo-somatic contact with the glandular cells or the neurosecretory axons in *Calliphora* (Normann, 1965) and *Locusta* (Cazal *et al.*, 1971) (Fig. 18). Two

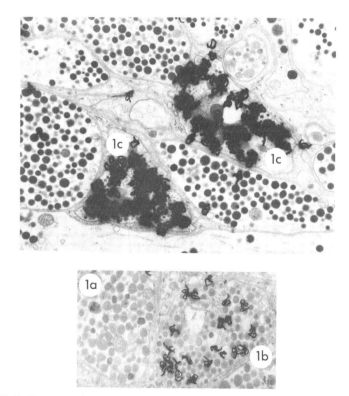

FIGURE 18. Demonstration of the presence of serotonin (1b) and dopamine axons (1c) in the neurohemal region of the corpora cardiaca in *Locusta migratoria,* located near peptidergic axons (1a) (labeled material incorporation experiments). (Courtesy of Lafon-Cazal.)

distinct amines, dopamine and serotonin, have been identified in the neural lobe in *Schistocerca* and *Locusta* (Klemm, 1971; Lafon-Cazal *et al.*, 1973; Lafon-Cazal and Arluison, 1976; Lafon-Cazal, 1976), and the presence of octopamine has been observed in the corpora cardiaca of both locusts and cockroaches (Evans, 1978).

Aminergic axons make synaptoid contacts but not true synapses. Thus it may be inferred that biogenic amines either enter into the bloodstream like other neurohormones or act *in situ*, modulating the release of polypeptidic neurohormones. In various organs, aminergic endings are also found beside the peptidergic ones; if both endings perform a regulatory function, it seems that a radical distinction cannot be made between biogenic amines and polypeptidic neurohormones.

1.4. CONCLUSIONS

Neurosecretory processes in insects have been the subject of many investigations performed using various methods. These studies contribute to the knowledge of neurohormones by permitting the distinction of several neurosecretory products, by classifying the functioning of the neurosecretory cells and neurohemal organs, while giving information on their location as well.

The neurohemal system of insects is distinguished by tremendous variety, combined with wide dispersion of neurosecretory neurons and neurohemal

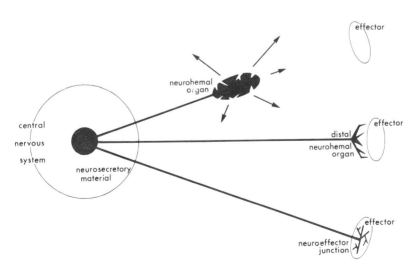

FIGURE 19. Major modes of release of neurosecretory products.

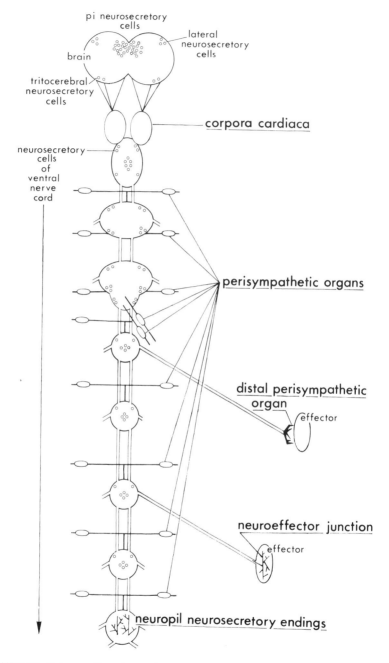

FIGURE 20. Release sites of neurohormones in insects: a recapitulation. pi, Pars inter-cerebralis.

sites, whose degree of proximity to the organs varies widely (Fig. 19). The early concept of a single site in the corpora cardiaca must now be supplanted by a different one: the neurohormones released from central neurohemal organs, such as the corpora cardiaca and the perisympathetic organs, enter into the bloodstream, but limited diffusion is frequently encountered in neurohemal areas located close to or within various organs (Fig. 20).

Thus, in place of an earlier concept postulating two opposing modes of action, that of the neurohormones transmitted by the circulatory pathway and that of the neurotransmitters released at the synapse, we now have a concept in which the differences between neurohormones and mediators diminish. It is considered that peptidergic neurohormones may, in some cases, be released in contact with effectors, whereas amines sometimes follow the circulatory pathway.

2

Control of Endocrine Gland Activity

2.1. MOLTING GLAND

The existence of a brain factor involved in the control of molting was discovered by Kopec (1922) in the gypsy moth *Lymantria dispar*, and demonstrated by Wigglesworth (1940) in the blood-sucking bug *Rhodnius prolixus*. Its role in activating the molting gland was shown by Williams (1946, 1947) in the great saturniid *Platysamia cecropia*, by Possompès (1953) in the blowfly *Calliphora erythrocephala*, and subsequently in various species. *In vitro*, moreover, molting glands have been shown to be stimulated by the brain (Agui, 1975).

Ecdysone, the hormone produced by the molting gland, was isolated by Butenandt and Karlson in 1954 from pupae of the silkworm *Bombyx mori*. We now know that ecdysone exists in two forms, α and β ecdysone. The former, which is secreted by the molting gland from cholesterol, is hydroxylated to β ecdysone or ecdysterone in the fat body.

The molting gland, the site of which varies according to species, has been designated under different names according to its location: thoracic gland, prothoracic gland, or ventral gland. The brain factor that causes its activation has also been given various names, the most widely used being brain hormone, prothoracicotropic hormone, thoracotropic hormone, and activation hormone. We shall refer to it as brain hormone.

It has been known for some time that the follicle cells of the ovary also secrete ecdysone, and investigations have been carried out to determine whether the brain controls their activity.

2.1.1. Molting Gland Secretion Control

The earliest investigators pointed out that the molting gland displayed discontinuous secretory activity, and this fact was substantiated recently by radioimmunoassays of hemolymph ecdysteroids. The secretion of brain hormone, which regulates the functioning of the molting gland, is also discontinuous.

The stimuli initiating brain hormone secretion are not always known. In *Rhodnius* (Wigglesworth, 1934), a single blood meal is taken at each intermolt period, and determines the swelling of the abdomen and the stimulation of the stretch receptors. From the latter, the nerve impulses are transported along the ventral nerve cord to the brain, where they trigger the brain hormone secretion process. In continuously feeding insects, e.g., *Locusta migratoria*, the intake of food also appears to stimulate the neurosecretory cells that produce brain hormone through the intermediary of nerve impulses passing through the frontal ganglion (Clarke and Langley, 1963). The existence of a photosensitive circadian clock has been demonstrated in the moths *Manduca sexta* and *Antheraea pernyi* (Truman, 1972): it appears to control brain hormone release, which is also dependent on the weight reached by the larva (Nijhout and Williams, 1974a) and on the disappearance of juvenile hormone from the hemolymph (Nijhout and Williams, 1974b).

The secretory activity of the brain can be identified according to the different behavioral sequences. In the wax moth *Galleria mellonella*, the completion of brain secretion occurs at the end of the cocoon-spinning phase. In the prepupa of *Manduca*, there are two short surges of brain hormone: the first takes place on day 4, namely, at the end of the feeding period, after a gut purge; it lasts 3–5 hr and induces passage to the wandering stage, characterized by the "exposed heart" due to deposition of the pigments and clearing of the tissues around the heart. The second, which occurs 2 days later, lasts 7 hr and determines the pupal molt (Truman and Riddiford, 1974) (Fig. 21).

Stimulation of the neurosecretory cells may involve the intervention of cAMP. In *Hyalophora* pupae taken out of the cold and reactivated, Rasenick *et al.* (1976) reported a significant increase in brain cAMP. Similarly, *Antheraea* pupae transferred from a short photoperiod to a long photoperiod, which interrupts diapause, reveal a considerable elevation of cAMP in the pars intercerebralis.

2.1.2. Molting Gland Activation

The molting gland exhibits morphological and cytological characteristics that differ from one species to another. Often formed of many small cells in primitive insects, it is made up of large polyploidal elements constituting diffuse organs in more evolved insects.

In the giant silkworm *Hyalophora cecropia*, the histological study of Herman and Gilbert (1966) revealed that the inactive cells are small, possess regular nuclei containing small nucleoli, and a sparse cytoplasm devoid of

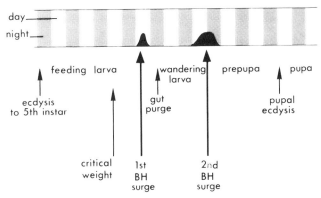

FIGURE 21. Timing of morphological and endocrinological events during the final instar and prepupal stage of the tobacco hornworm *Manduca sexta*. (Modified from Truman and Riddiford, 1974; Nijhout and Williams, 1974a.)

vacuoles. Activation is first reflected by an increase in nuclear and nucleolar volume, and then by the appearance of vacuoles first in the nucleus and then in the cytoplasm, whose volume increases. In the phase of maximum activity, the nuclei are extremely irregular and the cytoplasm contains many vacuoles.

Ultrastructural changes in the molting gland have been investigated by various authors. Despite a number of contradictions in the data presented, it appears that the main variations detectable by the electron microscope concern the endoplasmic reticulum and the mitochondria. According to Fain-Maurel and Cassier (1968), the same modifications are observed in *Locusta* during each intermolt period. The endoplasmic reticulum initially occurs in the form of ribosomes or polysomes, followed by the appearance of ergastoplasmic saccules that later increase in size. In the final phase, the ergastoplasm is replaced by a smooth endoplasmic reticulum and free ribosomes.

Activation of the molting gland triggers nuclear and cytoplasmic RNA syntheses (Krishnakumaran *et al.*, 1965; Oberlander *et al.*, 1965; Kobayashi *et al.*, 1968; Gersch and Stürzebecher, 1970; Weber and Emmerich, 1976). It is also accompanied by an elevation of the membrane potential, which gives rise to increased permeability (Gersch and Birkenbeil, 1973). The latter may facilitate the entry of cholesterol and the release of ecdysone, which occur at an interval of a few days according to Blazsek and Malá (1978).

Brain hormone, which is responsible for activating the molting gland, involves the action of cAMP. The latter is produced by molting gland homogenates and increases the quantity of ecdysone produced, *in vitro*, by the molting gland (Vedeckis *et al.*, 1974). Furthermore, the cAMP level in the molting gland rises during the brain hormone release period (*Manduca*, Vedeckis *et al.*, 1976).

2.1.3. Degeneration of the Molting Gland

The mechanisms that control imaginal degeneration of the molting gland appear to be complex and are difficult to identify. Degeneration of the molting gland may derive from a juvenile hormone deficiency, for corpora allata implants and the injection of juvenile hormone-rich extracts prevent its degeneration in *Antheraea polyphemus* (Gilbert, 1962) and in *Locusta* (Cassier and Fain-Maurel, 1970c). In *Rhodnius*, however, Wigglesworth (1954) demonstrated that the absence of juvenile hormone is not sufficient to cause the breakdown of the molting glands. In order to degenerate, they must undergo a hormonal stimulus during the imaginal molting period, of which the extracerebral source has not been identified.

In the cockroach *Nauphoeta cinerea*, Lanzrein (1975) confirmed the role of juvenile hormone, the absence of which programs the degeneration of the molting gland, but also reported a protective function performed by the corpora cardiaca as indicated previously by Bodenstein (1953a) in *Periplaneta americana*, and demonstrated the existence of a humoral factor inducing degeneration, which could be ecdysone.

A completely different scheme was postulated by Herman and Gilbert (1966) in *Hyalophora*. They suggested that degeneration of the molting gland could be compared to the degeneration of the intersegmental muscles, and could be due to a decrease of the nerve impulses.

2.1.4. Location of Brain Hormone

The location of the neurosecretory cells that produce brain hormone appears to vary according to the insects studied (Fig. 22). Following the discovery by Hanström (1938) that neurosecretory cells occur in the pars intercerebralis of *Rhodnius*, Wigglesworth (1940) experimentally showed that these neurosecretory cells were the origin of brain hormone. This was then confirmed in *Locusta* (Girardie, 1964), in the dragonfly *Aeshna cyanea* (Charlet and Schaller, 1975a,b), and in the bug *Panstrongylus megistus* (Furtado, 1977a). In *Hyalophora* pupae, however, brain action requires the 16 median cells of the pars intercerebralis and also the 10 lateral neurosecretory cells (Williams, 1948). The function of the lateral neurosecretory cells has been confirmed in *Locusta* (Girardie, 1974; Girardie and de Reggi, 1978) by using electrical stimulations followed by determination of hemolymph ecdysteroids. But these cells appear to be of lesser importance in this insect than the neurosecretory cells of the pars intercerebralis. In the young *Manduca* larva, on the other hand, the lateral cells seem to be the only ones involved (Gibbs and Riddiford, 1977). The precise source cells producing brain hormone in this insect were identified by Agui *et al.* (1979), who incubated the molting gland with different parts of the brain. Ecdysone production was determined, and it

Aeshna cyanea

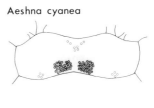

Hyalophora cecropia

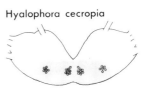

Manduca sexta

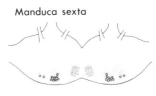

FIGURE 22. The source cells of brain hormone in three insects. [*Aeshna cyanea*, modified from Charlet *et al.*, 1974; Charlet and Schaller, 1975a; *Hyalophora cecropia*, modified from Williams, 1951; *Manduca sexta*, modified from Gibbs and Riddiford, 1977 (first results); Agui *et al.*, 1979 (recent results).]

Manduca sexta

was demonstrated that one particular neuron in each lobe of the brain is involved in brain hormone production. In aphids, one or two pairs of median protocerebral neurosecretory cells were reported to be involved in molting control (Steel, 1978b). In other insects, histophysiological investigations conducted on diapausing insects and on insects just terminating diapause demonstrated that the cells involved are neither type B cells nor all type A cells, but rather one of the categories of the A cells, the A' cells. These results obtained on lepidopterans were substantiated on the bug *Panstrongylus* (see Chapters 3 and 4).

Experimental investigations performed on *Panstrongylus* revealed the existence of a dual brain control over molting exerted according to two modes by two distinct neurosecretory cell types: an early direct control by type A neurosecretory cells and a delayed control exerted by type A' neurosecretory cells through the molting gland (Furtado, 1977a). It remains to be seen if such a dual control of molting by the brain exists in other insects.

2.1.5. Brain Hormone Release Site

Evidence concerning neurosecretory processes implies that brain hormone is released from the corpora cardiaca. However, it has been observed in some cases that corpora cardiaca implants are less active than pars intercerebralis implants. This may be explained by the fact that isolated corpora cardiaca are formed of separate endings and cannot receive release impulses either from the neurosecretory cells themselves or from any neurons controlling their activity; in addition, the release site of brain neurosecretory products is not always located in the corpora cardiaca and can be localized in other sites, namely, in the aorta wall in Heteroptera and around the corpora allata in Lepidoptera (see Chapter 1). The presence of brain hormone activity in the corpora allata was demonstrated in *Bombyx* (Ishizaki, 1969) and *Manduca* (Gibbs and Riddiford, 1977). *In vitro* experiments that determined the amount of ecdysone produced by the molting gland when kept in contact with various organs have demonstrated unequivocally that in *Manduca* the corpora allata are the source of brain hormone (PTTH) (Agui *et al.*, 1980).

The mechanisms regulating the release of brain hormone have been studied in some cases. By electrical stimulation of the nccI or nccII in *Periplaneta*, it was observed that brain hormone release depends exclusively on the nccII (Gersch *et al.*, 1970; Gersch, 1972). This does not necessarily imply that the synthesis site is located in the lateral parts of the protocerebrum, because the stimulated fibers may only regulate the release of neurosecretory products. The use of sympathico- and parasympathicomimetics and sympathico- and parasympathicolytics confirms this hypothesis and indicates the existence of adrenergic transmission.

The mode of transport of brain hormone to the molting gland has also been investigated. The results obtained suggest that brain hormone released into the hemolymph is conveyed by hemocytes. In fact, working on *Rhodnius*, Wigglesworth (1955) succeeded in delaying molting by the injection of substances that block the hemocytes, e.g., India ink, trypan blue, and iron saccharate. Brain hormone might also reach the molting gland directly via its nerves, for the presence of neurosecretory products has often been shown in these nerves as in the molting gland itself. Moreover, in *Calliphora*, the connections between the brain and Weismann's ring are indispensable to the functioning of the molting gland (Possompès, 1953), thus supporting the proposed hypothesis.

2.1.6. Brain Hormone and Ventral Nerve Cord

The molting gland is richly innervated by ventral nerve cord ganglia, sometimes by the subesophageal ganglion only, sometimes by the subesophageal and prothoracic ganglia, and sometimes by several thoracic ganglia. Neurosecretory fibers are present within the molting gland, in its nerves, and in some cases in the perisympathetic organs located in close contact with the molting gland (*Gryllus domesticus,* Thomas and Raabe, 1974; *Labidura riparia,* Raabe, 1972).

In *Periplaneta*, connection between the molting gland and the prothoracic ganglion activates its functioning *in vitro* (Marks and Reinecke, 1965), even though its disconnection *in vivo* was reported to have no effect on its secretory activity in the larva of the lemon butterfly *Papilio demoleus* (Srivastava *et al.*, 1977).

Some investigations utilized the cobalt chloride technique in an attempt to determine the source cells of the fibers that innervate the molting gland. In the wax moth *Galleria* (Granger, 1978; Singh and Sehnal, 1979), the neurons innervating the molting gland have been identified. They include some neurosecretory cells: a median pair located in the subesophageal ganglion, and two lateral pairs located in the prothoracic and mesothoracic ganglia. Hence, it is very likely that the functioning of the molting gland is regulated, apart from the cerebral brain hormone, by nerve impulses and neurosecretory products originating in certain ventral ganglia. This is suggested by experiments in which ligated, decapitated, and decerebrated insects are still able to resume molting (*Carausius morosus,* Dupont-Raabe, 1952b; *Bombyx mori*, Kobayashi *et al.*, 1960; *Antheraea polyphemus,* McDaniel and Berry, 1967; *Manduca sexta*, Judy, 1972; Fain and Riddiford, 1976; Nijhout, 1976) and by the demonstration of brain hormone activity in purified extracts of the ganglia of the cockroach *Periplaneta* (Gersch and Stürzebecher, 1967). However, the experiments performed on Lepidoptera by Gibbs and Riddiford (1977), Meola and Adkisson (1977), Alexander (1970) and Safranek and Williams (1980) do not substantiate these indications.

Gibbs and Riddiford (1977) investigated the presence of brain hormone in the ventral nerve cord ganglia of *Manduca* penultimate-instar larvae. Following cervical ligation of the larvae, injections of different nerve ganglia extracts were made into the hind part of the body. While the brain is highly active, the subesophageal ganglion, like the thoracic ganglia and the abdominal ganglia, seems inactive and possibly even inhibitory (Table VIII). The activity of the different ganglia of the ventral nerve cord was also assayed in *Manduca* after performing implantations on diapausing pupae (Safranek and Williams, 1980). Only the brain was able to initiate wing epidermis apolysis, used as a bioassay.

Experiments conducted by Malá *et al.*, (1977) determined the effect of removal of the brain and of the nerve ganglia innervating the molting gland in *Galleria*. They reported that growth was disturbed only by removal of the brain. The implantation of different ganglia in ligated last-instar larvae was also carried out. In these experiments, the presence of a ligature allowed molting to subsist in a high percentage of insects, although delaying it

TABLE VIII. Spiracle Apolysis Percentage in *Manduca* Larvae
Injected with Homogenates from Different Parts
of the Central Nervous System[a]

Injected organ	Donor stage	No. of organs injected	Spiracle apolysis percentage
—			17
Brain	Fourth-instar larva	1	23
		2	67
		4	78
		6	86
	Fifth-instar larva	1	54
Subesophageal ganglion	Fifth-instar larva	1	13
	Pupa	1	0
		2	8
		5	0
Thoracic ganglia	Fifth-instar larva	1	0
	Pupa	4	20
		7	0
Abdominal ganglia	Pupa	2	20

[a]Modified from Gibbs and Riddiford (1977).

considerably. Hence the effects of the experiments must be considered essentially as pertaining to the time of molting. This is significantly shortened by brain implants, slightly shortened by subesophageal ganglion implants, and essentially the same with prothoracic and mesothoracic ganglia implants. Furthermore, the percentage of insects undergoing molting is equal to that of the controls when the brain and subesophageal ganglion are implanted, while it is very low with the prothoracic and mesothoracic ganglia, which definitely show inhibitory activity (Table IX).

Inhibitory action of the subesophageal ganglion was also suggested by Alexander (1970) in experiments on the same species, where immobilization of the larvae in narrow glass tubes delayed pupation but was restored within normal intervals by removing the subesophageal ganglion.

Thus instead of demonstrating the existence of brain hormone activity in the ventral nerve cord, the experiments conducted on *Manduca* and *Galleria* appear to indicate the existence of an antagonistic factor. It is not excluded that such a factor may mask the presence of brain hormone in some cases.

2.1.7. Antagonistic Brain Factor

The existence of an inhibitory factor in the brain has also been suggested in some cases. Investigating the regeneration of legs in the presence of hormones *in vitro* in the cockroach *Leucophaea maderae*, Marks *et al.* (1968)

TABLE IX. Pupation of Decapitated *Galleria* Larvae
Implanted with Different Organs[a]

Implanted organs	No. of organs implanted	Percentage of pupated insects	Days between implantation and cuticle secretion
—		84	42
Brain	1–2	80	25
	3	87	17
	6	91	15
Brain + corpora cardiaca + corpora allata	1	63	15
	3	94	11
	5–12	100	10
Subesophageal ganglion	3	80	26
Prothoracic ganglion	3	11	35
Mesothoracic ganglion	3	16	30
Prothoracic and mesothoracic ganglia	3	6	27
Subesophageal and prothoracic ganglia	3	63	35
Subesophageal, prothoracic, and mesothoracic ganglia	3	32	21

[a]Modified from Malá *et al.* (1977).

reported that the prothoracic glands stimulate regeneration, but that their own activity, enhanced by the presence of the corpora allata or prothoracic ganglion, is depressed by the presence of the brain. Carlisle and Ellis (1968), who injected extracts of different parts of the brain in *Locusta migratoria* and *Schistocerca gregaria*, obtained a stimulatory action deriving from the cells of the pars intercerebralis and an inhibitory action originating in the lateral cells. Furthermore, by removing the brain at the beginning or around the middle of the intermolt period, they observed diametrically opposed cytological changes in the molting gland and suggested the possibility of a precocious inhibitory control and a late stimulatory control.

2.1.8. Feedback

The ecdysone and juvenile hormone titer of the hemolymph appears to exert a feedback control on the neurosecretory cells that synthesize brain hormone and on the molting gland (Fig. 23).

2.1.8.1. Action of Ecdysone on Brain Hormone Secretion

Removal of the prothoracic gland and injection of β ecdysone into the larvae of *Aeshna* and of the cabbage armyworm *Mamestra* induces changes in the appearance of the neurosecretory cells (Schaller and Charlet, 1970; Agui and Hiruma, 1977a). These changes are also found in the brains of diapausing

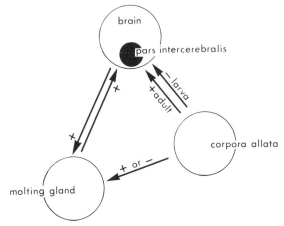

FIGURE 23. Reciprocal feedback exerted by the corpora allata, the molting gland, and the neurosecretory cells of the pars intercerebralis.

pupae incubated with β ecdysone (Agui and Hiruma, 1977b). They are accompanied by elevated brain activity detected in diapausing prothoracic glands incubated with pieces of integument. In *Leucophaea*, cultured brains accumulate neurosecretory products in the pars intercerebralis and release them upon the addition of β ecdysone (Marks *et al.*, 1972).

β ecdysone also stimulates the production of cAMP by the brain of *Manduca* (Vedeckis and Gilbert, 1973), a process that appears to be related to brain hormone secretion. By parabiosing 1-day fed *Rhodnius* and 8-day fed *Rhodnius*, in which the ecdysone titer is high, Steel (1978a) reported overloaded perikarya after 8 days in the former. This seems to indicate that the inhibition of the release processes of the brain neurosecretory products is due to ecdysone. Tritiated ecdysone injections show that this hormone exerts its action in the perikarya, where selective concentration occurs.

Furthermore, it appears that the molting gland can be stimulated by its own secretion, as reported by Williams (1952) and Siew and Gilbert (1971) in *Platysamia cecropia* and *Samia cynthia*, and by Kimura and Kobayashi (1975) in *Bombyx mori*.

2.1.8.2. Action of Juvenile Hormone on Brain Hormone Secretion

Negative feedback exerted by the corpora allata on the source cells of brain hormone has been demonstrated by experiments on diapausing larvae of several Lepidoptera (see Chapter 3) and on the last-instar larvae of *Manduca*, in which removal of the corpora allata induces premature release of brain hormone (Nijhout and Williams, 1974b). In the adult, however, a positive feedback mechanism was demonstrated. In *Schistocerca* and *Locusta*, removal of the corpora allata causes storage of type A neurosecretory products in the pars intercerebralis (Highnam *et al.*, 1963), and supernumerary corpora allata implants or the application of juvenile hormone clearly stimulates the synthesis and transport of neurosecretory products (McCaffery and Highnam, 1975a,b). Allatectomy also depresses the synthetic activity of the neurosecretory cells of the pars intercerebralis in the mealworm *Tenebrio molitor* (Mordue, 1965c), the Colorado beetle *Leptinotarsa decemlineata* (de Wilde and de Boer, 1969), and the blowfly *Calliphora erythrocephala* (Thomsen and Lea, 1968), while juvenile hormone injection or application of juvenile hormone analogues activates the neurosecretory cells in adult diapausing *Leptinotarsa* (Schooneveld, 1972; Schooneveld *et al.*, 1977).

Apart from its action on the neurosecretory cells of the brain, juvenile hormone seems to act directly on the molting gland by stimulating it. In fact, Williams (1959), Gilbert and Schneiderman (1959), and Ichikawa and Nishiitsutsuji-Uwo (1959) succeeded in terminating diapause and the second pupal molt in decerebrated pupae of the great saturniids by corpora allata implants. As we have seen above, lepidopteran corpora allata store brain neurosecretory products. Their effect upon molting comes, thus, from the brain hormone they contain and is quite normal. Several authors, however, obtained similar results by injecting juvenile hormone or juvenile hormone analogues (Krishnakumaran and Schneiderman, 1955; Bhaskaran *et al.*, 1980). It may well be, therefore, that the corpora allata exert a dual action upon the molting gland through brain hormone and also through juvenile hormone.

The molting gland, however, appears to respond to juvenile hormone only in certain periods (Hiruma *et al.*, 1978). The application of juvenile hormone analogues to *Mamestra* larvae or pupae ligated and implanted with molting glands induces molting only if the insects have attained a certain age, while brain implants are effective at all times. In the southern armyworm *Spodoptera littoralis,* juvenile hormone analogues applied to young and old last-instar larvae also appear to have a dual effect upon the molting gland, first a stimulatory one and later an inhibitory one (Cymborowski and Stolarz, 1979). The same results were obtained in *Manduca sexta* (Safranek *et al.*, 1980.)

2.1.9. Chemical Nature of Brain Hormone

The identification of brain hormone has spurred a good number of investigations in the past 20 years, which have succeeded in isolating active fractions generally considered proteins. However, the data presented by the various authors reveal certain discrepancies, and the chemical nature of brain hormone is still unknown (Table X).

TABLE X. Brain Hormone Purification

Reference	Brain hormone origin		Brain hormone assayed on	Characterization					Nature
	Insect	Organ		Water solubility	Heat stability	Destroyed by proteolytic enzymes	Fraction No.	Molecular weight	
Ichikawa and Ishizaki (1961, 1963); Ishizaki and Ichikawa (1967)	*Bombyx mori*	Brain	*Samia* decerebrated pupae	+	+	+	3	9,000 12,000 31,000	Protein
Nagasawa et al. (1979)	*Bombyx mori*	Head					1	4,400	Peptide
Yamazaki and Kobayashi (1969)	*Bombyx mori*	Brain	*Bombyx* decerebrated pupae		+	+	1	20,000	Protein
Nishiitsutsuji-Uwo (1972)	*Bombyx mori*	Head	*Samia* and *Bombyx* decerebrated pupae	+	+			5,000	Peptide
Williams (1967)	*Antheraea pernyi*	Brain	*Antheraea* decerebrated pupae	+	+	:			Mucopolysaccharide
Gibbs and Riddiford (1977)	*Manduca sexta*	Brain	*Manduca* ligated larvae	+	+	+			
Gersch et al. (1973)	*Periplaneta americana*	Brain	*Galleria* and *Periplaneta* molting gland	+			2	50,000 10,000–20,000	Protein

The first investigations of Kobayashi and Kirimura (1958) and Kobayashi *et al.* (1962), carried out with brains of the silkworm *Bombyx mori* pupae (8500 and then 220,000), succeeded in isolating 4 mg of a substance active on decerebrated *Bombyx* pupae, which was simply cholesterol. This unexpected result is still unexplained, but may derive from the fact that cholesterol is the precursor of ecdysone.

Ichikawa and Ishizaki (1961, 1963) obtained completely different results with aqueous extracts of brains of *Bombyx* pupae tested on decerebrated *Samia* pupae. The active factor is a water-soluble, heat-stable, nondialysable protein resistant to pepsin and trypsin but inactivated by pronase and nagarse. In subsequent work, Ishizaki and Ichikawa (1967) demonstrated the existence of three different proteins with molecular weights of 9000, 12,000 and 31,000.

Investigating the brain hormone of the giant silk moth *Antheraea pernyi*, Williams (1967) observed the same properties and found that of trypsin, chymotrypsin, pepsin, lysozyme, hyaluronidase, tyrosinase, ribonuclease, and pronase, only the latter resulted in inactivation of the hormone. By spectrophotometric analysis, no absorption was visible in the protein region. This author therefore questioned the protein nature of brain hormone and suggested that it could be a mucopolysaccharide.

Yamazaki and Kobayashi (1969) succeeded in purifying *Bombyx* brain hormone active at a dose of 0.02 μg. The substance is heat-stable and is inactivated by trypsin, pronase, and nagarse, but not by chymotrypsin or sialidase. Its molecular weight is about 20,000 and its isoelectric point is 8.35–8.65. It is different from the substance isolated by Ishizaki *et al.*, (1977) and could be a glycoprotein.

Nishiitsutsuji-Uwo (1972) also attempted purification using *Bombyx* heads, and stated that brain hormone was probably a peptide of molecular weight below 5000.

Suzuki *et al* (1975), also starting from *Bombyx* heads and using decerebrated *Samia* pupae as an assay, developed a simple purification procedure. The acetone powder of the heads was extracted with ethanol followed by sodium chloride. The active extract was heated and precipitated by ammonium sulfate followed by picric acid and acetone. If the minimum quantity of hormone required for imaginal development is expressed in *Samia* units, it may be observed that 1 *Samia* unit corresponds to 1 or 2 μg, and that each head contains 30–40 *Samia* units. Chromatography on different columns revealed the purified extract to contain two components.

In 1977, Ishizaki and co-workers subjected the purified hormone to different enzymes and chemical reagents. From the results obtained, they inferred that the NH_2 and COOH terminals are blocked, and that the tryptophan and disulfide bonds are indispensable to hormone activity, which is not the case for sulfhydryl groups, tyrosine, or carbohydrates.

Gibbs and Riddiford (1977) reported that *Manduca* brain hormone is soluble in water, heat-stable, and sensitive to pronase.

Gersch and co-workers purified the factors present in the brain, corpora cardiaca, corpora allata, and ventral nerve cord of the cockroach *Periplaneta* and found several physiological effects. The C and D factors were the first to be isolated (Gersch and Mothes, 1956), and from them four different substances were separated, the factors C_1, C_2, D_1, and D_2 (Gersch *et al.*, 1960). The presence of brain hormone in the D_1 fraction was demonstrated (Gersch, 1962) by injections into ligated abdomens of *Lymantria* and *Calliphora* implanted with Weismann's rings. Two distinct factors, AH I and AH II, inducing molting in decerebrated *Antheraea* pupae were separated from the D factor by Sephadex gel filtration using *Periplaneta* brains (Gersch and Stürzebecher, 1968). These factors exhibit different properties. AH I, with a molecular weight of about 50,000, stimulates RNA syntheses in the molting gland of *Periplaneta*, but does not exert this action on other tissues; AH II, with a molecular weight of 10,000–20,000, elevates the membrane potential of the molting gland of *Galleria*, and thus induces an increase in permeability (Gersch and Birkenbeil, 1973). Subsequent investigations showed that activation factor I enhances RNA syntheses *in vitro* (Gersch and Bräuer, 1974) and increases α ecdysone synthesis by the molting gland (Bräuer *et al.*, 1977).

Quite recently, Nagasawa *et al.*, (1979) reported a simple and highly reproducible procedure for purification of brain hormone. Starting with 96,000 adult *Bombyx* heads, a 28,500-fold purification was achieved (Fig. 24). Six nanograms of this purified preparation was shown to cause adult development in a brainless *Samia*. Gel filtration analyses showed the molecular weight to be 4400.

2.1.10. Stimulation of Ecdysone Production by the Ovary

Several investigations have shown that insect ovaries, particularly the follicle cells (Goltzené *et al.*, 1978), secrete ecdysone intermittently. It was of interest to determine whether this ecdysone secretion responded to the same control as the molting gland.

Ovarian ecdysone appears to have different secretion and action mechanisms in different species. In the locust, ovarian ecdysone remains in the ovary and enters into the oocytes when vitellogenesis is terminated. In the mosquito, on the other hand, it is secreted after a blood meal, enters the hemolymph, and stimulates vitellogenin synthesis by the fat body. In both cases, however, ovarian ecdysone secretion is controlled by a factor originating in the pars intercerebralis of the brain and which may be brain hormone.

In *Locusta* (Charlet *et al.*, 1979), cautery of the pars intercerebralis or removal of the corpora cardiaca suppresses the increase of ovarian ecdysteroids that normally occurs at the end of the oogenetic cycle. Further-

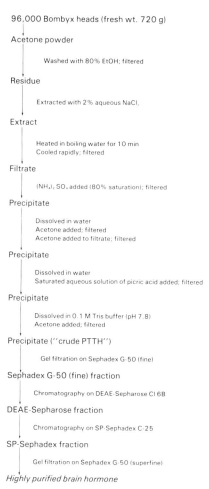

FIGURE 24. Procedure for partial purification of brain hormone (PTTH) from the head of *Bombyx mori*. (From Nagasawa *et al.*, 1979.)

more, the implantation of brain–corpora cardiaca complexes makes the ovarian steroid peak arrive sooner.

Working on the yellow fever mosquito *Aedes aegypti*, Hagedorn *et al.* (1979) incubated the ovary with different nerve ganglia, brain, or corpora cardiaca and determined the quantity of ecdysteroids synthesized. Whereas the ventral nerve cord ganglia were without effect, the brain proved to be highly active, as well as the corpora cardiaca, even when the donors were species as different as the cockroach *Periplaneta*, the butterfly *Papilio* and the housefly *Musca*. The presence of the active factor in the larva suggested that brain hormone might be involved.

2.1.11. Conclusions

Brain hormone, which controls molting gland activity, is the first neurohormone demonstrated in insects. It involves the action of cAMP and induces changes in the gland's permeability.

The location of the source cells of brain hormone has been investigated by several authors. The differing results obtained suggest that the location may vary according to species. In *Manduca,* the source cells have been precisely identified.

As for the occurrence of brain hormone in the ventral nerve cord, the investigations conducted on *Galleria* and *Manduca* do not support this hypothesis and suggest the possible existence of an inhibitory factor.

Brain hormone secretion is influenced by the juvenile hormone and ecdysone titer.

Various investigations have been conducted by different teams in an attempt to isolate brain hormone. It has not yet been accurately identified, but is believed to be a protein with a relatively high molecular weight.

2.2. CORPORA CARDIACA

Little is known about the mechanisms that control the activity of the glandular cells of the corpora cardiaca. However, it is probable that both brain neurosecretory products and biogenic amines are involved, as contacts have been observed in many species between glandular cells and neurosecretory endings, or aminergic endings containing small dense core granules (*Locusta,* Cassier and Fain-Maurel, 1970a; Cazal *et al.,* 1971; Rademakers and Beenakkers, 1977; *Calliphora,* Normann, 1965). The biogenic amine that they contain might be octopamine, which is very abundant in the glandular lobe (David and Lafon-Cazal, 1979). The neurons controlling the glandular cells were successfully localized in *Locusta* by the cobalt chloride method (Rademakers, 1977b) (Fig. 25). They form two clusters of 15 cells located laterally in the brain, whose axons transit through the nccII and innervate the glandular lobe. The lateral neurosecretory cells, on the contrary, innervate the neurohemal lobe of the corpora cardiaca and also the corpora allata.

The need for intact nerve connections with the brain for the functioning of the glandular cells of the corpora cardiaca was demonstrated in *Locusta* by severance of the nccI and II (Goldsworthy *et al.,* 1972a) and also by comparative ultrastructural study of the corpora cardiaca in periods of rest and flight and after transplantation (Rademakers, 1977a). Cutting of the nccI and II causes the disappearance of adipokinetic hormone from the hemolymph, for it cannot be released from the glandular lobe (see Chapter 9). After a flight of 60 min, the glandular cells of the corpora cardiaca display higher secretory

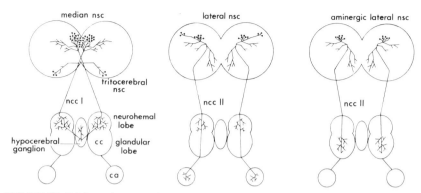

FIGURE 25. Origin and course of fibers leading to the neurohemal and glandular parts of the corpora cardiaca, to the corpora allata, and to the hypocerebral ganglion in the locusts *Schistocerca gregaria* and *Locusta migratoria.* nsc, Neurosecretory cells. (From histological data of Strong, 1966, and cobalt chloride filling data of Mason, 1973, and Rademakers, 1977b.)

activity, as can be judged by the state of the Golgi apparatus and the intensity of pinocytotic processes. These flight effects are nonexistent in transplanted corpora cardiaca (Rademakers, 1977a) and thus demonstrate the necessity of an intact innervation for their functioning.

2.3. CORPORA ALLATA

The corpora allata in larvae and in most adult females exhibit discontinuous activity that is related to specific phases of the insect's life cycle and to external environmental factors. During postembryonic development, the corpora allata display cyclic activity during each intermolt period. In the reproductive phase, the gonadotropic activity of the corpora allata varies during each vitellogenetic cycle, the variations being especially pronounced in viviparous species in which the corpora allata are inactive during gestation. Corpora allata activity is also affected by temperature, humidity, photoperiod, and diet, the functional dependence on these external factors occurring through the central nervous system, which receives and integrates the information.

2.3.1. Volume Changes and Activity of Corpora Allata

Early investigations on various species suggested that a correlation exists between corpora allata secretory activity and volume. In the adult female blood-sucking bug *Rhodnius prolixus* (Wigglesworth, 1936), the blood meal induces a considerable enlargement in the volume of the corpus allatum, resulting from an increase in size of the glandular cells, the cytoplasm of which becomes more abundant. At the same time, vitellogenesis is triggered. If the insect is then deprived of food, vitellogenesis is arrested and the size of the corpus allatum decreases significantly

In the phytophageous milkweed bug *Oncopeltus fasciatus* (Johansson, 1958), the corpus allatum also changes in size in relation to food intake and ovarian development. In young fasting females, the ovaries show little development and the corpus allatum is small, whereas in fed females, the ovaries display considerable growth and the corpus allatum increases sixfold.

In the viviparous cockroach *Leucophaea maderae*, the normal reproductive cycle includes a long gestation period during which vitellogenesis is interrupted. The volume of the corpora allata is large during vitellogenesis, diminishes considerably during gestation, and then again increases after parturition when vitellogenesis resumes (Engelmann, 1957; Scharrer and Von Harnack, 1958). The glandular cells of active corpora allata have abundant cytoplasm and well-developed cytoplasmic organelles (Golgi bodies, ergastoplasm, mitochondria). In inactive corpora allata, on the contrary, the cytoplasmic volume is small, the cell membranes are folded, and the Golgi bodies, ergastoplasm, and mitochondria show little development (Scharrer, 1964a, 1978).

A correlation between corpus allatum activity and volume has been reported in other insect orders (Isoptera, Orthoptera, Hymenoptera, Coleoptera, and Diptera). In some insects, however, this correlation does not seem to exist, suggesting that secretion and release processes are not necessarily interrelated (Mordue, 1965c).

2.3.2. Innervation of Corpora Allata

Because corpora allata activity is influenced by different factors, it is important to understand how information is transmitted to the corpora allata. The corpora allata are innervated by two pairs of nerves, the nca1, which exit from the corpora cardiaca, and the nca2, which originate in the subesophageal ganglion. While the starting point of the nca2 is perfectly clear, that of the nca1 is relatively obscure because they exit from the corpora cardiaca. It is well known that these nerves originate in the brain and generally contain nerve axons and neurosecretory axons, but in most cases it is unknown whether the neurosecretory axons originate in the nccI arising from the pars intercerebralis, in the nccII originating in the lateral neurosecretory cells of the protocerebrum (pars lateralis), or in the nccIII originating in the tritocerebrum.

The cobalt chloride technique affords an excellent tool for determining the source of the axons that innervate an organ; however, the evidence at the present time concerns only a small number of species. In *Schistocerca*, Mason

(1973) reported that the axons innervating the corpora allata originate only in the nccII and nccIII. In *Locusta*, however, nerve sections (Strong, 1965b), electron microscopy, and cobalt chloride filling show that they originate in both the nccI and the nccII (Cassier and Fain-Maurel, 1970b; Rademakers, 1977b). In the cockroach *Periplaneta* (Fraser and Pipa, 1977), the source cells of the ncal are found in both the pars intercerebralis and the lateral parts of the protocerebrum. Each corpus allatum is innervated by 50–60 cells in the contralateral pars intercerebralis, 8–12 cells in the contralateral pars lateralis, and 16–20 cells in the ipsilateral pars lateralis. In the larva and pupa of *Manduca sexta* (Nijhout, 1975), the corpora allata receive axons originating in the pars intercerebralis and in the neurosecretory cells of the pars lateralis. It appears, therefore, that the corpora allata are often innervated by the pars intercerebralis and pars lateralis (Fig. 26).

In those investigations dealing with the nca2, the source cells have been shown to be the medioventral neurosecretory cells of the subesophageal ganglion in *Periplaneta* (Pipa and Novak, 1979), *Locusta* (Chalaye, 1976), and *Schistocerca* (Mason, 1973).

2.3.3. Early Investigations

To understand the precise nature of corpora allata control, many studies have involved the removal of the pars intercerebralis, and the severance of the ncal, nccI, nccII, and nca2 (Tables XI and XII).

Severance of the ncal in the cockroaches *Leucophaea* and *Diploptera* and in *Locusta* has yielded the same results as severance of the nccI, signifying that the axons contained in the nccII play no part in control of the corpora allata. In *Diploptera*, severance of the nccII substantiates this hypothesis, because it has no effect on vitellogenesis. In *Schistocerca*, however, Strong (1965b), supported by Cassier (1970), reached a different conclusion: that severance of the nccII had the same effect as that of the ncal.

Severance of the ncal sometimes results in a volume increase in the corpora allata and in their physiological activation, which can be detected by juvenilizing or gonadotropic effects (cockroaches, bugs). These effects point to the inhibiting function of the brain. In other cases, on the contrary, this operation arrests the activity of the corpora allata, as in *Locusta* and *Schistocerca,* the earwig *Anisolabis maritima,* and the mealworm *Tenebrio molitor*. One is thus led to conclude that in these species, the brain exerts a stimulating rather than an inhibiting control on the corpora allata. Occasionally, severance of the ncal appears to have no effect on the corpora allata, as in the case of the Colorado beetle *Leptinotarsa decemlineata* and the tobacco hornworm *Manduca sexta*.

The role of the nerves originating in the subesophaegeal ganglion (nca2) has rarely been investigated (Table XII). Some authors have reported that the

Schistocerca gregaria

Locusta migratoria

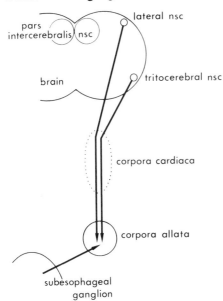

Periplaneta americana

Manduca sexta

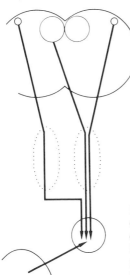

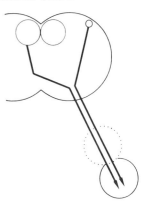

FIGURE 26. Corpus allatum innervation as revealed by cobalt chloride filling in four insects. Apart from their subesophageal innervation, the corpora allata are innervated by lateral and tritocerebral neurons in *Schistocerca gregaria* (Mason, 1973), by pars intercerebralis and lateral neurons in *Locusta migratoria* (Cassier and Fain-Maurel, 1970b; Rademakers, 1977b), by pars intercerebralis, lateral, ipsilateral, and contralateral neurons in *Periplaneta americana* (Fraser and Pipa, 1977; Gundel and Penzlin, 1980), and by pars intercerebralis and lateral neurons in *Manduca sexta* (Nijhout, 1975). nsc, Neurosecretory cells.

TABLE XI. Corpora Allata Disconnection: Brain Nerve Section

Insect	Reference	Level	Results	nccI ncaI role[a]
Leucophaea maderae	Scharrer (1952)	nccI	Corpora allata volume increase Juvenilizing activity increase Vitellogenesis stimulation	−
	Engelmann (1957)	nca1 or nccI	Vitellogenesis stimulation	−
Periplaneta americana	Fraser and Pipa (1977)	nca1	Juvenilizing activity decrease	−
Diploptera punctata	Engelmann (1959)	nca1 or nccI	Larval corpora allata activation	−
		nccII	No effect	0
	Stay and Tobe (1977)	nca1 + nca2	Corpora allata synthetic activity increase Vitellogenesis stimulation	−
Locusta migratoria	Staal (1961)	nca1	Corpora allata volume decrease Juvenilizing activity decrease Green pigmentation decrease	+
	Girardie (1966b)	nca1 or nccI	Vitellogenesis interruption	+
	Pener (1965)		Male sexual behavior and yellow pigmentation decrease	+
	Cassier (1970)	nccI	Egg-laying stimulation	−
		nccII	Vitellogenesis inhibition	+
	Johnson and Hill (1973)	nca1	Corpora allata growth arrested (weak synthetic activity)	+
Schistocerca gregaria	Strong (1965b)	nca1 or nccII	Corpora allata volume decrease Vitellogenesis interruption	+
	Tobe *et al.* (1977)	nca1	Corpora allata synthetic activity decrease	+
Tetrix undulata	Poras (1977a)	nca1	Vitellogenesis stimulation in diapausing females	−
Anisolabis maritima	Ozeki (1962)	nccI + nccII	Juvenilizing activity decrease	+
Tenebrio molitor	Mordue (1965c)	nccI	Vitellogenesis inhibition decrease	+
Leptinotarsa decemlineata	de Wilde and de Boer (1969)	nca (nca1 + nca2)	Ovarian diapause maintenance	0
Manduca sexta	Sroka and Gilbert (1971)	nccI + nccII	Vitellogenesis occurrence	0
Dindymus versicolor	Friedel (1974)	nca1	Corpora allata volume increase Vitellogenesis stimulation	−

(*Continued*)

TABLE XI. (Continued)

Insect	Reference	Level	Results	nccl ncal role[a]
Oncopeltus fasciatus	Johansson (1958)	nca1	Corpora allata volume increase Vitellogenesis stimulation	−
Rhodnius prolixus	Baehr (1973)	nca1	Vitellogenesis stimulation	−
Pyrrhocoris apterus	Hodková (1976)	nca1	Corpora allata volume decrease Vitellogenesis stimulation	−
Eurygaster integriceps	Panov (1977)	nca1	Corpora allata volume increase Vitellogenesis stimulation	−

[a] 0, No effect; +, stimulatory role; −, inhibitory role.

nca2 take part in controlling corpora allata activity, and that they have a stimulatory role, as in *Leucophaea*, or an inhibitory role, as in *Leptinotarsa*. It is, however, possible that the role of the nca2 is less important than that of the nca1, because their severance has no effect on the activity of the corpora allata in *Schistocerca* and *Locusta*. In *Diploptera* and *Periplaneta*, severance of both the nca1 and the nca2 has the same effect as severance of the nca1 alone, thus implying that the nca2 are not of fundamental importance.

Neurosecretory material has been observed histologically in the corpora allata and in their nerves in species belonging to the major orders. Electron microscopy studies have also been carried out, particularly in *Leucophaea* (Scharrer, 1964a) and the blowfly *Calliphora erythrocephala* (Normann, 1965;

TABLE XII. Corpora Allata Disconnection: Subesophageal Nerve Section (nca2)

Insect	Reference	Effector	nca2 role[a]
Leucophaea maderae	Engelmann (1957)	Vitellogenesis	+
Periplaneta americana	Fraser and Pipa (1977)	Supernumerary ecdysis	0
Locusta migratoria	Staal (1961)	Vitellogenesis	0
	Cassier (1966)	Vitellogenesis	0
	Chalaye (1975)	Vitellogenesis	0
Schistocerca gregaria	Strong (1965b)	Vitellogenesis Corpora allata volume	0
	Pener (1965, 1967)	Sexual behavior Male pigmentation	0
Leptinotarsa decemlineata	de Wilde and de Boer (1969)	Ovarian dispause	0

[a] 0, No effect; +, stimulatory role.

Thomsen and Thomsen, 1970) (Table VII). The results of these investigations indicated that two and sometimes three types of granules can be found in the neurosecretory material. In some cases, a correlation between the neurosecretory content of the corpora allata and their level of activity has been reported: in *Tenebrio* (Mordue, 1965b), neurosecretion is abundant in the brain and corpora allata of virgin females, but absent in mated and egg-laying females; in *Periplaneta* (Khan *et al.*, 1978a), the amount of neurosecretory material contained in the corpora allata decreases when these organs gain in size, both in the larva and in the adult.

These observations indicate that one or more neurohormones may intervene in the control of corpus allatum activity. Let us now examine the investigations that deal with the cauterization of the pars intercerebralis.

Like severance of the ncal, cauterization of the pars intercerebralis can cause stimulation of the corpora allata, although it generally inhibits the activity of these organs. This operation has been carried out chiefly on the adult female, whose responses were studied in relation to vitellogenesis. As we shall see later (Chapter 4), in some species the pars intercerebralis exhibits gonadotropic action independent of the corpora allata, and thus complicates the interpretation of the results observed after its cauterization. In fact, while cauterization that induces the stimulation of vitellogenesis indicates the presence of a factor inhibiting the corpora allata in the neurosecretory cells and the absence of an independent gonadotropic factor, on the other hand, cauterization that inhibits vitellogenesis cannot be directly interpreted, as this may be due to the suppression of a gonadotropic factor or to that of an allatotropic factor. The precise determination of what is occurring would require additional experiments involving the implantation of corpora allata into parsectomized insects. Two results may be obtained from such experiments: either the maintenance of vitellogenesis inhibition, which indicates that the response is caused by cerebral gonadotropic action, or a restoration of vitellogenesis, which indicates an allatotropic function of the pars intercerebralis.

The allatotropic function of the pars intercerebralis has been demonstrated in this way in the following species: the cockroach *Nauphoeta cinerea* (Barth and Sroka, 1975), the Colorado beetle *Leptinotarsa decemlineata* (de Wilde and de Boer, 1969), the tobacco hornworm *Manduca sexta* (Sroka and Gilbert, 1971), the flesh fly *Sarcophaga bullata* (Pappas and Fraenkel, 1978), and the blood-sucking bug *Rhodnius prolixus* (Baehr, 1973). The results obtained by cauterization of the neurosecretory cells of the pars intercerebralis of *Nauphoeta* and *Rhodnius* led investigators to a conclusion different from that drawn from ncal severance experiments. More extensive research is definitely needed in order to clarify the problem. Some such investigations have already been begun.

2.3.4. Detailed Investigations

2.3.4.1 Cockroaches

The nature and source of the inhibition transiting through the corpora allata nerves in cockroaches were investigated in detail by Engelmann and Lüscher (1957) in the viviparous cockroach *Leucophaea maderae*. Meticulous anatomical studies showed that, in addition to the neurosecretory axons issuing from the neurosecretory cells of the pars intercerebralis, the nccI contain axons arising from neurons located in the vicinity of the neurosecretory cells. While the bundles of neurosecretory axons originating in the right and left neurosecretory cells intersect in the posterior part of the brain, the axons originating in the nonneurosecretory neurons leave the brain through the corresponding ipsilateral nerves. Selective electrocoagulations of the neurosecretory cells of the pars intercerebralis and of the contiguous nerve cells yield opposite results: activation fails to occur in the former, while it occurs in the latter. Experiments involving unilateral severance of the nerve connections also show that the volume of the corpus allatum only increases on the operated side. This indicates that the inhibition exerted by the brain on the corpora allata is nervous and not neurosecretory, and that it does not originate in the neurosecretory cells themselves, but in the adjacent neurons.

Engelmann's investigations were resumed by Pratt and Tobe (1974) who developed an *in vitro* short-term radiochemical assay to determine the synthetic activity of the corpora allata. In their technique, the corpora allata were incubated for 3 hr with [methyl-^{14}C]methionine in order to synthesize labeled JH III. Radioactivity determination of the incubation medium by liquid scintillation after thin-layer chromatography was used to directly obtain the activity level of the corpora allata in different phases of the reproductive cycle and in various experimental conditions. Investigations conducted on the viviparous cockroach *Diploptera* (Stay and Tobe, 1977, 1978; Stay *et al.*, 1980; Tobe and Stay, 1980) showed that disconnection of the corpora allata activates them in the virgin female, which is in agreement with the foregoing results (Fig. 27), but it also appears that after complete denervation, the corpora allata of mated females retain a cyclic activity comparable to that of control insects, thus indicating that their functioning is also humorally regulated. Hence, in addition to inhibitory nervous control, the corpora allata are subject to a stimulatory humoral control, which may derive from the neurosecretory cells of the pars intercerebralis as was seen in the cauterization experiments of the pars intercerebralis that were followed by corpora allata implants or by ecdysterone produced by the ovary (Stay *et al.*, 1980).

Other studies, carried out by Fraser and Pipa (1977) on *Periplaneta*, confirmed the inhibitory role of the nca1 in the larva, in which their severance induces supernumerary molts.

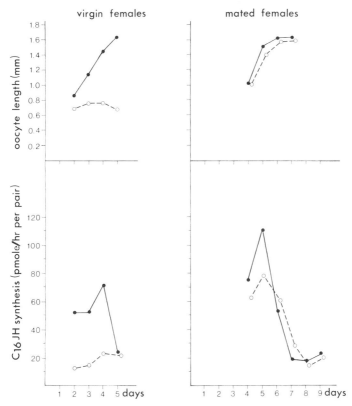

FIGURE 27. Effect of denervating corpora allata in virgin and mated *Diploptera* females on juvenile hormone synthesis rates and oocyte lengths. (——) Denervated; (------) sham operated. (Modified from Stay and Tobe, 1977.)

2.3.4.2. Locusts

We have seen above that severance of the ncal does not have the same effects in locusts as in cockroaches. In a histological and experimental investigation carried out on *Locusta migratoria*, Girardie (1966b, 1967) thoroughly studied the mechanisms of cerebral control of corpora allata activity in these insects.

The histological and ultrastructural study revealed three distinct types of neurosecretory cells in the pars intercerebralis: the numerous, small type A and B cells located in the central region of the pars intercerebralis, and the larger type Cj cells located around the type A and B cells.

Electrocoagulation performed with 15- to 40-μm electrodes allowed selective removal of the A–B cells and of the Cj cells. These two experiments produced contrasting effects. Removal of the A–B cells is equivalent to an implantation of corpora allata and causes supernumerary molts, the formation of green pigment in gregarious insects, and the stimulation of vitellogenesis. Removal of the Cj cells, on the contrary, has the same effect as removal of the corpora allata. It induces the appearance of diminutive adults, depigmentation, and the suppression of vitellogenesis, sexual behavior, and the yellow coloration of mature males.

Hence, the corpora allata are under a twofold cerebral control: inhibiting control exerted by the A–B cells, and stimulating control exerted by the Cj cells.

The mode of action may be different in both cases. According to Joly (1970), the inhibitory activity of the A–B cells is not exerted on the corpora allata but on the juvenile hormone itself. The secretion of the A–B cells appears to inactivate the hemolymph juvenile hormone by increasing the titer of specific esterases of this hormone (Retnakaran and Joly, 1976). Comparative ultrastructural examination of the corpora allata after removal of the Cj and of the A–B cells substantiates the hypothesis of indirect action of the latter on the corpora allata, which display a normal appearance in insects deprived of their A–B cells (Joly et al., 1969).

In experimental conditions, the action of the A–B cells is exerted humorally as was observed for the Cj cells, but it remains to be determined whether this would also be the case in natural conditions. With respect to the Cj cells, experiments show that the stimulation that they exert on the corpora allata can only occur if their nervous connections with the corpora cardiaca remain intact (Cazal et al., 1971).

Do the mechanisms revealed in Locusta exist in other locusts? Electrical stimulation of the pars intercerebralis of Anacridium aegyptium reactivates the disconnected corpora allata and terminates diapause, thus confirming the stimulatory and humoral function of the pars intercerebralis (Girardie et al., 1974).

In Schistocerca, the situation seems to be slightly different. We have already seen that severance of the ncal produces the same results as severance of the nccII. This suggests that the lateral protocerebral neurosecretory cells play a part in the regulation of corpora allata activity. According to Strong (1965b), these cells are allatotropic and their action is exerted by the neurosecretory material transported from the brain toward the corpora allata, via the nccII. Mason's investigation using cobalt chloride supports this view since in Schistocerca, the corpora allata receive axons originating in the nccII. The lateral neurosecretory cells also play a part in Locusta, where they appear to possess both the activity of the A and B cells and that of the Cj cells (Girardie, 1974).

2.3.4.3. *Pyrrhocoris apterus* and *Rhodnius prolixus*

In the linden bug *Pyrrhocoris apterus* (Hodková, 1976, 1977, 1979), reproductive activity is dependent on the photoperiod, an ovarian diapause occurring in short-day females. Reproductive diapause is broken by severance of the ncal and also by implantation of the corpus allatum, but implantation of the brain–corpora cardiaca–corpus allatum complex is far less active (Fig. 28).

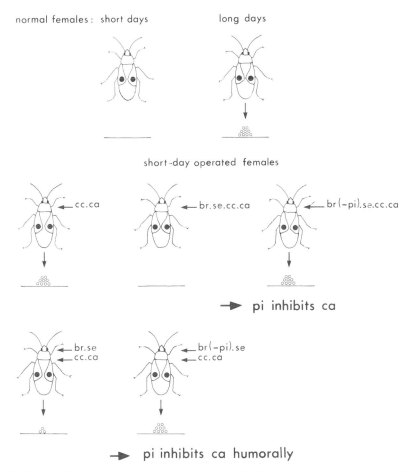

FIGURE 28. Experiments performed on the bug *Pyrrhocoris apterus* to elucidate the regulation of the corpora allata. Vitellogenesis and egg laying, suppressed in short-day insects, are restored in long-day insects. br, Brain; ca, corpora allata; cc, corpora cardiaca; pi, pars intercerebralis; se, subesophageal ganglion. ←, Implanted organs; ., associated organs; −, removed organs. (From data of Hodková, 1976, 1977, 1979.)

Thus the brain exerts an inhibitory action on the corpus allatum, comparable to that demonstrated in cockroaches. Moreover, the implantation of the corpora cardiaca–corpus allatum complex into females deprived of their own corpus allatum is far more effective in long-day females than in short-day females, as though the former were capable of stimulating the implant. The hypothesis of humoral stimulation of the corpus allatum may also be inferred from the nca severance experiments performed on long-day females, which failed to cause any decrease in fecundity.

A stimulatory and an inhibitory control exerted by the brain on the corpus allatum were also demonstrated in *Rhodnius* (Baehr, 1973, 1976), in which cauterization of the pars intercerebralis causes the appearance of adultoids when performed during the larval stage, and interrupts vitellogenesis in the adult. As shown above, these two effects disappear when the corpus allatum is implanted, and thus demonstrate the allatotropic action of the pars intercerebralis. This action, however, is a very early one, disappearing within 24 hours after the blood meal. Furthermore, severance of the ncaI stimulates vitellogenesis in fasting females, and thus confirms the inhibitory function of the brain on the corpus allatum).

2.3.4.4. *Leptinotarsa decemlineata*

The Colorado beetle *Leptinotarsa decemlineata*, raised in short-day photoperiod, enters an ovarian diapause that cannot be terminated by disconnecting the corpora allata (severance of the nca containing all the axons of cerebral and subesophageal origin in this species). Therefore, the corpora allata do not appear to be under nervous inhibitory control in this insect. A stimulatory hormonal control may exist, however, for the implantation of corpora allata in allatectomized females results in a stimulation or an inhibition of vitellogenesis, depending on the photoperiod conditions to which the insect is subjected (de Wilde and de Boer, 1969).

2.3.4.5. *Anisolabis maritima*

To estimate corpora allata activity in the dermapteran *Anisolabis*, several operations were performed in pupae and adults (removal of the brain, the subesophageal ganglion, and the prothoracic ganglion) followed by the implantation of their corpora allata into allatectomized pupae in order to estimate their activity from their ability to induce supernumerary molts. It appears that the activity of the corpora allata of decerebrate insects is greater than that of intact insects (Ozeki, 1979).

2.3.4.6. *Galleria mellonella* and *Manduca sexta*

The experiments of Granger and Sehnal (1974) and Sehnal and Granger (1975) on *Galleria mellonella* also reveal the existence of a twofold cerebral control on the corpora allata: neurohemal stimulatory control and nervous inhibitory control. The experiments were conducted on last-instar larvae. A brain implant during the first half of the larval instar induces supernumerary larval molts, indicating humoral activation of the corpora allata, for the juvenilizing effect of the brain disappears if the insects are allatectomized. Allatotropic activity seems to be located in the pars intercerebralis, because brains deprived of this region are no longer active. In the second half of the last larval instar, the corpora allata become insensitive to cerebral action and do not react to brain implants, but this insensitivity is removed by severance of their nerves. Hence it appears that, apart from its humoral activating function, the brain has an inhibitory nervous function.

In *Manduca*, the experiments performed demonstrated also that a transplanted brain may stimulate the corpora allata to induce supernumerary molts (Bhaskaran and Jones, 1980). Thus its stimulatory function is confirmed in this species.

2.3.5. Juvenile Hormone Titer and Cerebral Control

The juvenile hormone titer in the hemolymph affects the activity of the corpora allata. The implantation of supernumerary corpora allata reduces the activity of the host's corpora allata, which become smaller in *Carausius morosus* (Pflugfelder, 1939), *Drosophila melanogaster* (Bodenstein, 1947), *Oncopeltus fasciatus* (Johansson 1958), *Pyrrhocoris apterus* (Novák and Rohdendorf, 1959), *Locusta migratoria* (Staal, 1961), and *Anisolabis maritima* (Ozeki, 1965). The same result is obtained by the injection of juvenile hormone (*Leptinotarsa decemlineata,* Schooneveld *et al.,* 1977, 1979) or by the injection of juvenile hormone-rich extracts (*Periplaneta americana,* Nayar, 1962). According to Siew and Gilbert (1971), however, the reverse effect is observed in the eri silkworm *Philosamia cynthia* and the giant silk moth *Hyalophora cecropia.*

Unilateral removal of the corpus allatum causes hypertrophy of the remaining organ and an increase in its synthetic activity in the cockroach *Leucophaea* (Engelmann and Lüscher, 1957). In *Diploptera* (Stay and Tobe, 1978) and *Leptinotarsa* (Schooneveld *et al.,* 1979), the remaining corpus allatum doubles the amount of hormone synthesized.

One may enquire whether this is due to a direct response of the corpus allatum or to a cerebral response. By unilaterally removing one corpus allatum, Cassier showed that a cerebral response is involved, because com-

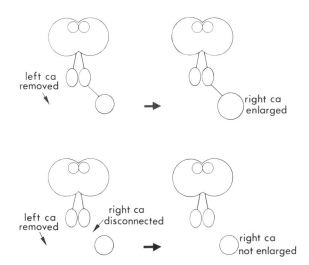

FIGURE 29. Corpus allatum activity controlled by the brain *in Locusta*. (Top) Right corpus allatum compensatory volume increase after removal of the left corpus allatum. (Bottom) Right corpus allatum increase suppressed by its disconnection. (From data of Cassier, 1966.)

pensating hypertrophy is observed only when innervation is intact (*Locusta*, Cassier, 1966) (Fig. 29). Positive feedback exerted by the corpora allata on the neurosecretory cells of the pars intercerebralis has been demonstrated in many species, but it remains to be determined whether this feedback is exerted on the cells that regulate the functioning of the corpora allata.

The corpora allata may directly react to certain factors. According to Engelmann (1965, 1968), the action of nutrient is exerted directly on the corpora allata of the cockroach *Leucophaea*, for the disconnected corpora allata respond by volume variations to hemolymph protein variations. Ecdysone may also have an effect on the activity rate of the corpora allata. Its injection enhances nuclear RNA synthesis in pupae of *Philosamia* and *Hyalophora* (Siew and Gilbert, 1971); in *Locusta*, it causes an increase in the juvenile hormone titer of the hemolymph (Joly and Joly, 1975), and in *Diploptera*, it causes a decrease in corpora allata activity (Stay *et al.*, 1980).

2.3.6. Regulation of Juvenile Hormone Esterases

The hemolymph juvenile hormone titer is controlled by the presence in the hemolymph of juvenile hormone-specific esterases synthesized in the fat body (see reviews in Gilbert, 1976), and also in the epidermis, but in amounts that differ according to the insect stage (*Manduca*, Mitsui *et al.*, 1979). It appears from the literature (Whitmore *et al.*, 1972; Kramer, 1978; Reddy *et al.*, 1979) that juvenile hormone-specific esterase activity might be correlated with juvenile hormone titer, but neurohormones also seem to be involved. The synthesis of juvenile hormone esterases is stimulated in *Locusta* by the A–B cells of the pars intercerebralis (Retnakaran and Joly, 1976) and recent studies performed in *Galleria* larvae show similar results, the juvenile hormone esterases appearing in ligated abdomens following brain or subesophageal implantation (McCaleb and Kumaran, 1980).

2.3.7. Conclusions

Corpora allata activity is under a twofold cerebral control, inhibitory and stimulatory.

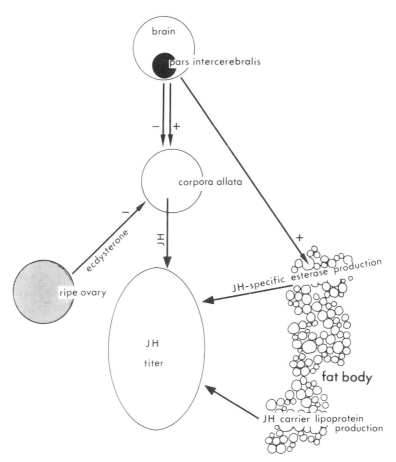

FIGURE 30. Schematic representation of juvenile hormone titer regulation. JH, Juvenile hormone.

The presence of neurosecretory material in the corpora allata and in its nerves raises the question of whether these organs are under nervous or neurosecretory control. In many cases, the effects engendered by disconnection of the corpora allata suggest the existence of nervous control, but modifications in corpora allata activity are also produced by neurosecretory cell implants. These and other experiments demonstrate the existence of stimulatory humoral control in many species.

The secretory material of the cerebral neurosecretory cells may exert its action humorally, but the normal mechanism appears to involve transport to the effector by a nerve pathway.

The precise locations of the inhibitory and stimulatory factors remain to be determined. The pars intercerebralis, the pars lateralis, and the neurosecretory cells of the subesophageal ganglion appear to be involved.

Neurohormones also play a part in regulating juvenile hormone-specific esterase production by the fat body.

The hemolymph juvenile hormone concentration thus depends on several distinct factors (Fig. 30).

3

Diapause

The term *diapause* covers processes that can occur at different developmental stages. Depending on the insect species, the adult, pupa, larva, and embryo may enter diapause, but these diapauses may be of varying intensity, obligatory or facultative.

Diapause is usually preceded by an accumulation of reserve materials, which gives rise to a hypertrophy of the fat body. It is characterized by the arrest of development or reproduction and by a decrease in water content, respiratory activity, etc. The brain becomes electrically inactive in some cases, its cholinesterase rate drops sharply, and the muscular system may degenerate.

Diapause is triggered by changes in environmental conditions (temperature, photoperiod, humidity, and diet) whose effects are exerted for periods of variable length, sometimes from one generation to the next. It usually occurs in late summer in temperate zones, and during the dry season in the tropics. The photoperiod is perceived either by the ocelli or directly by the brain, which is still capable of responding even when transplanted into the abdomen (e.g., giant silk moth *Antheraea pernyi*, Williams and Adkisson, 1964). In the cabbage white butterfly *Pieris brassicae*, the clypeus, which covers the brain, becomes much thinner during the period in which photoperiodic induction takes place (Claret, 1966). Light can also be perceived through the integument by means of epidermal pterobilin (*Pieris*, Vuillaume *et al.*, 1971).

3.1. IMAGINAL DIAPAUSE

Imaginal diapause is a reproductive arrest and, more specifically, the arrest of vitellogenesis. Its endocrine determinism has been demonstrated in many cases, of which the best known is that of the Colorado beetle *Leptinotarsa decemlineata*, investigated by de Wilde since 1954. Diapause in this insect is initiated by short days, light stimuli being transmitted to the brain, which regulates corpora allata activity. Diapause is preceded by a stage in which feeding increases and reproduction ceases. Significant metabolic changes take place in the proteins and glucides, which accumulate in the fat body (de Loof and Lagasse, 1970). On termination of this stage, the insect burrows into the ground for a period of several months, during which it remains motionless, its muscles degenerating (Fig. 31).

Imaginal diapause stems from juvenile hormone failure (de Wilde and de Boer, 1969; de Wilde *et al.*, 1968) and is terminated in both sexes by active corpora allata implants (in the grasshopper *Oedipoda miniata*, Broza and Pener, 1969, and in the grouse locust *Tetrix undulata*, Poras, 1975) and by treatment with juvenile hormone or its analogues (Bowers and Blickenstaff, 1966, on the alfalfa weevil *Hypera postica*; Schooneveld, 1972, on the Colorado beetle *Leptinotarsa decemlineata*; Burov *et al.*, 1972, on the bug *Eurygaster integriceps*).

The inactivity of the corpora allata may be due to cerebral inhibition (see Chapter 2). Various investigators, especially Siew (1965), Geldiay (1967), and Schooneveld (1969), report an overloading of the neurosecretory cells in diapausing insects. In addition, cauterization of the pars intercerebralis causes different aspects of the diapause syndrome to appear (de Wilde and de Boer, 1969), whereas electrical stimulation of the pars intercerebralis of the Egyptian grasshopper *Anacridium aegyptium* (Girardie *et al.*, 1974) causes an in- in the activity of the corpora allata and terminates diapause. However, the pars intercerebralis probably also plays a direct role in breaking diapause, both in the adult and in the larva and pupa.

3.2. PUPAL DIAPAUSE

Diapause in the pupa results from an ecdysone deficiency due to the lack of secretion of brain hormone. It is terminated by cerebral reactivation, as demonstrated on the giant saturniids *Platysamia* and *Antheraea* (Williams, 1946, 1947, 1952). Diapause in these species begins at the end of summer and lasts until spring, being interrupted by the warmth that follows the winter cold. In experimental conditions, the same result can be obtained by keeping the insect at low temperature for a short period, followed by a return to moderate temperature. Implantation of the brains of reactivated pupae into diapausing pupae induces the termination of diapause, as also occurs with the implantation of a molting gland or the injection of ecdysterone. Regulation occurring in this case is identical to growth regulation (see Chapter 2). Sometimes, however, the secretion of brain hormone proves inadequate to activate the molting gland. In the corn earworm *Heliothis zea*, brain hormone is released within 24 hr after molting and can only cause termination of diapause if the temperature is raised. Hence, activation of the molting gland appears to

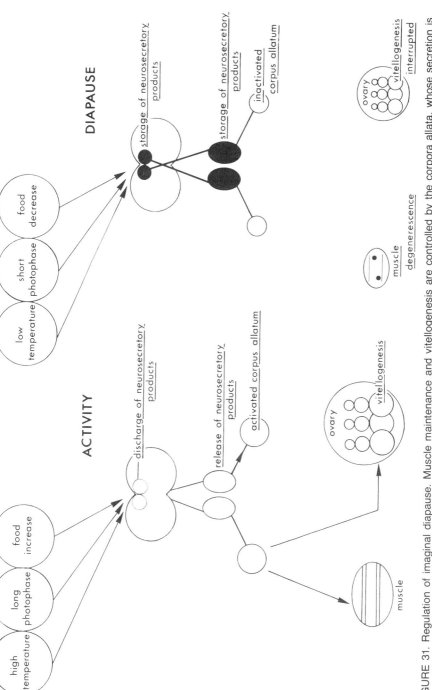

ACTIVITY

high temperature · long photophase · food increase

discharge of neurosecretory products

release of neurosecretory products

activated corpus allatum

ovary

vitellogenesis

muscle

DIAPAUSE

low temperature · short photophase · food decrease

storage of neurosecretory products

storage of neurosecretory products

inactivated corpus allatum

ovary

vitellogenesis interrupted

muscle degenerescence

FIGURE 31. Regulation of imaginal diapause. Muscle maintenance and vitellogenesis are controlled by the corpora allata, whose secretion is regulated by the brain. Environmental factors act upon this organ, which stimulates or inhibits the corpora allata according to the information received.

include two stages in this species: conditioning by brain hormone, and final activation by heat (Meola and Adkisson, 1977). Many lepidopterans (*Bombyx, Antheraea, Pieris, Manduca*) are capable of escaping diapause despite their decerebration (Kobayashi *et al.*, 1960; McDaniel and Berry, 1967; Maslennikova, 1970; Truman, 1972) and even if their corpora cardiaca are also removed (Judy, 1972). This derives from the fact that brain hormone is stored in the corpora allata in these species (as seen above). Moreover, juvenile hormone seems to exert a positive feedback upon the molting gland.

3.3. LARVAL DIAPAUSE

Instead of terminating diapause, as it sometimes does, juvenile hormone can exert the opposite effect and initiate diapause. This has been observed in larvae of many lepidopterans: the rice stem borer *Chilo suppressalis* (Fukaya and Mitsuhashi, 1957), the Indian meal moth *Plodia interpunctella* (Waku, 1960), the sugar cane borer *Diatraea grandiosella* (Chippendale and Yin, 1973; Yin and Chippendale, 1973), the European corn borer *Ostrinia nubilalis* (Yagi and Akaike, 1976), the burnet moth *Monema flavescens* (Takeda, 1978), and the dermestid beetle *Trogoderma granarium* (Nair, 1974).

The first investigation, conducted on *Chilo*, showed that the brain and the prothoracic gland definitely caused termination of diapause, but this resulted from the absence of the corpora allata (Fig. 32) (Fukaya and Mitsuhashi, 1957). The presence of juvenile hormone was also demonstrated in the diapausing larva.

Later investigations confirmed the role of the corpora allata in the initiation and maintenance of larval diapause. Diapause was triggered by the application of juvenile hormone analogues in *Diatraea* (Yin and Chippendale, 1973) and *Chilo* (Yagi and Fukaya, 1974), and its termination was caused by removal of the corpora allata in *Chilo, Ostrinia, Monema,* and *Diatraea* (Fukaya and Mitsuhashi, 1961; Yagi and Fukaya, 1974; Yagi and Akaike, 1976; Takeda, 1978; Yin and Chippendale, 1979).

Evaluation of the juvenile hormone titer in diapausing larvae confirms the existence of a high titer in comparison with developing insects (Yin and Chippendale, 1976, on *Diatraea*; Sieber and Benz, 1977, on the codling moth *Laspeyresia pomonella*), especially those in early (Yagi and Fukaya, 1974, on *Chilo*) and mid-diapause (Takeda, 1978, on *Monema*). Hormone determination by the *Galleria* test in nondiapausing and diapausing *Diatraea* larvae, before and during diapause, revealed interesting changes. Before and immediately after the onset of diapause, the juvenile hormone titer is high (4300 *Galleria* units/ml, 1230–2610 *Galleria* units/ml). It is maintained at 700–1500 *Galleria* units/ml up to day 110, and then starts to decline: 200 *Galleria* units/ml on day 130, 70 *Galleria* units/ml on day 190 (Yin and Chippendale, 1979). In

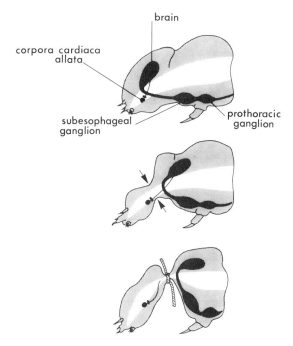

FIGURE 32. Method used to demonstrate the role of corpora allata in regulating larval diapause in *Chilo suppressalis*. A ligature is applied under strong pressure to the head on both the dorsal and the ventral sides and results in isolation of the corpora allata. (Modified from Fukaya and Mitsuhashi, 1961.)

nondiapausing larvae, the juvenile hormone titer, which is high before molting, falls to 140 *Galleria* units/ml in 6 hr and then remains at a very low level.

Experiments and determination of juvenile hormone titer in *Ostrinia* indicate that the hormone titers are low during diapause. Hence in this species, it appears that juvenile hormone acts only to initiate diapause and plays no part in its maintenance (Chippendale and Yin, 1979; Bean and Beck, 1980). Similar results have been obtained in the parasitoid wasp *Nasonia vitripennis* (de Loof *et al.*, 1979).

If the role of juvenile hormone has been demonstrated, its mode of action in diapause remains to be determined. Based on the appearance of type B neurosecretory cells of the pars intercerebralis during diapause, Mitsuhashi and Fukaya (1960) suggested the existence of a relationship between the activity of the corpora allata and the activity of these cerebral neurosecretory cells. Incubation of the brain of the cabbage armyworm *Mamestra brassicae* in the presence and absence of juvenile hormone analogues (Hiruma and Yagi, 1978) also showed that juvenile hormone acts on the cerebral neurosecretory cells by inhibiting their release. Inhibition of brain hormone release by juvenile hormone has already been postulated in nondiapausing species, in which development is lengthened by the injection of juvenile hormone and shortened by the removal of the corpora allata (Nijhout and Williams, 1974b).

If juvenile hormone acts by inhibiting the secretion of brain hormone, the endocrine situation in larval diapause is barely different from that of pupal diapause in which the absence of brain hormone determines the inactivity of the molting gland.

The hypothesis that larval diapause is terminated by ecdysterone or active molting glands was confirmed by Fraser (1958) in the fly *Lucilia caesar* by gland implants of a nondiapausing species, the blowfly *Calliphora erythrocephala*, and also by Sieber and Benz (1980) on the codling moth *Laspeyresia pomonella*. Moreover, Claret *et al.* (1978) observed in the endoparasitic wasp *Pimpla instigator* that the larva entering diapause has a very low circulating ecdysone titer.

The identification of the neurosecretory cells that store material during diapause and that release material at diapause termination is important, because these cells probably produce brain hormone. It is well known that the pars intercerebralis contains many distinct types of neurosecretory cells; thus, an overall examination is insufficient. Thorough investigations on Lepidoptera, despite the authors' use of different nomenclatures, show that not all of the type A cells are involved. Histological changes related to diapause and to its termination are almost always observed in a single type, usually called A' (sometimes B or II), in: the rice stem borer *Chilo suppressalis* (Mitsuhashi and Fukaya, 1960); the Far Eastern urticating moth *Euproctis flava* and the pine caterpillar *Dendrolimus spectabilis* (Mitsuhashi, 1963); the slug moth *Monema flavescens* (Takeda, 1972); the cabbage worms *Pieris rapae* and *brassicae* (Kono, 1973; Kind, 1978); *Mamestra brassicae* and the noctuid moth *Acronycta turnicis* (Kind, 1978; Kind and Vaghina, 1976; Hiruma and Agui, 1977); the cabbage armyworm *Barathra brassicae* (Kind, 1977). In the pink bollworm *Pectinophora gossypiella*, however, the A cells appear to act like A' cells (Raina and Bell, 1978). Furthermore, the lateral neurosecretory cells of *Mamestra* seem to show the same activity as the A' cells (Hiruma and Agui, 1977), but this is not so in the tobacco hornworm *Manduca sexta* (Borg and Bell, 1977).

3.4. EMBRYONIC DIAPAUSE

In the silkworm *Bombyx mori*, the number of annual generations varies according to strain and external conditions. Some univoltine strains have a single generation per year, and all their eggs enter diapause. Other strains, bivoltine or quadrivoltine, have two or four generations per year; the eggs they lay sometimes show rapid development and are sometimes diapausing. Embryonic diapause is determined by the temperature and photoperiod undergone by the mother at the embryonic or larval stage. A long photoperiod and a high temperature induce embryonic diapause.

The experimental investigations of Hasegawa (1951) and Fukuda (1952, 1963) demonstrated the existence of a diapause factor produced in the subesophageal ganglion whose release is controlled by the brain. The nature of cerebral control appears to be stimulating (Morohoshi and Oshiki, 1969) or both stimulating and inhibiting (Fukuda, 1953) (Figs. 33 and 34).

The subesophageal ganglion starts to release diapause hormone immediately after pupation (Hasegawa, 1952), before the onset of ovarian development. However, the command that it receives from the brain at this time results from information stored by the latter at the end of embryonic development, i.e., long before.

The ovary is the effector of diapause hormone. The glycogen content of

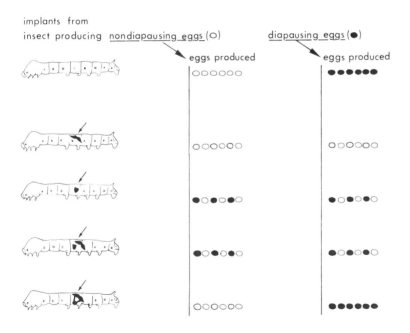

FIGURE 33. Experimental study of embryonic diapause regulation in the silkworm *Bombyx mori.* Implantation experiments are performed in larvae with organs either from larvae shown to produce nondiapausing eggs or from larvae shown to produce diapausing eggs. Only the complexes from the latter larvae are able to induce diapause. (Modified from Fukuda, 1963.)

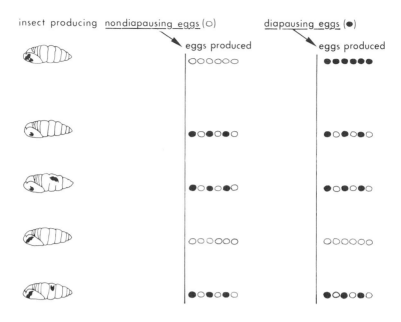

FIGURE 34. Experimental study of embryonic diapause in the silkworm *Bombyx mori*. Ablation and implantation experiments are performed either in pupae shown to produce nondiapausing eggs or in pupae shown to produce diapausing eggs. From top to bottom: control; brain removal; brain removal and reimplantation; subesophageal ganglion removal; removal and reimplantation of the subesophageal ganglion. (Modified from Fukuda, 1963.)

this organ is high in pupae that produce diapausing eggs. It drops sharply if the subesophageal ganglion is removed and increases considerably if extracts of this organ are injected (Yamashita and Hasegawa, 1966). In fact, diapause hormone acts on both the ovary and the fat body. Apart from its action on glycogen, it raises the level, in the eggs, of triglycerides (Ichimasa and Hasegawa, 1973), cholesterol (Ichimasa, 1976), and 3-hydroxykinurenin, an ommochrome precursor (Yamashita and Hasegawa, 1964). The latter action occurs *in vitro*, thus making it a convenient bioassay for diapausing egg determination (Yamashita and Hasegawa, 1976).

The cells producing the diapause factor were identified histophysiologically (Fukuda and Takeuchi, 1967) as two medioventral neurosecretory cells located in the subesophageal ganglion. They are unloaded and nearly invisible in pupae and adults that produce diapausing eggs, while their appearance is normal in the opposite case.

Embryonic diapause does not only occur in *Bombyx*, but is also encountered in other species such as the lymantriid moth *Orgyia antiqua* and the alfalfa plant bug *Adelphocoris lineolatus*, in which it is also determined by the mother. The source of the diapause factor is found in *Orgyia* in the subesophageal ganglion, whose removal prevents diapause (Kind, 1969), and in *Adelphocoris* in the thoracoabdominal nerve mass (Ewen, 1966). The neurosecretory cells of the subesophageal ganglion implicated in this secretion in *Orgyia* appear to be two posteroventral cells located as in *Bombyx* (Kind, 1965). In *Adelphocoris,* five pairs of lateral cells may be involved (Ewen, 1966).

Fukuda (1963) reported the presence of diapause hormone only in the subesophageal ganglion. However, more recent investigations reported that diapause hormone is present in the brain (Sonobe and Keino, 1975) and in the corpora cardiaca–corpora allata complex of *Bombyx* (Takeda and Ogura, 1976).

Diapause hormone is also found in other species: in the subesophageal ganglion of the gypsy moth *Lymantria dispar* and the giant silk moth *Antheraea pernyi* (Fukuda, 1951; Hasegawa, 1951), the noctuid moth *Phalaenoides glycinae* (Andrewartha *et al.*, 1974), and the cockroach *Periplaneta americana* (Takeda, 1977). It is also present in the brain, corpora cardiaca, and corpora allata of the armyworm *Leucania separata* (Ogura and Saito, 1973) and of *Periplaneta* (Takeda, 1977). These results suggest that diapause hormone has a different role in nondiapausing insects. It may be mentioned here that in the saturniid moth *Samia cynthia*, a hormone originating in the brain acts on the shape of the cocoon, which differs according to whether development includes a diapause or not (Pammer, 1966; Nopp-Pammer and Nopp, 1967).

Does diapause hormone act exclusively in embryonic diapause or does it play a part in larval and pupal diapauses? According to Fukuda (1968), the subesophageal ganglion does not act in pupal diapause, for it is incapable of initiating diapause in pupae of *Philosamia cynthia*, and its removal from the diapausing pupa does not prevent diapause. Nevertheless, some investigators, including Maslennikova (1973), support a different view, and it seems difficult to discount the existence of a diapause factor transmitted by the egg, as when larval or pupal diapause is determined by the conditions undergone by the mother, e.g., the wasp *Nasonia vitripennis* (Saunders, 1965).

Attempts to purify the hormone of the subesophageal ganglion of *Bombyx* have been made by Sonobe and Ohnishi (1971), and also by Hasegawa *et al.* (1972, 1974), Isobe *et al.* (1973, 1975), and Kubota *et al.* (1976).

Using 2,000,000 heads, different extraction solvents, and many chromatographies, 0.012 g of a substance was obtained, of which 6 μg corresponded to one DH unit. This unit is the quantity required for the production of 40% diapausing eggs in a female conditioned to produce nondiapausing eggs. The isolated substance is stable in bases and weak acids. It exists in two forms, DHA and DHB, one of which (DHB) is three times more potent.

A thorough analysis of DHA (Isobe *et al.*, 1975) showed that it is destroyed by proteolytic enzymes; it contains 14 different amino acids and two amino sugars, glucosamine, and galactosamine (Table XIII).

The molecular weight of DHA is 3300 ± 40; that of DHB is 2000 ± 200 (Kubota *et al.*, 1976). DHB is also peptidergic and contains 14 amino acids and no amino sugar. It has no free NH_2- or COOH-terminal groups (Kubota *et al.*, 1979). By using different extraction methods, Sonobe and Ohnishi (1971) also obtained two substances, both of which are thermostable and are destroyed by proteolytic enzymes. The molecular weight of the smaller is 5000–10,000.

3.5. CONCLUSIONS

Neurohormones play a pivotal role in the regulation of diapause, acting by different mechanisms.

In the embryo of *Bombyx*, a maternal diapause factor originating in the subesophageal ganglion acts on the pupal ovary and determines the evolution of the egg. It is well known that in *Bombyx* it causes an accumulation, in the eggs, of glycogen, triglycerides, and 3-hydroxykinurenin, but we do not know what relationship exists between these processes and diapause.

It is curious to note that the stimuli that condition the secretion of the embryonic diapause factor are perceived long before its release.

Purification experiments indicate the existence of two proteins or polypeptides, the smaller containing 14 identified amino acids.

TABLE XIII. Acidic Hydrolysate of DHA: Mole Ratios[a,b]

Lysine	0.8	Cystine	0
Histidine	Trace	Valine	1.7
Arginine	0.8	Methionine	Trace
Aspartic acid	1.3	Isoleucine	1.9
Threonine	1.0	Leucine	3.0
Serine	1.0	Tyrosine	0.9
Glutamic acid	2.1	Phenylalanine	0.9
Pyroline	3.0	Glucosamine	0.9
Glycine	2.1	Galactosamine	0.9
Alanine	2.2		

[a]From Isobe *et al.* (1975).
[b]Threonine is taken as a standard.

Diapause in the larva and pupa results from the absence of secretion of brain hormone, resulting in inactivity of the molting gland. In the pupa, brain activity is modified in accordance with environmental factors; in the larva, the brain is largely subject to juvenile hormone, which exerts a negative feedback effect.

Imaginal diapause is a reproductive diapause that is caused by a juvenile hormone failure due to brain inhibition of the corpora allata.

4

Reproduction

Reproduction takes place as the result of many processes that occur sequentially in time and are different in nature. The first of these is sexual differentiation, which includes the organization of the genital organs, the formation of exit ducts for sexual products, the development of secondary sexual characters, and subsequent spermatogenesis and oogenesis. Oogenesis in turn includes several successive phases—gonial mitoses, the entry of the gonocytes into meiosis, and the differentiation of the ovarian follicles within which vitellogenesis occurs and the chorion is formed. The male–female encounter leading to mating is preceded or accompanied by the release of pheromones and by special courtship behavior patterns. Mating sometimes takes place before vitellogenesis, sometimes follows it and precedes egg laying. It exerts a stimulatory role on these two processes through the secretions of the male accessory glands (possibly prostaglandins, Loher, 1979), which are transmitted to the female during mating. Egg laying does not generally occur at random; some females display a circadian egg-laying rhythm, while others select their laying site in accordance with the food requirements of the larva that will hatch from the egg.

Ecdysone and especially juvenile hormone act during certain phases of reproduction, but in these cases as in all functions, neurohormones play a pivotal role and are involved in all mechanisms related to reproduction.

4.1. SEX DETERMINATION

The sexual dimorphism of insects and the frequent cases of intersexuality due to the presence of parasites have suggested to investigators that the primary and the secondary sexual characters are under endocrine control.

Various experiments involving reciprocal grafts of gonads and imaginal discs of organs exhibiting strong sexual dimorphism, such as the wings and legs (moths, crickets), yielded negative results. Experiments conducted on the embryo of the Colorado beetle *Leptinotarsa decemlineata* (Richard and Charniaux-Cotton, 1970) were also negative, especially with respect to the action of neurohormones in sex determination. The destruction of presumptive cephalic areas in the 66th hour of embryonic life allows normal sexual differentiation to occur on the fourth day.

Positive results were obtained only on the glowworm *Lampyris noctiluca* (Naisse, 1965, 1969) (Fig. 35). In the adult stage, this insect displays highly pronounced secondary sexual characters. The wingless female possesses three pairs of abdominal photogenic organs, while the male, which is much smaller, is winged and has only one vestigial pair of photogenic organs.

Investigations carried out on the larvae of both sexes revealed a complex regulation of sexual differentiation, involving both a neurohormone and a factor originating in the male gonad.

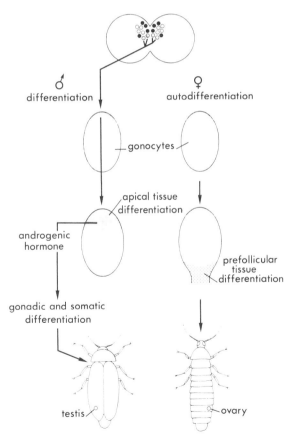

FIGURE 35. Sex determination in the glowworm *Lampyris noctiluca*. (From data of Naisse, 1965.)

During the first larval instars, the male and female gonads undergo the same transformations. The genital follicles, anlagen of the future testis and ovarian follicles, are formed by multiplication of the gonocytes and the mesodermal cells. At the end of the fourth instar, their differentiation begins in the male with the development of an apical mesodermal tissue that has a secretory appearance, which regresses at the end of the following instar, with the onset of spermatogenesis. This apical tissue determines the masculinization of the primary and secondary sexual characters. Its development occurs only in genetic males and depends on neurohormones originating in the pars intercerebralis. Its formation can be induced in female gonads implanted into castrated males, or in females in which the brain has been replaced by a male brain. Hence the male appears to possess a neurohormone that does not exist in the female, at least at the time when its action is necessary. The neurosecretory cells responsible for this action are apparently small-grained cells located in the pars intercerebralis that enter into activity in the male just before the development of the apical tissue.

4.2. GONIAL MITOSIS AND MEIOSIS

Elucidation of the initial stages of oogenesis, i.e., gonial mitosis and meiosis, was achieved in studies on the Brazilian blood-sucking triatomid *Panstrongylus megistus* (Furtado, 1976a–c, 1977a,b, 1979).

Study of the development of the female gonad during the final larval instar showed that the first evidence of transformation of the ovary is a mitotic crisis affecting the germinal and somatic cells, which begins 24–48 hr after the blood meal. This multiplication phase is followed by a differentiation phase, which includes the entry of the oogonies into meiosis and differentiation of the tropharium and the prefollicular tissue. This phase continues until the imaginal molt, when previtellogenesis begins. The pars intercerebralis of *Panstrongylus* contains four types of neurosecretory cells, A, A′, B, and D. The variations in the amount of their neurosecretory products observed during the final larval instar served to distinguish a relationship between the activity of the A cells and the mitosis mechanism, the activity of the A′ cells, and the appearance of meiosis (Fig. 36).

Experimental investigations were carried out to clarify the respective roles played by the various neurosecretory cells of the pars intercerebralis. The pars intercerebralis was totally electrocoagulated at two different times, before the mitotic phase and before the meiotic phase. In both cases, the operation prevented subsequent development of the ovary and caused its degeneration. The neurosecretory cells, therefore, exert a twofold regulation on the early development of the ovary by taking part in triggering mitosis and also in triggering meiosis.

In order to achieve a better understanding of these mechanisms, brain operations were repeated by destroying only one of the neurosecretory cell types. These experiments showed that selective removal of the type A cells does not cause the same repercussions as total removal of the pars intercerebralis; it only inhibits the mitotic crisis and allows meiosis to subsist. Meiosis, on the contrary, cannot occur in the absence of the type A' neurosecretory cells.

Radioimmunoassay of hemolymph ecdysteroids (Fig. 36) demonstrated the existence of a peak coinciding with the onset of meiosis. The operations confirmed the role of the molting gland in the determination of meiosis. The molting gland was removed at different times, and the results obtained showed that this gland does not take part in the determination of mitosis, but is essential to the triggering of meiosis. In order to understand the respective roles played by the brain and the molting gland in regulating meiosis, ecdysterone was injected into insects deprived of the pars intercerebralis. Normal ovarian development was shown to be restored.

Hence the brain plays a twofold role in controlling the initial phases of

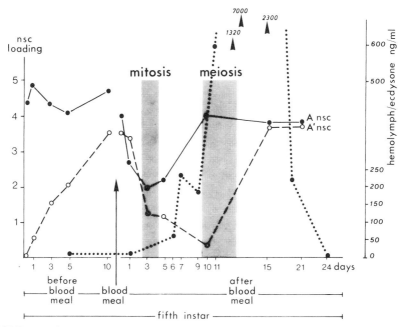

FIGURE 36. Variations in the amount of neurosecretory material in type A and A' neurosecretory cells (nsc), ecdysterone hemolymph titer, and ovarian development in the bug *Panstrongylus megistus*. (Modified from Furtado, 1979.)

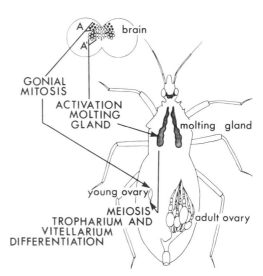

FIGURE 37. Schematic representation of mitosis and meiosis
control in the bug *Panstrongylus
megistus.*

oogenesis: an indirect role by stimulating the molting gland, and a direct role
by inducing mitoses (Fig. 37).

4.3. DIFFERENTIATION OF THE OVARIOLE—
PREVITELLOGENESIS

In vivo and *in vitro* investigations have shown that vitellarium differentiation, cytoplasmic growth of the oocytes, and previtellogenesis require the
presence of ecdysone. A brain factor is also necessary. In *Tenebrio molitor,*
cauterization of the pars intercerebralis causes a decrease in the size of the
germarium and inhibits previtellogenesis (Mordue, 1965b). Also, *in vitro* evidence of the role of a brain neurohormone in the initial growth of oocytes in
Tenebrio has been presented (Laverdure, 1972).

4.4. VITELLOGENESIS

4.4.1. Pars Intercerebralis

The role of the pars intercerebralis in controlling vitellogenesis has been
suspected for a long time, because changes in the load of the neurosecretory
cells had often been observed, related to reproduction and vitellogenesis.
    The first thorough investigations into this matter were conducted by

Highnam and Lusis (Highnam, 1962; Highnam and Lusis, 1962) on the locust *Schistocerca gregaria,* in which vitellogenesis is cyclic. The pars inter-cerebralis contains a large number of neurosecretory cells that exhibit cyclic modifications. Heavily loaded cells correspond to an inhibition of release, while a small load corresponds to an intense process of synthesis and libera-tion, as shown by experiments using [^{35}S]cysteine (Highnam and Mordue, 1970). The results of the different histophysiological observations are as fol-lows:

1. In females raised with males, the first batch of oocytes matures be-tween day 4 and day 14, and is accompanied by slight storage of type A neurosecretory products in the pars intercerebralis and the corpora cardiaca. On day 21, the oocytes are mature and considerable storage of neurosecretory material is observed in the pars intercerebralis and the corpora cardiaca. Egg laying takes place on day 28 and is fol-lowed by the release of neurosecretory material.
2. In females raised without males, oocyte development is slight or nil; a large accumulation of neurosecretory products is observed on day 14.
3. Electrical stimulation exerted on the ventral nerve cord, the brain, or the optic lobes in 14-day-old females raised without males induces the release of neurosecretory products and the development of the oocytes.
4. The same result occurs if the flask containing the insects is turned over repeatedly. This treatment carried out for 45 min induces release from the corpora cardiaca; continued for 1 hr over 3 days, it gives rise to development of the oocytes.
5. Finally, if females isolated for 17 days are placed in contact with males, mating occurs within 24 hr. The corpora cardiaca, still heavily loaded during mating, release their products in the subsequent 12-hr period, with the onset of vitellogenesis processes.

Thus in these diverse cases a clear correlation appears between the release of neurosecretory products and oocyte maturation.

Many other histophysiological investigations have been conducted and generally show, in species with a cyclic egg-laying pattern, a progressive load of the type A neurosecretory cells of the pars intercerebralis during vitellogenesis, and a release that coincides with egg laying. Storage was fre-quently observed during periods of vitellogenesis arrest, as for example during ovarian diapause and during the incubation of mature eggs or embryos in viviparous species.

In many insects, destruction of the neurosecretory cells of the pars inter-cerebralis causes inhibition of vitellogenesis (Table XIV). However, the re-sults exhibited by a single species are sometimes heterogeneous, for while some individuals react by arresting vitellogenesis, others are capable of carry-

TABLE XIV. Role of the Pars Intercerebralis in Vitellogenesis Regulation[a]

Insect	Reference	Effect on vitellogenesis of	
		pi removal	pi removal + ca, JH, or JHA
Leucophaea maderae	Engelmann and Penney, (1966)	+ or −[b]	
	de Bessé (1975)	+	
Nauphoeta cinerea	Barth and Sroka (1975)	−	+
Carausius morosus	Dupont-Raabe (1952b)	+	
Clitumnus extradentatus		+	
Locusta migratoria	Girardie (1966b); McCaffery (1976); Lazarovici and Pener (1978)	−	−
Schistocerca gregaria	Highnam (1962); Strong (1965a)	−	
Anacridium aegyptium	Geldiay (1967)	−	
Melanoplus sanguinipes	Gillott and Elliott (1976)	−	
Gomphocerus rufus	Loher (1966)	−	
Tetrix undulata	Poras (1977b)	+	
Tenebrio molitor	Mordue (1965b,c)	−	
Leptinotarsa decemlineata	de Wilde and de Boer (1969)	−	+
Solenopsis invicta	Barker (1978)	−	−
Danaus plexippus	Barker and Herman (1973)	−	−
Manduca sexta	Sroka and Gilbert (1971)	−	+
Aedes taeniorhynchus	Lea (1967)	−	−
Calliphora erythrocephala	Thomsen (1952)	−	−
Sarcophaga bullata	Wilkens (1968)	−	−
	Pappas and Fraenkel (1978)	−	+
Sarcophaga inzi	Chadha and Denlinger (1976)	−	
Paecilometopa spilogaster and _punctipennis_	Chadha and Denlinger (1976)	−	
Glossina fuscipes	N'Kouka (1977a)	+	
Musca domestica	Lea (1975)	−	−
	Sakurai (1977)	−	+
Drosophila melanogaster	Handler and Postlethwait (1977)	−	
	Boulétreau-Merle (1976)	−	

(Continued)

Table XIV. *(Continued)*

		Effect on vitellogenesis of	
			pi removal
		pi	+ ca, JH,
Insect	Reference	removal	or JHA
Oncopeltus fasciatus	Johansson (1958)	+	
Rhodnius prolixus	Baehr (1973)	−	+
Triatoma protracta	Mundall and Engelmann (1977)	+	

[a]Abbreviations used: ca, corpora allata; JH, juvenile hormone; JHA, juvenile hormone analogues; pi, pars intercerebralis.

[b]+ and − indicate that the pars intercerebralis exerts allatotropic and gonadotropic control, respectively.

ing it through, as in the cockroach *Periplaneta americana* (Engelmann and Penney, 1966) and the milkweed bug *Oncopeltus fasciatus* (Johansson, 1958). In other species, such as the cockroach *Leucophaea maderae* (de Bessé, 1975), the stick insects *Carausius morosus* and *Clitumnus extradentatus* (Dupont-Raabe, 1952b), the grouse locust *Tetrix undulata* (Poras, 1977b), the tsetse fly *Glossina fuscipes* (N'Kouka, 1977a), and the reduviid bug *Triatoma protracta* (Mundall and Engelmann, 1977), the pars intercerebralis appears to be useless for vitellogenesis.

It is well known that vitellogenesis is controlled by the corpora allata, and that the functioning of these organs is governed by the brain. Thus, cauterization of the pars intercerebralis cannot provide positive evidence of direct or indirect action: the demonstration of the direct role of the brain on vitellogenesis requires experiments in which the absence of the pars intercerebralis is offset by corpora allata implants (Table XIV).

These experiments are few in number, and uncertainty subsists for many insects: *Schistocerca* (Highnam, 1962; Strong, 1965a), *Anacridium* (Geldiay, 1967), *Melanoplus* (Gillott and Elliott, 1976), *Gomphocerus rufus* (Loher, 1966), *Tenebrio* (Mordue, 1965b,c), *Paecilometopa* (Chadha and Denlinger, 1976), *Drosophila* (Boulétreau-Merle, 1976; Handler and Postlethwait, 1977).

In others, it has been observed that the corpora allata offset the disorders caused by removal of the pars intercerebralis: in the cockroach *Nauphoeta* (Barth and Sroka, 1975), the coleopteran *Leptinotarsa* (de Wilde and de Boer, 1969), the lepidopteran *Manduca* (Sroka and Gilbert, 1971), and the bug *Rhodnius* (Baehr, 1973). These results thus show that the pars intercerebralis exerts an indirect role on vitellogenesis (Fig. 38).

In many insects, on the other hand, the pars intercerebralis exerts a "direct" action on vitellogenesis, and corpora allata implants fail to compensate for its destruction, as in the case of the locust *Locusta migratoria* (Girardie,

1966b; McCaffery, 1976), the butterfly *Danaus plexippus* (Barker and Herman, 1973), the mosquito *Aedes taeniorhynchus* (Lea, 1967), and the flies *Calliphora erythrocephala* (Thomsen, 1952), *Sarcophaga bullata* (Wilkens, 1968), and *Musca domestica* (Lea, 1975) (Fig. 38).

The gonadotropic action of the brain is especially pronounced in species such as mosquitoes, in which the corpora allata are only necessary in the young female, contrary to what occurs in other insects (Lea, 1969; Gwadz and Spielman, 1973). An egg development neurosecretory hormone (EDNH),

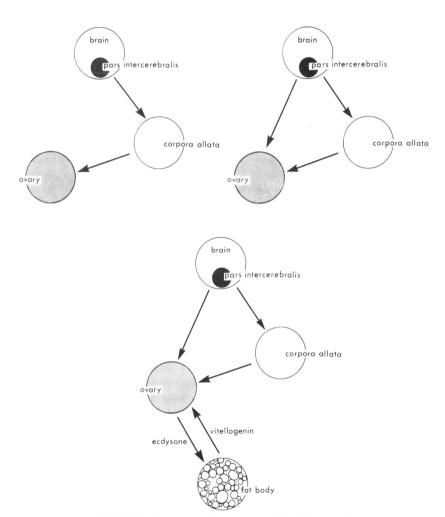

FIGURE 38. Several modes of control of vitellogenesis.

produced in the median neurosecretory cells of the brain and stored in the corpora cardiaca, is released into the hemolymph during a precise interval, which follows the filling of the midgut with ingested blood (Gillett, 1957; Lea, 1967, 1972). In *Aedes aegypti,* this neurohormone induces production by the ovary of "vitellogenin-stimulating hormone," which stimulates the pro-production of vitellogenins by the fat body (Hagedorn, 1974) and is nothing other than ecdysterone (Fallon *et al.,* 1974) (Fig. 38). Brain hormone or EDNH is located in the brain, and does not appear to exist in the ventral ganglia (Hagedorn *et al.,* 1979). It appears to be a stable molecule (Fuchs *et al.,* 1980).

In three insects, *Locusta, Sarcophaga,* and *Musca,* experimental data are contradictory. Whereas Lazarovici and Pener (1978), Pappas and Fraenkel (1978), and Sakurai (1977) indicate that the corpora allata can make up for the absence of the pars intercerebralis, Girardie (1966b), McCaffery (1976), Wilkens (1968), and Lea (1975) state that the pars intercerebralis is required for vitellogenesis to occur.

The discrepancies in the data are not easily reconciled. It should be noted from the outset that the brain neurosecretory products are mostly stored in the corpora cardiaca, and that these organs can play a substitute role in the absence of the pars intercerebralis. In *Carausius* and *Clitumnus,* removal of the pars intercerebralis does not arrest vitellogenesis (Dupont-Raabe, 1952b), but the simultaneous removal of the corpora cardiaca inhibits it completely (Mouton, 1971). Likewise, in *Aedes,* late removal of the neurosecretory cells is only effective if they are removed at the same time as the corpora cardiaca (Lea, 1972). The heterogeneity of the responses may also stem from differences in operating time and in observation time. In *Locusta,* for example (Goltzené and Porte, 1978), removal of the pars intercerebralis disturbs vitellogenesis only if it is performed during the first 5 days of imaginal life. In *Aedes,* it is effective only if performed shortly after eclosion (Lea, 1967). Histophysiological investigations on the fly *Sarcophaga argyrostoma* (Possompés *et al.,* 1967) show that the first oogenetic cycle is associated with a considerable release of neurosecretory products from the pars intercerebralis, which takes place before eclosion. When the observations deal exclusively with the first oogenetic cycle, the action of the brain can thus be easily overlooked if the operation is performed after the initial release of neurosecretory products.

In some insects, neurosecretory products enhance vitellogenesis without being indispensable to it, as observed by Lüscher (1968a) in the cockroach *Nauphoeta* decapitated and implanted with various organs. The corpora allata trigger vitellogenesis, but the corpora cardiaca implanted with them increase the number of growing oocytes. This effect appears to be induced by an oocyte-protection neurohormone that prevents their degeneration (Lüscher, 1968b).

What are the source cells of the gonadotropic factor in the pars inter-cerebralis? The only study carried out on this matter was conducted on *Locusta* (Girardie, 1966b) and showed that the median A–B cells of the pars intercerebralis were involved, whereas Cj cells control the activity of corpora allata.

Let us now consider how the brain factor acts in this process. Some authors have suggested that the brain exerts its action on vitellogenesis through protein metabolism (Table XV). The action of the pars intercerebralis upon hemolymph proteins is exerted, at least in some insects, through ovarian ecdysone, which triggers the fat body to produce them, as we saw above. This process, discovered in a mosquito, was also observed in *Drosophila melanogaster* (Postlethwait and Handler, 1979) and *Calliphora erythrocephala* (Thomsen *et al.*, 1980). In other insects, however, ecdysterone inhibits cor-pora allata activity and vitellogenin synthesis (*Diploptera*, Friedel *et al.*, 1980). In still others, protein synthesis is dependent on both pars inter-cerebralis and corpora allata, as for example in *Leucophaea* (Scheurer, 1969a). If decapitated insects receive a brain implant, certain hemolymph proteins increase, whereas the specific female protein increases only after corpora allata implantation (see Chapter 9).

Apart from its metabolic action, the pars intercerebralis also exerts a gonadotropic action. Hence in *Locusta*, the passage of certain hemolymph proteins into the oocyte is favored by the pars intercerebralis, while the passage of other hemolymph proteins is favored by the corpora allata (Bentz *et al.*, 1970).

The changes occurring in the follicular cells and the mode of action of juvenile hormone upon these cells have been thoroughly studied. During vitellogenesis, spaces appear between follicular cells due to a decrease in their volume (see Davey, 1978). Ouabain inhibits the development of this process (patency), thus suggesting that a Na^+–K^+ ATPase may be involved. Unfor-tunately, no such studies have been performed to date using brain instead of corpora allata.

4.4.2. Ventral Nerve Cord

Recent histophysiological investigations suggest that the neurosecretory cells of the ventral nerve cord act also in controlling vitellogenesis. As for pars intercerebralis, a correlation is frequently observed between the activity of certain neurosecretory cells and the gonadotropic cycle.

In *Locusta*, which exhibits successive vitellogenetic cycles separated by massive egg deposition, very clear-cut variations in neurosecretory product load are visible in two medioventral type A cells of the subesophageal ganglion (Fréon, 1964). In females reared with males, a progressive accumu-lation is observed in these cells during the vitellogenesis period, together with

TABLE XV. Role of the Pars Intercerebralis and Corpora Allata in Vitellogenesis and Protein Synthesis Regulation

Insect	Reference	Pars intercerebralis		Corpora allata	
		Protein synthesis	Vitello-genesis	Protein synthesis	Vitello-genesis
Periplaneta americana	Thomas and Nation (1966a)			⊕[a]	
	Bell (1969)			⊕	+
Leucophaea maderae	Engelmann and Penney (1966)	⊕/0		⊕	+
	Scheurer (1969a,b)	+		⊕	
Nauphoeta cinerea	Barth and Sroka (1975)				+
	Lanzrein (1974)			⊕	
Melanoplus san-guinipes	Elliott and Gillott (1977, 1978, 1979)	+		⊕	+
Locusta migratoria	Bentz *et al.* (1970); Goltzené-Bentz *et al.* (1972)		+	⊕	+
Schistocerca gregaria	Hill (1962, 1963)	+		+	+
	Engelmann *et al.* (1971)			+	
Gomphocerus rufus	Loher (1966)	+			
Tenebrio molitor	Mordue (1965c)	+		+	+
	Laverdure (1970)			⊕	
Leptinotarsa decemlineata	de Loof and de Wilde (1970)	⊕		⊕	+
	Dortland (1978, 1979)			0	
Apis mellifica	Ramamurty and Engels (1977)			±	
Hyalophora cecropia	Pan (1977)			0	
Antheraea polyphemus	Blumenfeld and Schneider-man (1968)			+	
Manduca sexta	Sroka and Gilbert (1971)			+	
Danaus plexippus	Pan and Wyatt (1976)			⊕	
Calliphora erythrocephala	Thomsen (1952)		+		+
Sarcophaga bullata	Wilkens (1968, 1969)	0	+	+	
Phormia regina	Mjeni and Morrison (1976)			⊕	
Drosophila grisea and *macroptera*	Kambysellis and Heed (1974)				+
Drosophila melanogaster ap[4]	Gavin and Williamson (1976)				+
Oncopeltus fasciatus	Kelly and Telfer (1977)			⊕	+
Rhodnius prolixus	Pratt and Davey (1972)				+
	Baehr (1974)			⊕	
Triatoma protracta	Mundall and Engelmann (1977)	0	0	⊕	
Pyrrhocoris apterus	Sláma (1964)				+

[a]⊕, Vitellogenin synthesis; +, protein synthesis or vitellogenesis; 0, no effect.

considerable release at the time of egg laying. In females reared without males, in which oocyte growth is very slight, storage occurs earlier and no release is observed. The variations in the neurosecretory products of the A cells of the subesophageal ganglion thus closely recall those observed by Highnam in the pars intercerebralis.

In the cricket *Gryllus (Acheta) domesticus,* vitellogenesis is not cyclic and neurosecretory material evolves in a very different manner (Huignard, 1964). Egg laying begins about 8 days after the imaginal molt, and then proceeds continuously. The activity of the type A neurosecretory cells of the subesophageal ganglion changes in accordance with genital activity: in normal females, the cells, which are heavily loaded during the first days of imaginal life, do not display any storage throughout the egg-laying period; if the insects are starved, vitellogenesis is arrested and storage processes recur; if the females are rendered incapable of laying through the absence of a suitable substrate for egg deposition, vitellogenesis continues for a certain period and then stops; the mature oocytes start degenerating and the neurosecretory cells resume the storage of their secretory products.

In the viviparous cockroach *Leucophaea maderae,* vitellogenesis is interrupted during gestation periods. Clear-cut variations are observed not only in the type A neurosecretory cells of the subesophageal ganglion, but also in the type A neurosecretory cells of all the ventral nerve cord ganglia. These cells are heavily loaded during gestation and release their material at each parturition (de Bessé, 1965).

Many other histophysiological investigations have been conducted. While no final conclusion can be drawn, it appears that the type A neurosecretory cells of the ventral nerve cord may play a part in controlling vitellogenesis. This is confirmed by castration experiments, which induce a change in the activity of the neurosecretory cells in *Leucophaea* (Scharrer, 1955), *Carausius* (Mouton, 1968), *Locusta* (Chalaye, 1975), and *Teleogryllus commodus* (Dürnberger *et al.,* 1978). As already mentioned, this operation also induces important changes in the corpora allata, as shown especially by Scharrer (1978).

Experimental data concerning the role played by ventral nerve cord neurohormones in controlling reproduction are rare, and this can be explained by the impossibility of removing scattered elements spread among many nerve ganglia. Delphin (1963) noted that removal of the last abdominal ganglion in *Schistocerca* prevents egg development, which is restored under certain conditions after its reimplantation. Thomas (1964) induced premature, abundant egg laying in *Acheta* by implanting subesophageal ganglia of females in sexual activity into females that had molted a few hours before the experiment. At the present time, however, it cannot be stated positively whether this endocrine action of the subesophageal ganglion implies an activation of the vitellogenesis process, or whether it points to a stimulation of egg laying similar to that observed by the same investigator in stick insects (see below).

4.5. OOSTATIC HORMONE (ANTIGONADOTROPIN)

It must be mentioned here that in several insects the ovary under certain conditions exerts a feedback control that results in an inhibition of vitellogenesis. This can be observed in the viviparous cockroach *Leucophaea* during embryo incubation (Engelmann, 1957) and in oviparous insects such as the fly *Musca domestica* (Adams *et al.*, 1968), the mosquitoes *Aedes sollicitans, Aedes aegypti, Aedes taeniorhynchus* (Meola and Lea, 1972), and the bugs *Iphita* (Nayar, 1958) and *Rhodnius* (Huebner and Davey, 1973) when ripe oocytes cannot be laid. Vitellogenesis inhibition is due to substances coming from the ovary, which were semipurified in *Musca* by Adams *et al.* (1970) and in *Rhodnius* by Liu and Davey (1974). These substances are low-molecular-weight proteins that seem to act through pars intercerebralis neurosecretory cells (Lea, 1975; Adams, 1976). In *Rhodnius* (Davey, 1978), the origin of the antigonadotropin seems to be outside the ovary in strands of connective tissue that contain many neurosecretory axons. Because no antigonadotropic activity is present in the brain or in the mesothoracic ganglionic mass, it may be that the antigonadotropin originates in peripheral neurosecretory neurons. This would be the first evidence of a physiological role of such peripheral neurons, which may, at the same time, function as stretch receptors and deliver active products.

4.6. SPERMATOGENESIS

Spermatogenesis control is not well understood in insects, especially when compared with oogenesis. Several investigations suggest that ecdysone is involved, but the nature of its involvement is not very clear (see Dumser, 1980). The regulation of spermatogonial mitosis and meiosis has been carefully investigated in *Samia cynthia* (Kambysellis and Williams, 1971), *Rhodnius* (Dumser and Davey, 1975), and *Locusta* (Dumser, 1980). It appears that ecdysone stimulates the mitotic index, but no neurohormonal regulation has yet been demonstrated. However, it is worthwhile mentioning that such a control was reported to occur in the myriapod *Lithobius forficatus* (Descamps, 1978).

4.7. ACCESSORY GLANDS, PHEROMONE PRODUCTION, SEXUAL BEHAVIOR

The functioning of male and female accessory glands, the production of pheromones, and sexual behavior depend upon the corpora allata and consequently upon the brain, which regulates their functioning. The direct role of neurosecretory materials of the pars intercerebralis on these different mechanisms has been clearly demonstrated in only a few cases: the sexual receptivity of *Leucophaea* (Engelmann and Barth, 1968), the female sexual behavior of *Drosophila* (Boulétreau-Merle, 1976), and the male sexual behavior of *Locusta* (Pener *et al.*, 1972) and *Teleogryllus* (Loher, 1974). The experiments of Riddiford and Williams (1971) on the great saturniid moths *Antheraea polyphemus* and *Hyalophora cecropia* revealed the action of the corpora cardiaca in the production of female pheromones, and in calling behavior.

In the tsetse fly, both the pars intercerebralis and the corpora allata are very important for the functioning of the milk glands (Ejezie and Davey, 1976; N'Kouka, 1977b). If one of these organs is removed, milk gland activity is reduced and the incubated larvae are very small (Fig. 39). The role of the pars intercerebralis was also demonstrated in histological studies. The type A neurosecretory cells display a cyclic activity in *Glossina austeni* (Ejezie and Davey, 1974, 1977) and *Glossina fuscipes* (N'Kouka, 1976).

4.8. OVULATION

Ovulation may involve a control that is distinct from that of egg laying. In most species, in fact, these two mechanisms do not succeed each other automatically and they may be separated by the period of embryo incubation in the case of viviparity, or by a more or less long period of egg storage when egg laying takes place exclusively in certain preferential sites.

Few investigations have been conducted into the regulation of ovulation. The only existing work is that of Chaudhury and Dhadialla (1976) on the tsetse fly *Glossina morsitans,* in which ovulation occurs normally after mating. It can be induced in virgin females by injection of the hemolymph of mated females, while virgin female hemolymph has no effect. The active factor may originate in the corpora cardiaca, for removal of the corpora cardiaca–corpus allatum complex inhibits ovulation, whereas removal of the corpus allatum alone does not. cAMP appears to be involved in neurohormone control of ovulation (Denlinger *et al.*, 1978), but it is possible that amines are also active, for reserpine inhibits ovulation in *Periplaneta* (Hentschel, 1972).

4.9. EGG LAYING

Egg-laying procedures vary widely among insects. Some species deposit their eggs at random, but most of them find a laying site that is favorable to the survival of their offspring. Thus, oviposition does not immediately follow

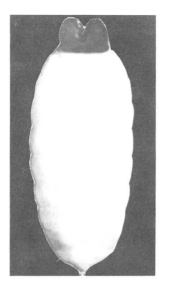

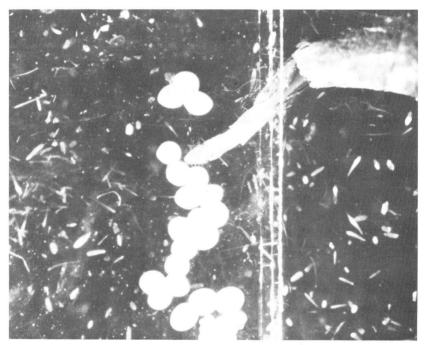

FIGURE 39. (Top) Regulation of milk gland secretory activity in the tsetse fly *Glossina fuscipes.* On the left, a normal third-stage pupa; on the right, two small third-stage pupae incubated by a fly whose milk glands possess reduced activity following pars inter-cerebralis removal. (From N'Kouka, 1977a.) (Bottom) A method for counting eggs in the wax moth *Galleria mellonella.* (Courtesy of Mesnier.)

egg maturation. In many cases, it occurs after mating, and the brain plays an important coordinating role in its occurrence.

The existence of endocrine regulation of egg laying has been suggested in some species based on experiments employing hemolymph injections. Mokia (1941) showed that the blood of a mated *Bombyx* female contains a substance that induces oviposition in an unmated female. Nayar (1958), on the bug *Iphita limbata,* and Okelo (1971), on *Schistocerca,* achieved premature egg laying by injecting the nearly mature female with the hemolymph of ovipositing females.

Endocrine control of egg laying was confirmed by experiments conducted *in vitro* on isolated oviducts. Oviduct contractions were increased by brain extracts in *Tenebrio* (Köller, 1954) and *Carausius* (Enders, 1955), by pars intercerebralis extracts in *Rhodnius* (Davey, 1965) and *Locusta* (Girardie and Lafon-Cazal, 1972), and by ventral nerve cord extracts in *Locusta* (Girardie and Lafon-Cazal, 1972). A detailed study of egg-laying mechanisms was performed in the stick insect *Carausius,* which lays eggs regularly throughout its imaginal life (Thomas, 1968, 1969, 1979) (Figs. 40 and 41).

Total removal of the brain or the subesophageal ganglion or a simple disconnection of these two organs definitively suppresses egg laying, whereas ovulation subsists in all cases, causing a significant accumulation of eggs in the oviducts (Figs. 40 and 41).

The injection of extracts of the brain, the subesophageal ganglion, the thoracic ganglia, or the first abdominal ganglia into insects rendered incapable of laying by one of the above-mentioned operations causes egg laying in all cases. This demonstrates the existence of an oviposition factor present throughout the central nervous system, whose production and release are controlled by the cephalic nerve centers. Its action is essentially exerted on the genital valves, which retain an egg during the day and eject it at nightfall.

The transection of the last abdominal connectives also inhibits egg laying, but this derives from the existence, in addition to endocrine control, of nervous control exerted by the last abdominal ganglion, which regulates the passage of the oocytes from the lateral oviducts into the genital chamber, in response to nerve impulses generated by inputs from the sensory hairs of the genital valves.

In another stick insect, *Clitumnus extradentatus,* the coordinating role of the brain and the presence of an active neurohormone throughout the ventral nerve cord were also shown, but nervous control of the last abdominal ganglion was absent. This is explained by the different position of the egg in the genital chamber, which in *Clitumnus* does not have sensory hairs (Mesnier, 1972a).

Egg-laying regulation was also analyzed in detail in a lepidopteran, *Galleria mellonella* (Mesnier, 1972b). In this species, massive egg laying takes place within a few days after mating. The female, provided with a long

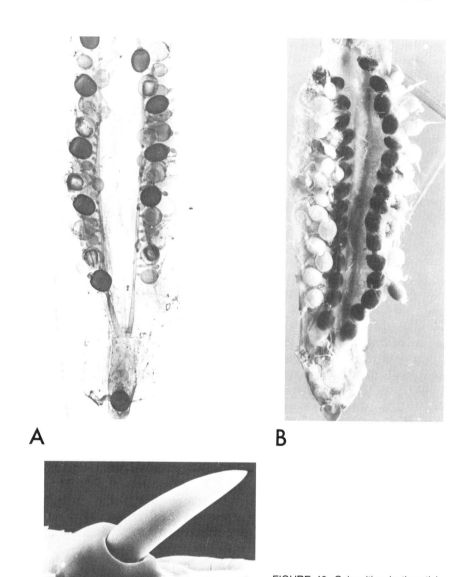

A

B

C

FIGURE 40. Oviposition in the stick insect *Carausius morosus.* Ovaries in a normal female (A) and in an operated one (B): oviposition is arrested and eggs accumulate in the oviducts. (C) A sensory hair of the egg chamber. (C, courtesy of Thomas, 1979.)

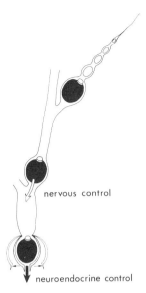

nervous control

neuroendocrine control

FIGURE 41. Schematic representation of egg-laying control in the stick insect *Carausius morosus.* (From data of Thomas, 1979.)

ovipositor, lays selectively in the slits that the ovipositor detects by its sensory hairs (Fig. 39). Experimental investigations are carried out by providing the females with laying areas created by thin strips cemented to glass slides, thus making it possible to count the eggs laid. As in the stick insect, the experiments showed that egg-laying is subject to twofold control, nervous and neuroendocrine. The last abdominal ganglion plays a fundamental nervous role. Its removal inhibits egg laying by immobilizing the ovipositor, as in *Bombyx* (Yamaoka and Hirao, 1977). The brain is also necessary as a command center. Like the ventral nerve cord ganglia, it is the source of an oviposition-stimulating hormone that triggers egg laying.

The role of a substance from the central nervous system in inducing egg laying has been confirmed in two orthopterans. In the locust *Schistocerca,* implantation of the last abdominal ganglion causes egg laying in the young female (Delphin, 1963), and brain implantation restores egg laying, which is inhibited by its removal (Okelo, 1971). In the grasshopper *Melanoplus,* brain extracts stimulate egg laying (Friedel and Gillott, 1976). In *Triatoma protracta,* removal of the brain or transection of the ventral nerve cord suppresses oviposition (Mundall, 1978).

One may well ask which is the active part of the brain. Many investigations point to the pars intercerebralis. Its implantation triggers premature egg laying in *Iphita* (Nayar, 1958); its removal causes a decrease in egg laying in *Rhodnius* (Davey, 1967), total arrest in *Locusta* (Lazarovici and Pener, 1978), and inhibits parturition in the viviparous bug *Stilbocoris natalensis* in which its reimplantation restores normal mechanisms (Furtado, 1971b). In the stick insects, however, removal of the pars intercerebralis allows normal egg laying to proceed (Dupont-Raabe, 1952b), possibly indicating that the corpora cardiaca or the ventral nerve cord play a substitute role.

No experimental work has been devoted to the identification of the brain and ventral nerve cord neurosecretory cells that regulate egg laying. However, we do have some histological data. In *Stilbocoris* and *Glossina,* the type A cells of the pars intercerebralis do not appear to be involved in this mechanism (Furtado, 1971b; N'Kouka, 1976). In the ventral nerve cord, however, it is precisely the A cells that appear to act (Thomas and Mesnier, 1973).

Do the neurohemal organs, corpora cardiaca, and perisympathetic organs store the neurohormones that trigger egg laying? The activity of the corpora cardiaca has been demonstrated *in vivo* in *Schistocerca* (Highnam, 1962; Delphin, 1963), *Hyalophora* (Truman and Riddiford, 1971), and *Galleria* (Mesnier, 1972b), and *in vitro* in *Locusta* (Girardie and Lafon-Cazal, 1972). The activity of the perisympathetic organs has been demonstrated *in vivo* in *Galleria* (Mesnier and Provansal, 1975). However, extracts of corpora cardiaca and perisympathetic organs of stick insects are inactive on stick insects *in vivo,* although they are active on *Galleria,* which is inexplicable.

4.10. CONCLUSIONS

Reproduction is intimately linked to environmental, internal, and nutritional factors. It also depends on the proximity of the sexes, on the release of pheromones, and on mating. Coordination between the different stimulations received and the progress of the successive phases of reproduction is carried out by the brain. Apart from its nervous role, this organ acts humorally in regulating the main stages of female sexual life. It also controls the functioning of the corpora allata, which are endocrine glands of prime importance in reproduction.

Neurohormones control sex determinism, at least in *Lampyris*. They trigger gonial mitoses and activate the molting gland, which induces meiosis.

Vitellogenesis is independent of neurosecretory material in some species but dependent on it in others. Neurohormones act both on protein metabolism and on the follicular cells that surround the oocyte.

Sexual behavior, activity of the accessory glands, and egg laying also depend on neurohormones. But in the latter case neurohormones are produced throughout the central nervous system.

On the whole, the source cells of the different factors are not accurately identified and the neurohormones have not been purified. Therefore, it is not known whether one or more factors are involved.

5

Functioning of the Visceral Muscles

Investigations initiated some 30 years ago revealed the existence in insects of factors acting on the contractility of the visceral muscles. The earliest work of Koller (1948) showed that head and brain extracts of different species induced the acceleration of contractions of the Malpighian tubules (Fig. 42). Subsequent investigations dealt with the heart, Malpighian tubules, gut, and oviducts, and were almost always carried out *in vitro*. Most of them dealt with the heartbeat of the cockroach *Periplaneta americana*. The active factors were discovered in the corpora cardiaca and the central nervous system, and in the effectors themselves (Fig. 43).

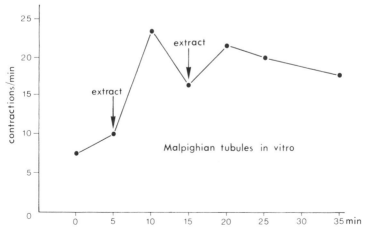

FIGURE 42. Effect of brain homogenate on the contraction frequency of Malpighian tubules in the water beetle *Dytiscus marginalis*. (Modified from Koller, 1948.)

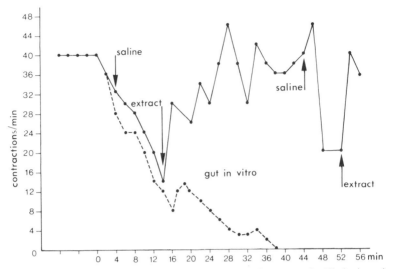

FIGURE 43. Effect of gut extract on gut contraction frequency in *Pieris brassicae*. (Modified from Koller, 1948.)

5.1. THE VISCERAL MUSCLES: INNERVATION AND FUNCTIONING

Visceral muscles in insects are striated like the skeletal muscles, but they differ in many respects from the latter. They contract in a spontaneous rhythmic way and display myogenetic functioning, which persists in the absence of any innervation, although the latter may be highly developed in certain cases.

5.1.1. Heart, Diaphragm, Alary Muscles

The heart is formed of two parts: the heart proper, which occupies the abdominal portion of the body, and the aorta, which extends to the head. The heart consists of ventricles surrounded by circular muscles and bearing lateral openings or ostioles. One pair of alary muscles is inserted laterally on each ventricle and is associated with connective tissue making up the dorsal diaphragm. The pumping process performed by the heart results from its own contraction and that of the alary muscles.

Innervation of the heart in insects has rarely been investigated and appears to be highly variable. In the cockroach *Periplaneta orientalis* (Alexandrowicz, 1926), the stick insect *Carausius morosus* (Opoczyński-Sembratowa, 1936), and the locust *Locusta migratoria* (Roussel, 1972), it includes several components. Two lateral heart nerves originating in the head run along the aorta and the heart. They have been given different names: lateral heart nerves (employed here), cardiac nerves, lateral cardiac nerve cords. The lateral heart nerves are formed of nerve fibers of different origins. Some of them originate in the corpora cardiaca, others, via the segmental nerves, in the ventral nerve cord ganglia, while others come from nerve cells placed on or near the lateral heart nerves.

The function of these different components has been elucidated in part by investigations conducted on *Periplaneta* (Johnson, 1966b; Miller, 1968; Miller and Thomson, 1968; Smith, 1969; Miller and Usherwood, 1971; Miller, 1973). Although the heart of this insect displays a basic myogenetic heartbeat, it has complex regulation, both nervous and neurosecretory. The impulses originating in the central nervous system control the functioning of the heart neurons, which in turn control the contractions of the myocardium. Some of the heart neurons are neurosecretory. Moreover, neurosecretory axons originate in the ganglia. They belong to two categories, one with small granules, the other with large granules. A portion of their secretory products is released into the connective membrane of the lateral heart nerves, but apart from this typical neurohemal functioning, the small-grained fibers innervate the heart neurons and form neuromuscular junctions with the myocardium, which thus displays a double innervation (Fig. 44).

The alary muscles also exhibit complex innervation originating in the lateral heart nerves and the segmental nerves. In *Carausius* and probably in *Periplaneta*, they are also innervated by two neurons located in each segment near the segmental nerves.

The lateral heart nerves and the heart neurons have been identified in some species but appear to be lacking in others. It is possible that they escape the attention of investigators because of their small size.

5.1.2. Malpighian Tubules, Oviducts, Digestive Tract

The Malpighian tubules often display considerable mobility owing to the contraction of the thin muscle strips that form spirals around them. They appear to be innervated only in some insects.

The oviducts, which are not always innervated, have like most visceral muscles the inherent ability to contract spontaneously and rhythmically. Their contractions are of importance for sperm transport to the spermatheca and for ovulation and oviposition.

The contractility of the digestive tract is rather more complex. Its different parts, the foregut, midgut, and hindgut, are independent and possess different innervations. The foregut and the anterior portion of the midgut are richly innervated by the nerves from the ganglia of the anterior sympathetic nervous system (frontal, hypocerebral, ingluvial, and proventricular ganglia).

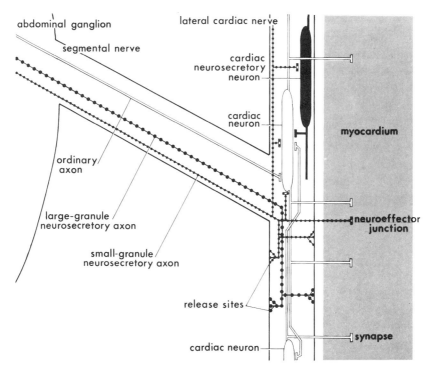

FIGURE 44. Diagram of left-side innervation of an abdominal heart chamber in *Periplaneta*. A segmental nerve joins the lateral nerve adjacent to the heart. The heart is innervated by intrinsic cardiac neurons and by descending neurosecretory axons from the central nervous system. The nonneurosecretory cardiac neurons are functionally interconnected either by synaptic contacts or by ephaptic junctions between axons. The activity of the cardiac neurons does not alter the activity in the cardiac neurosecretory neurons, which implies independent functions for these two types of nerve cells. Neural units from the central nervous system affect the activity of both neurosecretory cardiac neurons and cardiac motor neurons. (Modified from Miller, 1973.)

The innervation of the hindgut originates in a different site, the last abdominal ganglion. Numerous sensory receptors are present in the pharynx, foregut, and hindgut; there are tension receptors and also osmotic receptors.

5.2. SOURCE OF MYOTROPIC FACTORS, MULTIPLICITY

The presence of myotropic factors has not been investigated in a very large number of species, but positive results have been obtained in cockroaches, stick insects, locusts, *Corethra* larvae, and the bug *Iphita limbata*.

Although some investigations were carried out *in vivo*, most studies were made on semi-isolated organs such as the heart associated with the dorsal integument, the alary muscles, and the ventral diaphragm.

The effects observed are mainly concerned with the frequency of contractions of the effector, whether it is the heart, Malpighian tubules, gut, or oviducts. At the same time, however, an increase in the tonus and an increase in the amplitude of the contractions have often been reported.

Investigations into the myotropic factors of insects raise several problems, particularly concerning the number of active factors and their source and release sites.

The presence of a myotropic factor in the corpora cardiaca of the cockroach *Periplaneta* was clearly established in early investigations (Cameron, 1953; Davey, 1961a,b, 1962a). Davey (1963) noted that induced hyperactivity brings about a decrease in the myotropic factor content of the corpora cardiaca and causes the appearance of a powerful heart stimulator in the hemolymph, which is not found in decapitated insects or in those deprived of the corpora cardiaca and subjected to the same treatment.

Some investigators assume that the corpora allata also exhibit myotropic activity. This has been observed by Pilcher (1971) on the isolated Malpighian tubules of *Carausius*, by Morohoshi and Ohkuma (1968) on the heart of *Bombyx*, and by Roussel (1969a) who carried out considerable work on the *in vivo* heartbeat regulation in *Locusta*. According to Roussel, the most important organ for maintaining the heartbeat is the corpus allatum, and this has been confirmed by the cardioacceleratory action of the synthetic juvenile hormones JH II and JH I (Roussel, 1975a,b). However, other authors do not substantiate this effect of the corpora allata on the visceral muscles, and this point requires further investigation.

Myotropic factors have been identified in the central nervous system by several investigators who assayed the activity of brain, subesophageal ganglion, and ventral nerve cord ganglia extracts on the heart, and sometimes the activity of other organs. According to the data of Ralph (1962b), Rounds (1963), and Rounds and Gardner (1968) on *Periplaneta*, a cardioaccelerator appears to be present throughout the central nervous system. Gersch (1958) reached the same conclusion concerning *Corethra* (Fig. 45). The Malpighian tubules of *Carausius*, the foregut of *Corethra* (Gersch, 1955), and the oviducts of *Locusta* (Girardie and Lafon-Cazal, 1972) also respond to ventral ganglia extracts (Table XVI).

The work of Rounds (1963) on the cockroach *Blaberus giganteus*, in which extracts were prepared in alcohol–water mixtures of varying proportions, demonstrated the presence of cardiodecelerators in addition to cardioaccelerators, distributed like the latter throughout the central nervous system. In the cockroach *Blaberus craniifer*, Banks (1976) also observed a cardioac-

TABLE XVI. *In Vitro* Effects of Various Neuroglandular

Target organ	Donor	Recipient	Reference
Heart	*Periplaneta americana*	*Periplaneta*	Cameron (1953)
			Davey (1961a, 1962a)
			Ralph (1962b)
			Rounds and Gardner (1968)
			Gersch (1974b)[c]
	Blaberus giganteus	*Blaberus*	Rounds (1963)
	Blaberus craniifer	*Blaberus*	Banks (1976)[d]
	Locusta migratoria	*Locusta*	Cazal (1967)
	Schistocerca gregaria	*Schistocerca*	Mordue and Goldsworthy (1969)
	Carausius morosus	*Carausius*	Raabe *et al.* (1966)
	Clitumnus extradentatus	*Clitumnus*	
		Periplaneta	
	Locusta migratoria	*Locusta*	
		Periplaneta	
	Periplaneta americana	*Periplaneta*	
	Calliphora erythrocephala	*Calliphora*	Normann (1972)
	Corethra plumicornis	*Corethra*	Gersch (1958)
	Periplaneta americana	*Corethra*	
Malpighian tubules	*Periplaneta americana*	*Periplaneta*	Cameron (1953)
	Carausius morosus	*Dytiscus*	Koller (1954)
	Carausius morosus	*Carausius*	Enders (1955)
			Pilcher (1971)
Foregut	*Periplaneta americana*	*Periplaneta*	Cameron (1953)
	Periplaneta americana	*Corethra*	Gersch (1955)
	Corethra plumicornis	*Corethra*	
	Locusta migratoria	*Locusta*	Cazal (1969)
		Blabera	
	Blabera fusca	*Locusta*	
Hindgut	*Periplaneta americana*	*Periplaneta*	Cameron (1953)
			Davey (1962c)
	Carausius morosus	*Tenebrio*	Koller (1954)
	Carausius morosus	*Carausius*	Enders (1955)
	Locusta migratoria	*Locusta*	Cazal (1969)
		Blabera	
	Blabera fusca	*Locusta*	
	Aspongopus janus	*Iphita*	Nayar (1956b)
Oviducts	*Carausius morosus*	*Carausius*	Koller (1954)
	Carausius morosus	*Carausius*	Enders (1955)
	Locusta migratoria	*Locusta*	Girardie and Lafon-Cazal (1972)

[a]Abbreviations used: ab gg, abdominal ganglion; ca, corpora allata; cc, corpora cardiaca; po, perisympathetic organs; se gg, subesophageal ganglion; th gg, thoracic ganglion.
[b]+, ++, +++, Slight, medium, strong stimulation; −, inhibition; — + —, extract performed with different ganglia.

Organs on Visceral Muscle Contractions

| | | | | | | | | | | | | | Jast | |
| | | | se | th1 | th2 | th3 | ab1 | ab2 | ab3 | ab4 | ab5 | ab | |
ca	cc	Brain	gg	gg	gg	gg	gg	gg	gg	gg	gg	gg	po
0	+	0	0										
	+												
0	+++	++	++	++	+	+	0	0	+	+	+	++	
	+	+	+	————	+	————			————	+	————		
				−	−	−						+	
		+	+	————	+	————	————————	+	————————		
		−	−	————	−	————	————————	−	————————		
	++	−	+	−	−	−	0	+	+	+	+	++	
	+												
0		0											
													+
													+
													+
													+
													+
													+
	0												
+	+	+	+	————————————————	+	————————————————		
+	+	+	+	————————————————	+	————————————————		
0	+	+											
	+	+											
	+	+											
+	+	+	+	+								+	
	−												
				————————————————	+	————————————————		
				————————————————	+	————————————————		
	+												
	+												
	+												
	+												
+	−	−											
+	−	−											
	+												
	+												
	+												
	+	0											
		+											
−	−	+											
+	+	+	+					————————	+	————————		

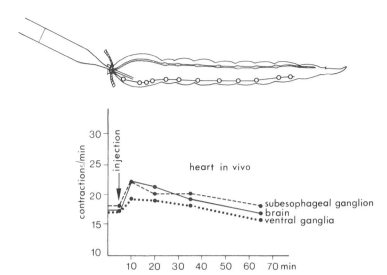

FIGURE 45. Effect of extracts from different parts of the central nervous system on the heartbeat in a decapitated *Corethra plumicornis* larva. (Modified from Gersch, 1958.)

celerator effect of the corpora cardiaca and the abdominal ganglia and a slight decelerator effect of the brain, the thoracic ganglia, and the first abdominal ganglion. Performing stimulations of the different thoracoabdominal ganglia *in vitro* on *Periplaneta*, Gersch (1974b) obtained the release of cardioaccelerators from the sixth abdominal ganglion, and the release of cardiodecelerators from the thoracic ganglia (Fig. 46). Furthermore, attempting to determine the presence of two types of factors in the ventral nerve cord at different periods of the circadian cycle, Rounds (1963) observed that cardioaccelerators and cardiodecelerators exist in *Blaberus* throughout the central nervous system, but that they are not stored in the same place at the same time: brain and thoracic ganglia store decelerators at noon and accelerators at 11 PM, while the subesophageal and the abdominal ganglia store accelerators at noon and decelerators at 11 PM.

While the presence of antagonistic factors appears probable, many distinct myotropic factors also seem to exist. In some cases, opposite results are obtained on different effectors by using the same extract, or on the same effector by using different extracts (Table XVII). Hence, for example, brain and corpora cardiaca extracts do not have the same action on the heart of *Periplaneta*: some investigators state that brain action is weak, nil, or even diametrically opposed to that of corpora cardiaca; the same extracts also act in opposite ways on the oviduct of *Carausius*.

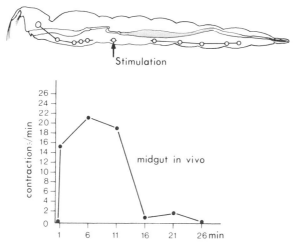

FIGURE 46. Effect on the *Corethra plumicornis* midgut of electrical stimulation of the second abdominal ganglion in an insect deprived of the first and third abdominal ganglia. A myotropic factor is released. (Modified from Gersch, 1955.)

5.3. RELEASE SITES OF MYOTROPIC FACTORS

5.3.1. Corpora Cardiaca and Perisympathetic Organs

The corpora cardiaca seem to perform a twofold function. On the one hand, they release a substance coming from the pars intercerebralis of the brain by the nccI pathway (Holman and Marks, 1974), as indicated by experiments in which functional brain–corpora cardiaca complexes of the cockroach *Leucophaea maderae* are maintained *in vitro*, or in other *in vitro* experiments where electrical stimulations were applied to the different nerves of the corpora cardiaca and the effect of the bathing fluid then assayed on the heartbeat of *Periplaneta* (Kater, 1967, 1968; Gersch *et al.*, 1970). On the other hand, the secretory cells of the corpora cardiaca appear to synthesize a factor distinct from the brain factor, for in locusts, the glandular lobe displays a much greater activity than the neurohemal lobe (Davey, 1963; Cazal, 1969).

Apart from their innervation by the brain, the corpora cardiaca are innervated by the anterior sympathetic nervous system. The ingestion of glucose, which induces a strong heartbeat acceleration in *Periplaneta,* serves to demonstrate these pathways (Davey, 1962a,b). The impulses from the sensory receptors of the labrum are transported through the frontal ganglion and the recurrent nerve, from which they reach the corpora cardiaca.

The cardioaccelerator of the ventral nerve cord is released at least in part

TABLE XVII. Effect of Neuroglandular Extracts on Various Target Organs

Donor	Recipient	Reference	Extract origin[b]	Target organ[a]				
				Heart	Malpighian tubules	Foregut	Hindgut	Oviduct
Periplaneta americana	Periplaneta	Cameron (1953)	Brain	0	+			
			cc	+	+			
		Ralph (1962b)	Brain	++			+	
			cc	+++		-		
Blaberus cranifer	Blaberus	Banks (1976)	Brain	-				
			cc	++				
Carausius morosus	different coleopterans and Carausius	Koller (1954)	Brain		+		-	+
			cc		+		-	
Carausius morosus	Carausius	Enders (1955)	Brain	+	+		-	+
			cc		+		+	-
			ca					-

[a]0, No effect; +, ++, +++, slight, medium, and strong stimulation; -, inhibition.
[b]cc, Corpora cardiaca; ca, corpora allata.

by the perisympathetic organs (Raabe *et al.*, 1966), which contain a large quantity of it in *Periplaneta, Carausius,* and *Locusta* (Fig. 47).

5.3.2. Distal Neurohemal Organs and Neuroeffector Junctions

Not all the myotropic factors are released in the traditional neurohemal organs, the corpora cardiaca and the perisympathetic organs. Distal neurohemal organs such as the lateral heart nerves and the stomodeal and proctodeal nerves also take part. Moreover, neurosecretory products are conveyed to the effectors, which, like their nerves, contain neurosecretory fibers with electron-dense granules and display strong myotropic activity. As shown in Table XVIII, myotropic activity has been found in many organs provided with visceral muscles—the Malpighian tubules, foregut, and hindgut—and in the nerves that innervate them—the lateral heart nerves in *Periplaneta* (Johnson and Bowers, 1963), the stomodeal and proctodeal nerves in *Periplaneta* and *Leucophaea* (Brown, 1967; Holman and Cook, 1970, 1972). These nerves display very strong activity in comparison with that of the organs they innervate and that of the nerve ganglia (Brown, 1967). It may be inferred, therefore, that these nerves simultaneously perform a neurohemal role and a transit role, for the activity of the hindgut disappears after severance of the proctodeal nerves (Brown, 1967).

In *Periplaneta,* the presence of electron-dense granules has been observed in the segmental and lateral nerves of the heart (Johnson and Bowers,

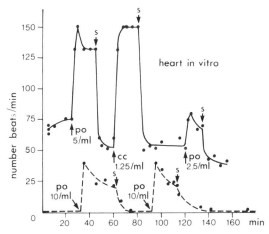

FIGURE 47. Effect of corpora cardiaca and perisympathetic organ extracts on the heartbeat in *Periplaneta americana*. Dashed lines represent the responses of an arrested heart. cc, Corpora cardiaca; po, perisympathetic organs; s, saline. (Modified from Raabe *et al.,* 1966.)

TABLE XVIII. Myotropic Activity of Some Organs and Their Nerves

| Donor | Recipient | Reference | Extract origin | Target organ[a] | | |
				Heart	Malpighian tubules	Hindgut
Periplaneta americana	Periplaneta	Johnson and Bowers (1963)	Heart lateral nerves	+		
		Brown (1967)	Foregut			+
			Hindgut			+
			Stomodeal nerves			+++
			Proctodeal nerves			+++
Leucophaea maderae	Leucophaea	Holman and Cook (1970, 1972)	Proctodeal nerves			+
			Hindgut			
Carausius morosus	Carausius	Pilcher (1971)	Malpighian tubules		+	
			Midgut		+	
Vanessa jo	Tenebrio and other coleopterans	Koller (1948)	Gut		0	+

[a]0, No effect; +, slight activity; +++, strong activity.

1963; Johnson, 1966b; Miller, 1968), in the nerves innervating the alary muscles (Adams *et al.*, 1973), in the proctodeal nerves (Holman and Cook, 1972), and in the effectors themselves, heart, alary and hyperneural muscle, midgut, and rectal papillae (Oschman and Wall, 1969; Wright *et al.*, 1970; Miller and Adams, 1974).

In other species, the presence of neurosecretory products has also been reported in the nerves or in the effector organs themselves: the proctodeal nerves of the larva of the coleopteran *Oryctes nasicornis* (Nagy, 1978), the hearts of the blowfly *Calliphora erythrocephala* (Johnson, 1966a; Normann, 1972) and the aphid *Myzus persicae* (Bowers and Johnson, 1966), the foreguts of the locust *Schistocerca gregaria* (Klemm, 1972) and the mosquito *Culex tarsalis* (Burgess, 1973), the hindguts of the coleopteran *Blaps mucronata* (Fletcher, 1969) and the aphid *Myzus persicae* (Johnson and Bowers, 1963), and the recta of the termite (Noirot and Noirot-Timothée, 1976) and the blowfly *Calliphora* (Johnson, 1966a; Gupta and Berridge, 1966).

Part of the neurosecretory products conveyed by the nerves is probably released into the hemolymph near the effectors; another part is released within the organ at the neurosecretomotor junctions. These have been observed in the heart and the gut of *Periplaneta* (Johnson, 1963; Johnson and Bowers, 1963; Brown, 1967) and in the rectum of *Calliphora* (Gupta and Berridge, 1966), where two types of granules have been observed, from 100 to 150 and from 100 to 300 nm, thus raising the problem of a potential role played by amines in addition to the polypeptidic neurohormones (see Section 1.3).

The substances represented by the electron-dense granules revealed by the electron microscope do not only originate in the neurosecretory cells of the central nervous system. Multipolar neurosecretory cells also exist in contact with or in the vicinity of the rectum and the heart, along the nerves that innervate these organs such as the proctodeal nerve or the lateral heart nerves. Their release sites are not only located against the connective membrane of the nerve (Nagy, 1978, proctodeal nerve of *Oryctes*), but also against the muscles themselves (Reinecke *et al.*, 1978, proctodeal nerve, ileum, and colon of *Manduca sexta*).

To conclude, it thus appears that the visceral muscles of insects possess a particularly sophisticated regulation that includes nervous control as well as neurohemal control of peripheral and central origin.

The neurohormone release sites are highly diversified. They are located in the central neurohemal organs—corpora cardiaca and perisympathetic organs, in the distal neurohemal organs located near the effectors, and in contact with the effectors themselves in the neuroeffector junctions.

A peripheral control exerted by sensory neurons seems to exist because they convey information concerning internal conditions. On the other hand, central control appears to be necessary. The nervous and neurohormonal regulations are not superimposed, because the latter has a far broader range of action than the former. The respective roles played by the centralized release site, the peripheral release sites, and the neuroeffector junctions are more difficult to distinguish, and the explanation is perhaps to be found in the degree of innervation of the organs analyzed. When innervation is nil or poor, only the centralized organs can intervene. If innervation is rich, other release modes probably take over. With respect to the lateral heart nerves, which are important neurohemal organs in the cockroach, it is not at all sure whether the neurosecretory products they contain are related only to regulation of the heartbeat. The lateral nerves may store several factors that, by being released in the vicinity of the heart, are vigorously propelled into the general bloodstream.

5.4. REALITY OF THE EXISTENCE OF MYOTROPIC FACTORS

Some authors have questioned the existence of myotropic factors, basing their observations on the absence of any effect of the corpora cardiaca on the heartbeat of some species, especially *in vivo* as in *Locusta* (Roussel and Cazal, 1969) where the removal or implantation of these organs and of the pars intercerebralis is without effect (Roussel, 1969b, 1970). On the other hand, other facts confirm the existence of these factors. Hence, in the larva of *Corethra,* antiperistaltic movements of the midgut are stimulated by excitations applied to the ventral nerve cord ganglia, even if the latter are totally disconnected (Gersch, 1955). Furthermore, cardiac acceleration has been observed *in vivo* in *Periplaneta* after the injection of corpora cardiaca extracts (Hertel, 1971), and it has been shown that this occurs naturally when the insect is fed with glucose (Davey, 1962a). Stress also elevates the level of myotropic factors in the hemolymph (Rounds, 1975).

Variations in the motility of the different organs, and in particular the heart, have been often reported in relation to age, sexual state, food intake, activity, and environmental factors; thus, it is probable that endocrine regulation exists, and this has been demonstrated definitively by the isolation of one of the active factors, proctolin.

5.5. SEPARATION AND IDENTIFICATION OF ACTIVE FACTORS

The purification and separation of myotropic factors have been attempted by several groups using different methods, so that the data are abundant but sometimes difficult to reconcile (Table XIX).

Most of the investigations used the semi-isolated heart of *Periplaneta* as

TABLE XIX. Effects of Some Fractions Isolated from Neurohemal and Nervous Tissue upon Various Targets

Insect	Factor origin	Factor	MW	Process					
				Heartbeat		Hindgut contractions	Spontaneous nerve activity	*Carausius* pigment movements	Blood trehalose concentration
				Frequency	Amplitude				
Locusta migratoria and *Schistocerca gregaria*	cc glandular lobe[a]	B		++	–				+
Periplaneta americana	cc neurohemal lobe[a]	C		+	+				
	Central nervous system[b]	C¹		+	–		+	+	
		C²		+	0			0	
		D¹	2000	+	+		+	–	
		D²		+	0			0	
	cc or head[c]	1a		+					
		1b		+					
		2a	1795	+					
		2b	1287	+					
Leucophaea maderae	Rectum[d]	Proctolin	500–700			+			
	Head[e]	HSN[f]	400–600			+			

[a]Data from Mordue and Goldsworthy (1969).
[b]Data from Gersch *et al.* (1960) and Gersch and Richter (1963).
[c]Data from Traina *et al.* (1976).
[d]Data from Brown and Starratt (1975).
[e]Data from Holman and Cook (1979).
[f]HSN, hindgut—stimulating neurohormone.

an assay for active factors and the corpora cardiaca as the donor organ, but the hindgut was also used as an effector and also as a donor organ. The authors agree that the myotropic substances are heat-stable peptides the activity of which is slightly increased by boiling.

As for the number of active factors, the results are fairly divergent. Ralph (1962b) and Rounds (1963), using aqueous and organic solvents and alcohol–water mixtures of varying proportions, respectively, found several accelerators and also decelerators. By employing chromatography, Brown (1965) separated six to nine factors of which the three most active are the factors P_1, P_2, and P_3, the first acting on the heart, the second on the hindgut, and the third on both organs.

Mordue and Goldsworthy (1969) compared the effects of each lobe of the corpora cardiaca on diuresis and on the heartbeat in *Schistocerca* and on the hemolymph trehalose level in *Periplaneta*; and then the effects of two fractions of each of these lobes separated by chromatography. The fractions B and C display effects comparable to those of extracts of glandular and neurohemal lobes of the corpora cardiaca. There are slight differences in their action. Fraction B increases heartbeat frequency but reduces its amplitude; fraction C is less active on heartbeat but increases its amplitude. A hypertrehalosemic effect appears to be associated with both factors, but diuretic activity seems to be present essentially in fraction C. The number of factors isolated is thus small, and they appear to display several activities.

The research team of Natalizi, Frontali, and Traina (1966, 1970, 1974, 1976), employing Sephadex gel filtration, carried out a comparable investigation, and attempted to isolate the cardioaccelerators and to determine whether they are distinct from the hypertrehalosemic factors and from the factor stimulating spontaneous nerve activity (see Chapter 7). At the outset of their investigation, hypertrehalosemic activity appeared to be associated with cardioaccelerator activity, but finally, owing to improvements in purification methods, four cardioaccelerators were discovered, distinct from the hypertrehalosemic factors. Two of them have been subjected to amino acid analysis. Taking leucine as a unit of molar proportions, and postulating the existence of a single tryptophan residue, molecular weights of 1795 and 1287 were calculated for two peptides composed respectively of 12 and 17 amino acids. Factor 2b contains lysine, aspartic acid, serine, glutamic acid, proline, glycine, alanine, valine, and leucine; factor 2a contains the same amino acids plus threonine and phenylalanine (Natalizi and Frontali, 1966; Natalizi *et al.,* 1970; Traina *et al.,* 1974, 1976).

The factors of the central nervous system have stimulated less research than those of the corpora cardiaca. They have been investigated by Krolak *et al.,* (1977), who isolated three substances from the last abdominal ganglion of *Periplaneta* that are active on the heart. One appears to be acetylcholine. The second, which increases the heartbeat amplitude, is a peptide of molecular weight 3500–7500, and which may be identical to one of the factors to be discussed next.

The main investigations concerning the factors present in the central nervous system have been those of Gersch and his co-workers. Three substances active on the heart of *Periplaneta*, factors C, D, and acetylcholine, were first found in the central nervous system and in the endocrine glands of this insect (Unger, 1957). It subsequently appeared that factors C and D each decomposed into two factors (Gersch *et al.*, 1960), giving four factors, factors C_1, C_2, D_1, and D_2, which increased heartbeat frequency but differed in their action on heartbeat amplitude, which was increased by factor D_1 reduced by factor C_1, and unaffected by factors C_2 and D_2. In a subsequent work, Gersch and Richter (1963) showed that factors C_1 and D_1 acted not only on the heartbeat but also on pigment movements in the stick insect, and on the spontaneous nerve activity of the phallic nerve in the cockroach. It should be noted that factors C_1 and C_2 are present in the central nervous system and in the corpora cardiaca, like factors D_1 and D_2; but they are also reported to be present in the corpora allata.

The results of the foregoing investigations suggest that many factors exist, and that there is a correspondence between some of them. Hence Brown's factor P_3 may be identical to Gersch's factor C, factor 2 of Traina *et al.*, and factor B of Mordue and Goldsworthy; and Brown's factor P_1 may be identical to Gersch's factor D, factor 1 of Traina *et al.*, and factor C of Mordue and Goldsworthy.

The most thorough purification was finally achieved by using the effector itself, in this case the hindgut, as the source of active factor. This succeeded in isolating and identifying a basic pentapeptide, proctolin (Brown, 1975; Brown and Starratt, 1975; Starratt and Brown, 1975), capable of stimulating contractions of the hindgut in the cockroach *Periplaneta*. Proctolin has a molecular weight of 500–700. Of chymotrypsin, trypsin, carboxypeptidases A and B, leucine aminopeptidase, only the latter was shown to inactivate proctolin. Its formula is:

Arg-Tyr-Leu-Pro-Thr

It was synthesized in 1977 by Starratt and Brown.

The proctolin unit, defined earlier as the quantity present in the rectum, was finally calculated, and corresponds to 0.86 ng. Its activity threshold on the proctodeal muscle *in vitro* is about 10^{-9} M.

Besides its action on the hindgut, proctolin stimulates cockroach heartbeat (Holman and Cook, 1979), and initiates a rise in muscle tonus or an increase in the frequency and amplitude of individual phasic contractions in the oviducts of the horsefly *Tabanus sulcifrons* (Cook and Meola, 1978), suggesting that it may be a unique visceral muscle stimulatory factor (Holman and Cook, 1972). However, further investigations performed in *Leucophaea* show that proctolin is not the only myotropic insect factor and that a second one, the hindgut-stimulating neurohormone (HSN), occurs in the head and probably in the central nervous system (Holman and Cook, 1979).

Hence the question of the number of active factors remains unresolved, and the demonstration of interspecific activity by many investigators fails to help solve this problem, although it shows that one or more active factors are present in the different species (Table XX). Corpora cardiaca and brain extracts of *Carausius* are active on the contractility of the Malpighian tubules of different coleopterans, and especially of *Dysticus* (Koller, 1954). Corpora cardiaca extracts of *Locusta* stimulate contractility of the different parts of the digestive tract of *Blabera* and vice versa (Cazal, 1969). Extracts of perisympathetic organs of *Periplaneta, Carausius, Clitumnus,* and *Locusta* are active on the heart of these four species (Raabe *et al.,* 1966). The different parts of the central nervous system of the coleopteran *Melolontha* stimulate motility of the digestive tract in *Corethra* and act in the same manner on the heart of *Corethra* and *Periplaneta* (Kirchner, 1962). A substance separated from the head of the honeybee is active on the heart of *Periplaneta* (Natalizi and Frontali, 1966).

Proctolin proves to be present in seven species of different orders, *Periplaneta americana, Locusta migratoria, Gryllus pennsylvanicus, Musca domestica,* the coleopteran *Sitophilus granarius,* the blood-sucking bug *Rhodnius prolixus,* the honeybee *Apis mellifica,* and the lepidopteran *Malacosoma americana* (Brown, 1977).

5.6. MODE OF ACTION OF MYOTROPIC FACTORS

The main question posed with respect to the mode of action of myotropic factors is related to their release mechanism, which, as we have seen above, can occur in the central neurohemal organs, in distal release sites located on the heart or gut nerves, or in contact with the muscles themselves.

Davey's investigations (1961a,b) led this author to postulate an indirect action for myotropic peptides which stimulate the production of indolalkylamine, probably tryptamine, by the pericardial cells. This concept is based on the use of an indolalkylamine inhibitor, the diethylamide of 2-bromolysergic acid, and on the blocking of the activity of the pericardial cells by dyes or by India ink.

The action of the pericardial cells in controlling cardiac acceleration has sometimes been questioned, but it appears that active extracts act at least partly in an indirect manner. They could exert their action on the nervous system of the heart through the intermediary of acetylcholine neurons, for the

TABLE XX. Interspecific Activity of Myotropic Extracts[a]

Reference	Donor		Effector	
	Insect	Organ	Insect	Organ
Koller (1954)	*Carausius morosus*	cc, brain	Coleopteran	Malpighian tubules
Cazal (1969)	*Locusta migratoria*	cc	*Blabera*	Foregut and hindgut
	Blabera fusca	cc	*Locusta*	Foregut and hindgut
Raabe *et al.* (1966)	*Clitumnus extradentatus* and *Locusta migratoria*	po	*Periplaneta*	Heart
Kirchner (1962)	*Melolontha melolontha*	cns	*Corethra*	Heart and foregut
		cns	*Periplaneta*	Heart
Natalizi and Frontali (1966)	*Apis mellifica*	Head	*Periplaneta*	Heart

[a]Abbreviations used: cc, corpora cardiaca; cns, central nervous system; po, perisympathetic organs.

application of neurohormone D is only effective in the presence of this neurotransmitter (Richter and Stürzebecher, 1969; Richter and Gersch, 1974; Gersch *et al.*, 1974). These neurons may be the heart neurons, because corpora cardiaca extracts induce a spectacular rise in their spontaneous electrical activity (Smith, 1969; Richter and Stürzebecher, 1969; Hertel, 1971; Richter, 1973).

It also appears that myotropic substances can act directly on the myocardium, for factor D depresses the membrane potential (Richter, 1973) and acts both on the innervated heart and on the denervated heart (Hertel, 1975). According to Brown (1975), proctolin acts like a neurotransmitter, whereas Holman and Cook (1979) consider that it behaves as a modulator. In a recent investigation, it was shown that like octopamine, it induces a rhythmic depolarization when used in small concentrations and acts postsynaptically in the body-wall muscles of *Lucilia sericata* (Irving and Miller, 1980). These questions are difficult to resolve, but the existence of a synthetic proctolin provides hope for rapid progress through the use of labeled antibodies.

Although acetylcholine and various amines exert an effect similar to that of the myotropic substances, several investigators have shown that the myotropic factors of the nervous system and of the effectors are distinct from the former.

The role of cAMP in the effects exerted by myotropic factors is debatable. In fact, the Malpighian tubules of *Carausius* and the heart of *Periplaneta* respond neither to cAMP nor to aminophylline (Pilcher, 1971; Collins and Miller, 1977), although the hindgut of *Leucophaea* does respond to cAMP (Cook *et al.*, 1975).

5.7. CONCLUSIONS

Myotropic factors active on the contractility of the heart, the digestive tract, the Malpighian tubules, and the rectum are present in insects.

They are synthesized in the neurosecretory cells of the central nervous system and are released into the traditional neurohemal organs, the corpora cardiaca and the perisympathetic organs, into the nerves such as the aortic, proctodeal, and stomodeal nerves, and also into the effectors themselves.

Several factors probably exist, and one of them, proctolin, a pentapeptide, has been identified and synthesized.

The mode of action of myotropic factors is not yet completely clear, but they may act either on peripheral neurons or directly on the effector.

Morphological and Physiological Color Change

The mimicry of insects (their resemblance in shape and color to their background), is often striking, but what is even more striking is the existence of internal mechanisms that enable a given species to adapt to environmental conditions in the course of its life. Color change can be achieved by two entirely different processes: pigment synthesis (morphological color change) or rapid movements of pigment granules within the cells containing them (physiological color change).

6.1. MORPHOLOGICAL COLOR CHANGE

Morphological color change is displayed by many species: dragonflies, mantids, grasshoppers, locusts, stick insects, beetles, and lepidopteran larvae and pupae.

Morphological color change usually consists in matching the color of the insect to its background. However, pigment modifications also occur in relation to environmental factors such as temperature, photoperiod, humidity, gregariousness (locusts and crickets), and maternal influences (Fig. 48). Pigment changes may be associated with a particular type of development such as diapause. The O_2 and CO_2 content of the surrounding atmosphere also has a bearing and possibly plays a part in the color changes associated with gregariousness.

In most cases, the compound eye is responsible for mimicry. The removal of the optic lobes (Atzler, 1931) or their disconnection from the brain (Moreteau-Levita, 1972a) makes adaptation impossible. In some species, however, color change is independent of the eye and results from the direct action of light on the integument.

Insect colors are widely varied. They are partly due to selective reflection of light radiation by the integumentary surface, but they result mainly from

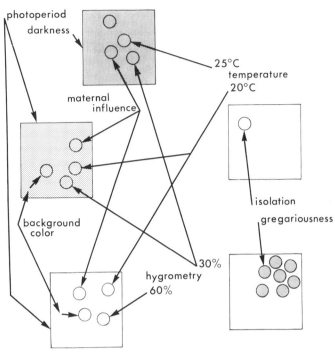

FIGURE 48. Effect of environmental conditions, gregariousness, and maternal influences on the morphological color change in insects. Open circles, light-colored insects; dark circles, dark-colored insects.

the color of the pigments present in the cuticle or in the epidermal cells, and sometimes in the internal tissues.

The pigments of the integument belong to many biochemical families. The cuticle contains melanins and pterins. The former, generally black, may be brown or yellow. Pterins are often white or yellow, sometimes red. The epidermal cells contain tetrapyrrolic pigments, carotenoids, and ommochromes. Tetrapyrrolic or bile pigments (biliverdin chromoproteids or other bile pigments) are green; upon oxidation, they become yellow, brown, or black. Carotenoids (β-carotene or astaxanthin) are yellow and orange. Ommochromes are represented by ommines and ommatines, brown xanthommatine, red dihydroxanthommatine, and colorless 3-hydroxykinurenin.

6.1.1. Locusts, Grasshoppers, Stick Insects

Pigment variations and their regulation are quite comparable among locusts, grasshoppers, and stick insects.

In the migratory locust *Locusta migratoria,* pigmentation depends essentially on the phase. Gregarious insects have a dark body color tending to black, while solitary insects have a light body color ranging from green to brownish-yellow, depending on the background. The pigments involved are brown ommochromes, carotenoids, green, yellow, and brown tetrapyrrolic pigments, and melanins.

In the red locust *Acrida turrita,* the grasshopper *Oedipoda cerulescens,* and the stick insect *Carausius morosus,* the insect harmonizes its pigmentation with that of the background. The color range is broader in *Oedipoda* (yellow, red, gray or black) and *Carausius* (green, beige, brown, or black) than in *Acrida* (green to yellow). The pigments involved are essentially bile pigments, carotenoids, and ommochromes (xanthommatine and ommine) (Bückmann and Dustmann, 1962). Pterins and melanins are of relatively lesser importance. In *Carausius,* coloration depends on many factors; elevated temperatures and darkness favor the formation of dark colors by the synthesis of ommochromes and melanins. High moisture content, illumination, and low temperatures stimulate the appearance of light-colored forms. Moreover, the color of the offspring depends on the color of the mother (Raabe, 1961).

6.1.1.1. Light Body Color

Green body coloration depends on the corpora allata in many grasshoppers (Rowell, 1967), the locusts *Locusta* (Joly, 1951; Joly *et al.,* 1956; Staal, 1961) and *Acrida* (Joly, 1952), and *Carausius* (Raabe, 1961; Dustmann, 1964). In *Acrida* and *Locusta*, the implantation of these organs results within 24 hr in pigmentation of the blood, which changes from yellow to green. The integumentary change occurs more slowly and only becomes apparent after molting. It involves two distinct processes: the elimination of melanins and ommochromes, and the synthesis of biliverdins. The action of the corpora allata appears to be direct, because if these organs are implanted outside the blood stream, between the integument and the fat body, a green patch appears above the implant (Joly, 1954).

Maintenance of the green pigmentation also depends on the corpora allata. Their removal results, after the next molt, in a gregarious pigmentation in *Locusta* (Joly, 1951) and in a beige pigmentation in *Carausius* (L'Hélias, 1955; Raabe, 1961). In the latter, however, the implantation of the corpora allata or the disconnection of these organs in green or beige insects leads, paradoxically, to the formation of ommochromes and the appearance of a black body color (Pflugfelder, 1938; L'Hélias, 1955; Raabe, 1961; Dustmann, 1964). This effect is probably due to some interaction between juvenile hormone and a darkening neurohormone (see Section 6.1.1.2).

Other correlations exist between pigmentation and corpora allata. In locusts, these organs induce yellow body color at sexual maturity (Loher,

1961; Pener, 1965; Pener *et al.*, 1972). In crickets, they are responsible for lightening the body color, which accompanies gregarious life (Roussel, 1966; Fuzeau-Braesch, 1968).

The pigmentation can be modified by operations dealing with the molting gland, but the results are contradictory and hence difficult to interpret. In the desert locust *Schistocerca gregaria*, partial removal of the molting gland in solitary green larvae gives rise to a yellow body color spotted with black (Ellis and Carlisle, 1961), whereas the implantation of this gland in the stick insect *Carausius* causes browning of the green and beige insects by ommochrome synthesis (Raabe, 1961). As we shall see below, ecdysone also has an effect on the epidermal ommochromes in the puss moth *Cerura vinula*.

6.1.1.2. Dark Body Color

The action of one or more neurohormones in the synthesis of dark pigments, ommochromes and melanins, appears to be established. The source of the active factor, however, differs from one species to another. In *Locusta* and *Oedipoda*, the pars intercerebralis and corpora cardiaca contain a darkening factor (Staal, 1961; Girardie and Cazal, 1965; Moreteau-Lévita, 1972b,c) originating in the Cj cells of the pars intercerebralis (Girardie, 1967), and also in the lateral cells of the protocerebrum (Girardie and Cazal, 1965). In *Carausius*, the implantation of the pars intercerebralis or the entire brain is without effect, whereas the implantation of the subesophageal ganglion causes darkening (Raabe, 1966). It is possible that the darkening factor originates in two medioventral cells of the subesophageal ganglion, for their appearance is quite different depending on body color (Fig. 49). In *Schistocerca* first-instar larvae, melanization is under the influence of a factor issuing from the metathoracic ganglion (Padgham, 1976a,b).

6.1.2. Lepidopteran Larvae and Pupae

Color changes are frequently encountered in lepidopterans, often occurring in the pupa but sometimes in the larva. They may be associated with diapause or with other developmental stages. Environmental factors such as light, background, temperature, and diet frequently determine the color change.

The saturniid larva *Hestina japonica* (Osanai and Arai, 1962) stops feeding in autumn, buries itself under dead leaves, and changes color from green to brown, the epidermal tetrapyrrolic pigments being replaced by an ommochrome (Osanai, 1966). In spring, the reverse occurs during the last larval instar, which takes place after the larva climbs back into the tree.

Pupae of the butterflies *Papilio xuthus* and *protenor* exhibit a variety of colors ranging from green to brown, which is also related to the type of

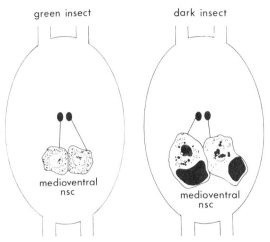

green insect   dark insect

medioventral
nsc

medioventral
nsc

FIGURE 49. Neurosecretory cells (nsc) producing the factor stimulating ommochrome synthesis in *Carausius morosus*. (Modified from Raabe, 1966.)

development, with or without diapause. Furthermore, the nutritional host plant induces a green coloration even in darkness. The pigments are essentially mesobiliverdin, carotenoids, and melanins (Ohnishi and Hidaka, 1957).

The coloration of the pupa of the white cabbage butterflies *Pieris rapae* and *brassicae* depends on the light intensity and background color perceived during a short period preceding pupation, and varies from light green to spotted black. The epidermal pigments are mesobiliverdin, yellow carotenoids, and reddish-brown ommochromes, while the cuticle is either transparent or melanized (Ohtaki and Ohnishi, 1967).

The pupa of the silkworm *Bombyx mori*, which varies from black to amber yellow, changes color in response to genetic factors and to the temperature experienced by the larva from the end of the spinning phase until pupation. The cuticle melanins responsible for the black pigmentation are deposited at low temperature in the strain bp (T); at 30°C they are absent from the integument, which assumes an amber color (Hashiguchi, 1960; Hashiguchi *et al.*, 1965).

Crowding sometimes affects body coloration, which is dark brown in the gregarious larvae of the armyworms *Leucania separata* and *Spodoptera exempta*, but whitish-yellow in the insects of these species living in isolation.

The puss moth *Cerura vinula* represents a special case because its prepuparial pigment changes are independent of environmental conditions and correspond to physiological stages. The green larvae bearing a light-brown dorsal pattern turn uniformly dark red when they stop feeding just before pupation. The dihydroxanthommatine that is formed results from the reduction of xanthommatine. It is first located in the epidermis and then spreads to the fat body and gut. Subsequently it is converted to colorless 3-hydroxykinurenin (Linzen and Bückmann, 1961).

6.1.2.1. Green Pigments

Green pigment synthesis in Lepidoptera appears to depend on the secretory activity of the corpora allata, as in Orthoptera. The injection of juvenile hormone into *Pieris* larvae induces the appearance in the pupae of a green or white patch (Hidaka and Ohtaki, 1963). If performed at the onset of the sensitive period, complete lightening of body color is observed (Ressin, 1980). The implantation of corpora allata or the injection of juvenile hormone into *Bombyx* larvae gives rise to a change in coloration from light gray to brownish-yellow (Kiguchi, 1972).

6.1.2.2. Dark Pigments

The synthesis of dark-brown or black pigments, ommochromes, and especially melanins appears to depend on one or more neurohormones produced or released into different parts of the ventral nerve cord. This has been demonstrated by the removal and reimplantation of organs, cutting of connectives, and ligatures performed on different regions of the body.

In *Papilio xuthus* and *protenor,* a series of thorough experimental investigations (Hidaka, 1956, 1961) revealed the control mechanisms governing the color change from green to brown. Ligatures were placed on young prepupae after the critical molting period, thus enabling pupation to take place in both parts of the body. However, ecdysis could not take place because of the ligature, and it was necessary to extract the pupa from its old cuticle. Wherever the ligature was placed, the green pigmentation persisted, but the ligature prevented browning of the hind region. Hence the darkening factor probably comes from the anterior region of the body. The removal of different nerve ganglia and the interruption of their connectives show that the brain, the subesophageal ganglion, and the prothoracic ganglion play a major role in the browning process (Table XXI and Fig. 50). By implanting different ganglia into ligated larvae, which are normally transformed into half-brown, half-green pupae, it may also be observed that the brain, the subesophageal ganglion, and the prothoracic ganglion are active only if interlinked. One may thus ask whether the three ganglia play the same role, or whether some of them only intervene by triggering and transmitting a nerve command, while the others secrete and/or release the active factor. If the nerve ganglia are removed at different times, it is observed that the critical period of browning occurs earlier for the brain and for the subesophageal ganglion than for the prothoracic ganglion, which may be the final source of the active factor.

TABLE XXI. *Papilio* Pupal Pigmentation
following Several Operations[a]

	Pigmentation	
	Green	Brown
Ganglion removed		
Control		+
Brain	+	
Subesophageal ganglion	+	
Prothoracic ganglion	+	
Mesothoracic ganglion		+
Metathoracic ganglion		+
First abdominal ganglion		+
Nerve section region (/)		
Brain/subesophageal ganglion	+	
Subesophageal ganglion/prothoracic ganglion	+	
Prothoracic ganglion/mesothoracic ganglion		+
Mesothoracic ganglion/metathoracic ganglion		+
First abdominal ganglion/second abdominal ganglion		+

[a]From data of Hidaka (1956, 1961).

Green *Pieris* larvae transferred to constant darkness change into brown pupae, but if they are ligated between the thorax and the abdomen, the anterior portion of the body turns brown, while the posterior portion remains green. As in *Papilio,* the brain, subesophageal ganglion, and prothoracic ganglion must be present and have intact connections in order to allow synthesis of the dark pigments (Ohtaki, 1960).

Experiments were also conducted on *Bombyx* prepupae of the strain bp (T), which are transformed into black pupae when raised at 20°C. In this case, it is the complex formed by the brain, subesophageal ganglion, and the three thoracic ganglia that is required for synthesis of the black melanic pigment, which cannot develop if one of these ganglia is removed. Synthesis is restored by the implantation of all five ganglia. By removing one of the five at different times, the role of each can be clarified, as was done by Hidaka on *Papilio.* It appears that in *Bombyx,* the metathoracic ganglion plays the part performed by the prothoracic ganglion in *Papilio* (Hashiguchi, 1960, 1962, 1964).

The results obtained in *Leucania, Spodoptera,* and *Hestina* larvae are rather different. A melanization factor is present in *Leucania* in the brain–corpora cardiaca complex, and also in the subesophageal ganglion (Ogura *et al.,* 1971; Ogura, 1975), but the nerve connectives are not a prerequisite for the functioning of the two nerve ganglia (Table XXII). The subesophageal ganglion is active when implanted in solitary larvae or *in vitro* (Ogura and Mitsuhashi, 1978, 1979). In *Spodoptera,* the study of which is just beginning,

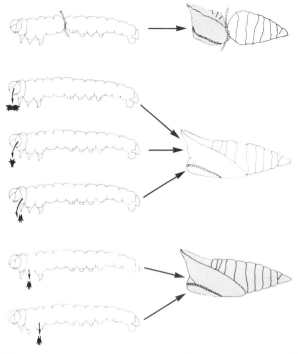

FIGURE 50. Effect of removal of various nerve ganglia on color change in *Papilio xuthus* and *protenor*. (From data of Hidaka, 1961.)

the subesophageal ganglion also contains a darkening factor (Yagi, 1980). In *Hestina*, which darkens by the formation of ommochromes, ligatures reveal an active factor released between the third and the seventh abdominal segments (Table XXIII) (Osanai and Arai, 1962).

Apart from a darkening factor, many studies on *Pieris* (Ohtaki, 1963; Oltmer, 1968; Bückmann, 1971) and also on *Vanessa* (Bückmann, 1960) indicated the existence of a factor that inhibits melanin synthesis. This hypothesis is based on the finding that dark ligated insects are unable to lighten the posterior portion of the body. The source of this inhibitory factor is unknown at present.

Color changes encountered in *Cerura* (Bückmann, 1956a,b, 1959) are rather different from the cases described above. They do not correspond to a true chromatic adaptation, but rather to physiological stages that precede molting. When the larvae stop feeding, their green, dorsally brown body coloration becomes uniformly red. Ecdysone, whose secretion is stimulated by brain hormone, plays a prominent role, demonstrated by ligatures and injections of pure hormone. It is interesting to observe that the timing and the

TABLE XXII. Pigmentation of Isolated
Abdomen of Light-Colored
Leucania following
Different Implantations[a,b]

Implanted organ	Pigmentation obtained
Ligated control	Light
Brain–cc–ca	Very dark
Subesophageal ganglion	Very dark
Brain	Weakly darkened
cc–ca	Weakly darkened
Prothoracic ganglion	Light
Mesothoracic, metathoracic, and abdominal ganglia	Light

[a]From data of Ogura (1975).
[b]Abbreviations used: ca, corpora allata; cc, corpora cardiaca.

amount of hormone required are different in both cases. Very small doses are necessary for epidermal reddening (60 *Calliphora* units), whereas pupal molting requires large doses (3000 *Calliphora* units); the molting critical period occurs 6 days later than that of epidermal color change. Hence it appears that, depending on its concentration, ecdysone induces reactions of different types. In *Pieris brassicae* larvae, the deposition of epidermal ommochromes and the excretion of xanthommatine are related to the hemolymph ecdysone titer (Bouthier and Lafont, 1976; Beydon *et al.*, 1980). The action of ecdysone is exerted not on tryptophan, which in the absence of ecdysone is converted to 3-hydroxykinurenin and eliminated, but rather on 3-hydroxykinurenin, causing its conversion to ommochromes (Bückmann, 1974).

TABLE XXIII. Body Pigmentation of *Hestina* Larvae following Ligation at Different Levels[a,b]

Head	Thorax			Abdomen									
	1	2	3	1	2	3	4	5	6	7	8	9	10

[a]From data of Osanai and Arai (1962).
[b]Vertical lines indicate the site of the ligature.

6.1.2.3. Purification of a Melanizing Factor

Purification of the melanizing factor of *Bombyx*, MRCH (melanization and reddish coloration hormone), was undertaken using brain–subesophageal ganglion complexes and especially heads of *Bombyx* pupae, the bioassay being the isolated abdomen of *Leucania*. The isolated substance is thermostable and is destroyed by proteolytic enzymes. The purification obtained, while very advanced, is not total. Using 3000 heads, or 23 g of material, 60 mg was isolated (Suzuki *et al.*, 1976).

The blackening factor in *Bombyx* pupae was purified by Hashiguchi *et al.*, (1965) using the thoracic ganglia of this species. The material was extracted with methanol, vacuum dried, and taken up with water or ether. Only the aqueous fraction is active, and 0.005 ml of this fraction, equivalent to 15 ganglia, induces black pupal pigmentation.

Unfortunately, no comparison was made between the factors isolated respectively by Suzuki and Hashiguchi, and thus it cannot be concluded that the same factor is involved. It may be noted, however, that MRCH exerts its action in *Spodoptera* solitary larvae (Yagi, 1980).

6.1.2.4. Mode of Action of MRCH

When injected into ligated abdomens of *Leucania* larvae, several cyclic nucleotides have been shown to cause a strong darkening of the body color, which suggests that cAMP acts as a second messenger in MRCH stimulation of melanization (Matsumoto *et al.*, 1979).

6.1.3. Summary

Little is known about the control of epidermal and cuticular pigment metabolism in insects, and this is understandable in view of the large number of pigments involved.

While juvenile hormone triggers the production of light pigments, one or more neurohormones intervene in the synthesis of dark pigments in Orthoptera, Phasmida, and Lepidoptera (Table XXIV and Fig. 51).

In certain cases, the isolated brain or isolated subesophageal ganglion is active, but in Lepidoptera the synthesis of dark pigments occurs only in the presence of several ganglia connected to the brain. The functioning of these complexes, which extend to the prothoracic ganglion and even the metathoracic ganglion, can be interpreted in different ways. While it is probable that the brain controls the secretion and the release processes, the precise location of these processes is far more difficult to pinpoint. It is conceivable that synthesis and release take place in each ganglion, but it is also possible

TABLE XXIV. Origin of the Darkening Factor in Some Insects

Insect	Inducing factors	Color change	Synthesized pigments	Organs involved in pigment synthesis[a]								
				cc	Brain	se gg	th1	th2	th3	ab1	ab2	ab3–ab7
Locusta[b]	Gregariousness	Green to brown–black		O	O							
Schistocerca		Darkening	Melanins						O			
Oedipoda[b]	Background color	Blackening			O	O						
Carausius	Heat and darkness	Green to black	Ommochromes		O	O						
Hestina	Diapause	Green to brown	Ommochromes									O–O–O–O
Leucania	Crowding	Yellow to brown	Melanins	O—O		O						
Papilio	Host plant	Green to brown	Melanins		O	O	O					
Pieris	Light background	Green to ± black	Melanins		O	O	O					
Bombyx	Genetic factors, temperature	Yellow to black	Melanins		O	O	O	O	O			

[a]Abbreviations used: ab, abdominal ganglion; cc, corpora cardiaca; se gg, subesophageal ganglion; th, thoracic ganglion; O, organ involved; O——O, organs involved whose connections cannot be interrupted.

[b]The pigments involved are poorly known; they might be melanins, ommochromes, or oxidized tetrapyrrolic pigments.

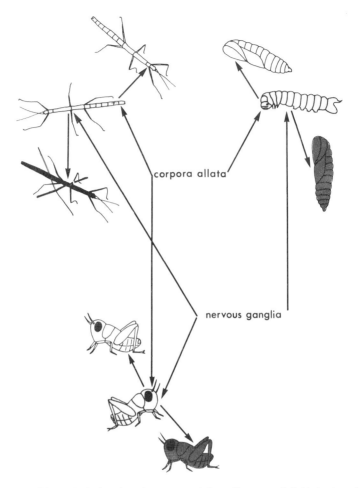

FIGURE 51. Morphological color change regulation. Green and light body color are induced by the corpora allata, and darkening is caused by neurosecretory cells of the central nervous system.

that the brain alone synthesizes the factor, which is transported along the ventral nerve cord, where it is then released in one or more sites.

6.2. PHYSIOLOGICAL COLOR CHANGE

Physiological color change, which is fairly rare in insects, occurs by two different processes: by means of chromatophores, which create a subcutaneous carpet for the entire body or a portion thereof, or by the surface migration of dark pigments within the epidermal cells. The latter process is clearly apparent in the stick insect *Carausius morosus*, which displays a circadian color change; it also exists in the grasshopper *Kosciuscola tristis*, which darkens with falling temperature (Key and Day, 1954a,b), and in various dragonflies (O'Farrell, 1964).

6.2.1. The Phantom Midge, *Corethra*

The presence of chromatophores is rare in insects and is only well known in the larva of the phantom midge *Corethra,* in which large chromatophores containing black ommochromes cover the dorsal face of two pairs of anterior and posterior tracheal bladders. These amoeboid chromatophores exhibit reversible reactions. On a black background, they form a continuous coating of jointed hexagonal cells. On a light background, they contract, assuming a rounded form, and display movement. The bladder surface appears to be dotted with small dark patches dispersed on a transparent background, paralleling that seen on the insect (Fig. 52).

Based on ligature and eye blinding experiments, Teissier (1947) showed that the chromatophore function is humorally controlled and ordered by the head. Organ extract injections (Dupont-Raabe, 1949, 1957; Gersch, 1956) and implants (Hadorn and Frizzi, 1949) show that a chromactivating factor is

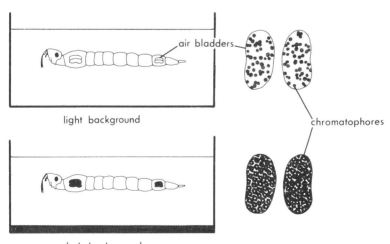

FIGURE 52. Physiological color change in *Corethra plumicornis* larvae. Contraction and migration of the chromatophores covering the air bladders. (Modified from Dupont-Raabe, 1957.)

located throughout the central nervous system. It causes adaptation to a black background, namely, the expansion of the chromatophores.

Hormone release from the ventral ganglia is confirmed by ligature; the stress caused by the latter is followed by a slight expansion of the posterior chromatophores (Dupont-Raabe, 1949). An electrical stimulation applied to the ventral ganglia, separated by a ligature from the anterior region, also causes the posterior chromatophores to expand (Gersch, 1956).

By using progressive dilutions of different nerve ganglia extracts, it can be demonstrated that the brain is more active than the ventral ganglia (Table XXV) (Dupont-Raabe, 1949, 1957). Implant experiments lead to the same conclusion (Hadorn and Frizzi, 1949). At equal concentrations, the corpora cardiaca exhibit lower activity than the brain.

One may question whether a substance causing chromatophore contraction exists. If a larva placed on a black background is ligated, the chromatophores of the posterior bladders contract. Hence it appears (Teissier, 1947) that the contraction stage represents a state of rest of the chromatophore, which it reaches by progressive transformations, in the absence of

TABLE XXV. Comparison of Activity
of Different Nervous and Glandular Organs[a]

Donor insect	Donor organ	Corethra and Carausius mean color change with increasing dilutions[b]		
		Corethra chromatophores		
		1/3	1/30	1/300
Corethra	Periaortic ring (corpora allata + molting gland)	1.6		
	Corpora cardiaca	3.8		
	Optic lobes	3		
	Brain	4.8	4.7	4.1
	Ventral ganglia	4.2	1.5	
		Carausius epidermis		
		1	1/10 1/20	1/100
Carausius	Brain	5	4.9	4.9
	Subesophageal ganglion	4.8	5	4.7
	Mesothoracic ganglion		4.3	1.6
	Connectives	4.6	1.7	
	Second abdominal ganglion	4.2	1.3	
	Corpora cardiaca	3.6	3.1	

[a]Modified from Dupont-Raabe (1957).
[b]The *Corethra* chromatophore expansion and *Carausius* body darkening are estimated according to a five-stage scale that enables us to calculate the mean reaction. 5: Maximum expansion in *Corethra* and darkening in *Carausius;* 3: intermediate stage; 1: chromatophore contraction in *Corethra* and lightening of the body color in *Carausius*.

darkening hormone. By isolating the segment containing the bladders, or the bladders themselves, contraction of the previously expanded chromatophores also occurs (Dupont-Raabe, 1949, 1957; Hadorn and Frizzi, 1949). This reaction occurs more slowly than expansion and is slower than the contraction observed after a ligature. This fact suggested to Hadorn and Frizzi the existence of an antagonistic factor, also postulated by Kopenec (1949), because the chromatophores of neck-ligated insects kept against a white background were less contracted than those of unoperated insects. However, this result can also be explained by a secretion of the darkening factor by the posterior nerve ganglia.

6.2.2. Dragonflies

Color change in dragonflies depends on temperature and involves two phases: a light phase in which the pigment granules gather at the base of the epidermal cells, and a dark phase in which they are distributed throughout the cell.

The ligature and removal of different ganglia performed on the damselfly *Austrolestes annulosus* (Véron, 1973) show that a darkening factor is located throughout the central nervous system, and in particular in the terminal abdominal ganglion. A direct response is nevertheless demonstrated, together with a nycthemeral variation in response to temperature. These facts recall the results obtained by Giersberg (1928) on *Carausius*, in which dark and light striations appear after keeping the insect under rings containing hot water. It is probable that the epidermal cell is under endocrine control, but still retains the capacity to respond directly to certain factors.

6.2.3. Stick Insects

The stick insect *Carausius morosus* displays a circadian color change that affects the entire body and occurs at dusk in varying degrees, depending on the pigmentation of the epidermis. The green insects remain unchanged, the beige insects turn brown, and the gray insects become completely black (Fig. 53). The black pigments play a prominent role, migrating from the base of the epidermal cell to the surface along a fixed system of microtubules (Berthold, 1980). The orange-yellow pigments spread in the median zone of the cell, and the green pigments fail to exhibit any movement at all. The transparent cuticle makes the color change especially clear. Humidity (Schleip, 1910; Giersberg, 1928) and cold also cause changes in body color, the tendency being to darken.

The humoral control of circadian color change was demonstrated by Giersberg (1928) and Janda (1936). The earliest experiments aimed to find the source of the humoral factor were conducted by Giersberg (1928), Atzler

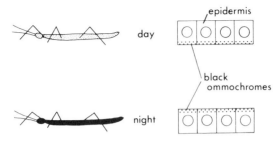

FIGURE 53. Physiological color change in the stick insect *Carausius morosus*. Circadian color change from gray to black is due to pigment migration within epidermal cells. (Modified from Dupont-Raabe, 1957.)

(1931), and Janda (1936). These workers revealed the importance of the brain and the lack of participation of the anterior sympathetic nervous system, corpora allata, and corpora cardiaca in controlling the color change. Two subsequent investigations provided further knowledge of the matter (Dupont-Raabe, 1951, 1956c, 1957; Mothes, 1960).

Like the subesophageal ganglion, the brain plays a prominent role. Its removal or simple disconnection eliminates any possibility of darkening, whereas this does not occur when the optic lobes or other nerve ganglia are removed (Dupont-Raabe, 1951, 1956c, 1957).

Injections of organ extracts into insects incapable of darkening by lack of a brain show that an active factor is present throughout the central nervous system, as in *Corethra* and some dragonflies. Its concentration is not the same in the different ganglia. The brain and subesophageal ganglion are very active, including a maximum reaction with an injection equivalent to 1/1000 organ; the mesothoracic ganglion and the second abdominal ganglion lose their activity respectively at doses of 1/100 and 1/20 (Dupont-Raabe, 1956b) (Table XXV).

With respect to the corpora cardiaca, the situation is completely different. These organs contain abundant amounts of a factor that shows medium activity on *Carausius* pigments. We shall see below that this is a different factor, which is provided with high activity in relation to crustacean chromatophores.

6.2.3.1. Intracerebral Location

The source of the chromactivating factor was investigated in the brain of *Carausius* by the selective removal of different parts of this organ, and by the injection of their extracts. It is somewhat difficult to compare the results obtained by Dupont-Raabe (1954, 1956c, 1957) with those obtained by Mothes (1960). In the former case, the results were evaluated according to the persistence of color change in the whole insect or after its reactions to raw extracts; in the latter case, the factors C_1 and D_1, isolated by Gersch *et al.*, (1960), were investigated in different parts of the brain by bioassays on integument pieces *in vitro*. Factor C_1 always causes darkening, while factor D_1 causes darkening in low concentrations and lightening in high concentrations.

Different divisions of the brain made by Dupont-Raabe (1954) lead to the conclusion that the chromactivating factor is located only in the tritocerebrum (Figs. 54 and 55). Divisions made by Mothes (1960) confirm the presence of factor C_1 in the deuto-tritocerebral part of the brain, but indicate that factor D_1 is present in the pars intercerebralis. However, the selective removal of this part of the brain, even if combined with removal of the corpora cardiaca, in no way interferes with color change (Dupont-Raabe, 1957) and renders questionable the role of the pars intercerebralis in color change. Removal of the tritocerebrum, on the contrary, definitely prevents color change at dusk. Neurosecretory cells, identified in the tritocerebrum (Dupont-Raabe, 1954, 1956a), probably produce the chromactivating factor.

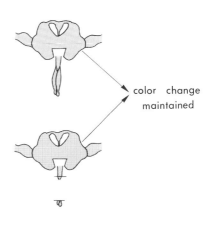

color change
maintained

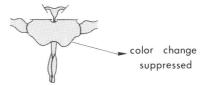

color change
suppressed

FIGURE 54. Demonstration of the source of the chromactivating factor by selective removal of different parts of the brain in the stick insect *Carausius morosus*. (Modified from Dupont-Raabe, 1957.)

FIGURE 55. Localization of the chromactivating factor within the brain of the stick insect *Carausius morosus* by extract injection experiments. (White) inactive brain parts; (dark) active parts; (gray) slightly active parts. (Modified from Dupont-Raabe, 1957.)

6.2.3.2. Release Site

Corpora cardiaca do not appear to release the cerebral chromactivating factor, because their removal, even when accompanied by that of the pars intercerebralis, allows normal color change to subsist. Furthermore, the determination of factors C_1 and D_1 in the corpora cardiaca shows that they fail to exhibit any circadian change, unlike the brain and the ventral nerve cord ganglia (Mothes, 1960).

The chromactivating factor of the ventral nerve cord also appears to be absent from the perisympathetic organs (Raabe *et al.*, 1966). Thus it is almost certain that other neurohemal structures are involved in the release of this factor.

6.2.3.3. Secretory Processes

The chromactivating-factor secretory processes were investigated by Mothes (1960), who evaluated their concentration in the hemolymph and in the different nerve ganglia during the circadian cycle. The determination was based exclusively on the *in vitro* assay of factors C_1 and D_1.

The amount of chromactivating factor contained in the central nervous system varies considerably during the circadian cycle (Fig. 56). The variations are, however, quite different in the brain and the ventral nerve cord. The brain is loaded by day and reaches a maximum in the late afternoon, namely just before the time when the hormone is released into the blood to induce color change. In the ventral nerve cord, on the contrary, storage occurs by night. Mothes postulated that this was due to diurnal release, but this seems questionable because the insects are light by day. It is proposed that the absence of the darkening factor within the ventral nerve cord during the day is more probably due to the fact that the chromactivating factors synthesized in the ganglia are stored by day in neurohemal organs. Considering what occurs in the hemolymph, it may be observed that factors C_1 and D_1 are highly abundant during the night. Their concentration, low by day, increases rapidly at dusk, the time when hormone release undoubtedly occurs.

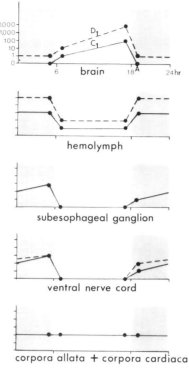

FIGURE 56. Circadian modifications of the C_1 and D_1 neurohormone content of the hemolymph and various organs in the stick insect *Carausius morosus*. (Modified from Mothes, 1960.)

6.2.3.4. Lightening Factor

Factor D_1 is lightening *in vitro* at high concentrations, but this does not necessarily imply physiological action. Simultaneous fluctuations in factors C_1 and D_1 in the hemolymph and in the ganglia appear to discount this possibility.

6.2.4. Interspecific Activity

Carausius organ extracts, which are active on *Carausius*, are also active on *Corethra*, and extracts of *Corethra* heads cause darkening in *Carausius* (Dupont-Raabe, 1949, 1957). It may therefore be assumed that the factor involved is the same in both cases. This factor is also present in species lacking color change. Although it has not been found in some insects, such as the water beetle *Dytiscus marginalis*, the wax moth *Galleria mellonella*, and the blowfly *Calliphora erythrocephala* (Dupont-Raabe, 1957), it is found in many other species such as the cockroaches *Blabera fusca* and *Blatta orientalis*, the grasshopper *Tettigonia viridissima*, the bug *Notonecta glauca*, and the mosquito *Culex apicalis*. Organ extracts from these species cause the expansion of *Corethra* chromatophores and the migration of epidermal pigments in *Carausius*. Thus it is probable that, apart from its action on pigments, the chromactivating factor is involved in another physiological regulation. It is interesting to note that the location of the factor is always the same. Absent from the pars intercerebralis, it is located in the lateral parts of the protocerebrum and in the tritocerebrum, where neurosecretory cells appear to be widespread (Raabe, 1963a,c).

6.2.5. Crustacean and Insect Hormones

Hanström's pioneering work (1937, 1940) revealed the existence of a substance in the heads of different insects that acts on crustacean chromatophores, causing concentration of the yellow and red chromatophores of the prawns *Leander adspersus* and *serratus* and the black chromatophores of the shrimp *Crangon vulgaris*, and which also acts on crabs by inducing expansion of the melanophores. This factor is present in the corpora cardiaca (Brown and Meglitsch, 1940; Thomsen, 1943; Dupont-Raabe, 1952a). Its source is, therefore, quite different from that of the insect chromactivating factor, which is found in the central nervous system.

The corpora cardiaca factor that is active on crustaceans is found in many but not all insects. It is present in Dictyoptera, Phasmoptera, Orthoptera, and Coleoptera, absent or possibly slightly active in Apterygota, Odonata, Trichoptera, Hymenoptera, Lepidoptera, and the Diptera assayed. Thomsen calculated the minimum fraction of corpora cardiaca capable of inducing a maximum reaction of *Leander* erythrophores. It is 1/100 for the greenhouse camel cricket *Tachycines asynamorus*, 1/64 for *Carausius*, and only 1 for *Blatta* and *Dytiscus*. Recent experiments revealed a similar heterogeneity between the species with respect to the concentration of chromactivating factor in the corpora cardiaca (Herman *et al.*, 1977) (Table XXVI).

6.2.6. Purification

The separation and comparison of the chromactivating factors of the brain and corpora cardiaca of *Carausius* on the one hand, and of the sinus gland of the prawn *Leander serratus* on the other, was carried out by electrophoresis (Carlisle *et al.*, 1955; Knowles *et al.*, 1955). These experiments revealed the existence of two distinct factors, A and C. Factor A, present in the corpora cardiaca and the sinus gland, causes the contraction of red and yellow chromatophores in *Leander*, contraction of black chromatophores in *Crangon*, and has a slight effect on *Carausius*. Factor C,

TABLE XXVI. Crustacean Red Pigment-
Concentrating Hormone Content
in Insect Corpora Cardiaca[a]

Insect	*Leander* units/organ[b]
Schistocerca gregaria	920/cc
Locusta migratoria	340/cc
Dytiscus sp.	120/cc
Tachycines asynamorus	28/cc
Tenebrio molitor	13/head
Periplaneta americana	2/cc
Danaus plexippus	0.8/brain + cc + ca
Calliphora erythrocephala	0.1/head

[a]After Herman *et al.* (1977).
[b]ca, Corpora allata; cc, corpora cardiaca.

which has no effect on *Leander*, nevertheless acts on *Crangon* and exerts a strong action on the black pigments of *Carausius*. It is present in the brain of *Carausius* and absent from the corpora cardiaca and the sinus gland of *Leander*. Both factors, which may be related, withstand boiling and are destroyed by proteolytic enzymes (Knowles *et al.*, 1956) (Table XXVII).

Gersch and co-workers also conducted thorough investigations in order to purify and separate insect neurohormones as reported above. Two factors, the neurohormones C and D, and then C_1 and D_1, were separated from the central nervous system of *Periplaneta* and *Carausius* (Gersch *et al.*, 1957, 1960). While they are both active on pigment movements in *Carausius* and *Corethra* (Mothes, 1960; Gersch, 1956), they also act on the heartbeat and endogenous nerve activity of insects, and on crustacean chromatophores. Purification was carried out first on 320 central nervous systems of *Periplaneta* (Gersch, 1958), resulting in 2 mg of active factor. The second purification, using an amount of nerve substance 10 times larger, led to the production of 50 μg of a crystallized protein, the factor D (Gersch *et al.*, 1960), with a molecular

TABLE XXVII. Effect of Insect and Crustacean A and C Factors on
Pigment Migration in Insects and Crustaceans[a]

		Target[b]		
			Crustaceans: Chromatophores	
		Insects:		
Source of factor	Factor	*Carausius* epidermis	Red and yellow (*Leander*)	Black (*Crangon*)
Carausius brain	C	+++	0	++
Carausius corpora cardiaca	A	+	+++	+++
Leander sinus gland	A	+	+++	+++

[a]From data of Knowles *et al.* (1955).
[b]0, No effect; +, ++, +++, light, medium, and strong effects.

weight of 2000 (Gersch and Stürzbecher, 1967). The neurohormones C and D, which are destroyed by proteolytic enzymes, are polypeptidic in nature.

It was shown recently (Cheeseman *et al.*, 1977; Carlsen *et al.*, 1979) that the corpora cardiaca of locusts contain one or more adipokinetic factors, the molecular structure of which closely approaches that of a chromactivating hormone in crustaceans (RPCH). Thus it is probable that the adipokinetic factor imparts to extracts of corpora cardiaca their activity on crustacean pigments.

One may well ask why these factors seem to be lacking in certain species. They may be deprived of adipokinetic hormone, or they may possess an adipokinetic hormone that has a different formula. As for the substance of the central nervous system that is active on insect pigments, its structure is presently unknown, and the question of whether it acts in the regulation of processes other than physiological color change remains unanswered.

6.3. CONCLUSIONS

Insects, which are often mimetic, undergo color change by the differential synthesis of epidermal and cuticular pigments or, more rarely, by pigment movements. The former process is frequently encountered in Orthoptera and Lepidoptera, while the latter occurs in stick insects by the migration of dark pigments in the epidermal cells, and in *Corethra* larvae by the expansion and contraction of chromatophores.

In all cases, neurohormones produced in the central nervous system play a leading part: they induce the synthesis and migration of ommochromes and the synthesis of melanins. However, it is unknown whether the active factors are the same in the various cases, although purifications have been carried out by many investigators.

The neurohormones do not originate only in the brain. The ventral nerve cord ganglia are also important, but they do not all perform the same function. Some of them intervene in the release of a melanization factor in many Lepidoptera. As for the regulation of pigment movements, all the ventral ganglia appear to be involved in the synthesis and release of the chromactivating factor. However, the brain and subesophageal ganglion play a leading part. They both contain a large quantity of hormone but are, however, incapable of secreting it if their connectives are interrupted. According to Mothes, the brain plays the leading role in light perception, but the ventral nerve cord exhibits an independent rhythmicity, which possibly explains the need for intact connectives between these two components of the same system.

Physiological color change is independent of the molting gland and corpora allata, but ecdysone and juvenile hormone exert an influence on pigment syntheses.

Behavior and Rhythmic Phenomena

The behavior and polymorphism of insects depend on environmental factors and on seasonal and circadian rhythms. Locomotor activity is also influenced by physiological states (imminent molting, sexual maturity) that are dependent on hormones. For example, in lepidopteran prepupae, a decrease in juvenile hormone or a surge of ecdysterone causes behavioral changes such as arrested feeding and decline in locomotor activity, or, on the contrary, provokes the wandering stage. During the period of sexual activity, juvenile hormone triggers the behavioral sequences that precede copulation and oviposition. In addition, juvenile hormone stimulates the alimentary flight and migratory flight in several insects. Metabolic neurohormones are involved in flight and in activity, providing the substrates for muscle functioning, but certain neurohormones exert a specific action in the regulation of behavior and polymorphism.

These phenomena are difficult to investigate and it must be emphasized that except for the data concerning the eclosion hormone our knowledge on the subject is rather limited in spite of the vast accumulation of data.

7.1. FLIGHT

Experiments performed on locusts and some other insects show the necessity of certain neurohormones for flight to occur. Removal of the neurosecretory cells of the pars intercerebralis causes the disappearance of nocturnal flight (Hinks, 1967) in the noctuid moths *Noctua pronuta* and *Agrostis ipsilon*; and in the blowfly *Calliphora erythrocephala*, the corpora cardiaca are indispensable to the achievement of flight (Vejbjerg and Normann, 1974).

In the locust *Schistocerca gregaria*, the neurohemal part of the corpora cardiaca performs an important role in sustained flight, which is prevented by the disconnection or removal of the corpora cardiaca and is only restored after their implantation or the implantation of their neurohemal part, the glandular part being ineffective (Michel, 1972, 1973). The electrocoagulation of the pars intercerebralis also inhibits flight, but its implantation fails to restore it (Michel and Bernard, 1973). Thus, according to Michel, flight in *Schistocerca* is stimulated by a substance produced in the pars intercerebralis and activated in the corpora cardiaca.

In *Locusta migratoria*, removal of the pars intercerebralis disturbs flight only in young insects, and it is suggested that its action is exerted in part by the corpora allata, for the injection of juvenile hormone analogues diminishes the effects caused by destruction of the pars intercerebralis (Goldsworthy *et al.*, 1973, 1977). Nevertheless, the glandular part of the corpora cardiaca is indispensable to flight, which is depressed by their removal. The ultrastructural study of the corpora cardiaca in *Locusta* has confirmed the secretory activity of the glandular cells during flight (Rademakers, 1977a; Rademakers and Beenakkers, 1977). The rough endoplasmic reticulum of these cells displays sharp flight-related variations in another locust, *Schistocerca* (Michel and Lafon-Cazal, 1978).

By comparing the results obtained on the two species of locust (Table XXVIII), it may be observed that the pars intercerebralis is more or less essential, and that the corpora cardiaca, always necessary, act either through their neurohemal region or through their glandular part. The locusts are known to use two distinct metabolites for flight. Trehalose is used for short flights or at the onset of migratory flight, and diglycerides are consumed during the migratory flight. As we shall see in Chapter 9, the release of these two metabolites from the fat body is subject to different controls. The release of trehalose is triggered by the hypertrehalosemic hormone from the neurohemal part of the corpora cardiaca, while the release of diglycerides depends on the adipokinetic hormone produced in the glandular lobe. Thus it may well be

TABLE XXVIII. Effect of the Removal of Corpora Cardiaca and Pars
Intercerebralis on Flight Activity

Organ removed	Insect	Reference	Flight
Corpora cardiaca	*Schistocerca gregaria*	Michel (1972, 1973)	Suppressed
	Locusta migratoria	Goldsworthy *et al.* (1973)	Weak
	Oncopeltus fasciatus	Rankin and Riddiford (1977, 1978)	Normal
	Calliphora erythrocephala	Vejbjerg and Normann (1974)	Suppressed
Pars intercerebralis	*Schistocerca gregaria*	Michel (1972, 1973)	Suppressed
	Mature *Locusta migratoria*	Goldsworthy *et al.* (1977)	Normal
	Young *Locusta migratoria*	Goldsworthy *et al.* (1977)	Slowed down
	Noctua pronuba and *Agrotis ipsilon*	Hinks (1967)	Suppressed

that the discrepancies between the results are due to differences in the flight metabolites used by the locusts in the experimental conditions and to differences in the neurohormones involved in the release of these metabolites. In fact, the experiments were conducted differently in each species. The assays performed by Michel on *Schistocerca* concerned the spontaneous tendency to sustained flight, while Goldsworthy and co-workers, working on *Locusta*, evaluated the flight obtained after the insects were stimulated. Thus two distinct metabolic neurohormones are involved in flight regulation. It remains to be seen if another "behavioral" neurohormone, stimulating the activity, might be also involved. In the milkweed bug *Oncopeltus fasciatus*, such a factor does not seem to occur (Rankin and Riddiford, 1977, 1978).

7.2. CIRCADIAN RHYTHMS AND ACTIVITY LEVEL

Many insects display behavioral circadian rhythms, with a rest period and an active period or with changes in the activity level (Table XXIX). In cockroaches, activity begins in the early hours of darkness, and slows down in the second part of the night and throughout the day.

Experiments performed on the cockroach *Periplaneta americana* demonstrate the action of a humoral factor regulating activity, and show that it originates in the type A cells of the subesophageal ganglion. The experiments consist of parabioses between a normal insect, whose activity is recorded, and a legless insect joined dorsally to the first, which plays the role of donor from an endocrine standpoint. Parabiosis between a normal insect and an insect that has lost its rhythm by being kept permanently in light, restores a normal activity rhythm in the latter. This effect is also obtained by implanting the subesophageal ganglion of a normal donor, whereas brain, corpora allata, or corpora cardiaca implants do not have this effect (Harker, 1956, 1960). This work on *Periplaneta* was resumed by later investigations, which failed to confirm the role played by the subesophageal ganglion in the regulation of the circadian rhythm, or in the control of the locomotor activity level in *Periplaneta* (de Roberts, 1966; Brady, 1967a,b) and *Leucophaea maderae* (Nishiitsutsuji-Uwo *et al.*, 1967). Thus the role of neurohormones in regulation of circadian rhythm remains questionable.

The role of the brain has also been debated. It appears certain that a biological clock controlling circadian rhythm is located in this organ, probably near the pars intercerebralis, because various brain lesions or extensive removal of the pars intercerebralis causes the circadian rhythm to disappear. The nervous or endocrine mode of cerebral action is also the subject of controversy. The endocrine hypothesis is supported by the demonstration of a circadian activity rhythm in the neurosecretory cells of the pars intercerebralis (Cymborowski and Flisinska-Bojanowska, 1970) as well as in the type A cells of the ventral nerve cord (de Bessé, 1965), and by experiments in which the

TABLE XXIX. Activity Rhythm and Activity Level following Several Surgical Procedures

Surgical procedure	Insect	Reference	Locomotor activity[a]	
			Rhythm	Level
Decapitation	Periplaneta americana	Brady (1967b)		++
	Leucophaea maderae	Nishiitsutsuji-Uwo and Pittendrigh (1968)		++
Corpora cardiaca removal	Periplaneta americana	de Roberts (1966)	Normal	
		Brady (1967a)	Normal	
	Carausius morosus	Dupont-Raabe (1956b)	Normal	
		Eidmann (1956)	Normal	
Brain removal	Carausius morosus	Dupont-Raabe (1956b)	0	+
		Eidmann (1956)	0	
	Romalea microptera	Fingerman et al. (1958)	Normal	
Pars intercerebralis removal	Carausius morosus	Dupont-Raabe (1956b)	Normal	
	Acheta domesticus	Cymborowski (1970a, 1973a)	0	
	Teleogryllus commodus	Sokolove and Loher (1975)	0	
Pars intercerebralis + lateral neurosecretory cell	Leucophaea maderae	Nishiitsutsuji-Uwo et al. (1967)	0	

Operation	Species	Reference	Normal	Variable
esophageal connectives	*Leucophaea maderae*	Nishiitsutsuji-Uwo and Pittendrigh (1968)	Normal	Variable
	Carausius morosus	Dupont-Raabe (1956b)	0	
		Eidmann (1956)	0	
Subesophageal ganglion removal	*Leucophaea maderae*	Nishiitsutsuji-Uwo and Pittendrigh (1968)	0	++
	Carausius morosus	Dupont-Raabe (1956b)	0	++
		Eidmann (1956)	0	++
	Romalea microptera	Fingerman *et al.* (1958)	0	
Transection of connectives between the sub-esophageal ganglion and the prothoracic ganglion	*Periplaneta americana*	Brady (1967b)	0	++
	Leucophaea maderae	Nishiitsutsuji-Uwo and Pittendrigh (1968)	0	++
	Carausius morosus	Dupont-Raabe (1956b)	0	++
		Eidmann (1956)	0	++
Transection of thoracic connectives	*Leucophaea maderae*	Nishiitsutsuji-Uwo and Pittendrigh (1968)	Variable	
	Carausius morosus	Dupont-Raabe (1956b)	Normal/0	
Transection of abdominal connectives	*Leucophaea maderae*	Nishiitsutsuji-Uwo and Pittendrigh (1968)	Normal	Normal/+

[a]0, Lost; ++, +, more or less increased; a strong increase or a strong decrease of locomotor activity makes the circadian rhythm imperceptible.

implantation of a gel containing actinomycin D, a protein-synthesis inhibitor, in the pars intercerebralis makes the insects arrhythmic (Nishiitsutsuji-Uwo *et al.*, 1967, on *Leucophaea*), but the nervous hypothesis is supported by experiments in which selective cauterization of the pars intercerebralis yields variable results.

In the house cricket *Acheta domesticus* (Cymborowski, 1970a), removal of the pars intercerebralis upsets the locomotor rhythm and causes hyperactivity. Parabioses similar to those made by Harker on *Periplaneta* cause the activity rhythm to reappear, and brain implantation in insects deprived of the pars intercerebralis fails to restore the normal situation, but periodically reduces locomotor activity. The type A cells of the pars intercerebralis display rhythmic activity (Fig. 57), as in the cockroach (Cymborowski, 1970b; Dutkowski *et al.*, 1971). Histological and ultrastructural studies show that they are in a storage phase during the nocturnal activity period, and in a release phase during the day (Cymborowski and Dutkowski, 1969, 1970a,b). RNA and protein synthesis also follow a rhythmic pattern, as do the incorporation of labeled cysteine (Cymboroski, 1973b) and the acetylcholinesterase level (Cymborowski *et al.*, 1970), which drops with the rise in nocturnal activity. Serotonin determinations made in the brain and the hemolymph also reveal cyclic modifications, but no correlation exists between the serotonin content of the hemolymph and the circadian rhythm. However, a close correlation exists between the latter and the serotonin content of the central body, an important connection center of the brain. Serotonin injected into the hemolymph does not alter locomotor activity (Cymborowski, 1970c, 1973a), but injected into the brain it causes intense hyperactivity after a 24-hr interval. Thus it appears that there is a cyclical serotonin secretion in the central body that controls the activity of the type A neurosecretory cells of the pars intercerebralis, which, in turn, inhibits insect activity (Muszyńska-Pytel and Cymborowski, 1978a, b).

Investigations on other species show that, depending on each case, either the brain or the subesophageal ganglion may control the activity rhythm. In the cricket *Teleogryllus commodus*, removal of the pars intercerebralis makes the insects arrhythmic (Sokolove and Loher, 1975). In the grasshopper *Romalea microptera*, brain removal allows the locomotor rhythm to continue, but the latter is prevented by removal of the subesophageal ganglion (Fingerman *et al.*, 1958).

A circadian activity of the neurosecretory cells of the pars intercerebralis has also been reported in several insects (see Chapter 1), but it does not necessarily imply their intervention in the regulation of circadian processes.

In the stick insect *Carausius morosus*, which displays both circadian color change and circadian activity rhythm with a nocturnal activity period, investigations were carried out by Dupont-Raabe (1956b) and Eidmann (1956). The triggering mechanisms of these two processes are different; whereas neu-

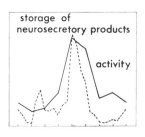

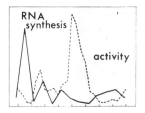

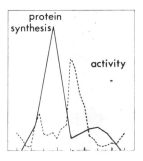

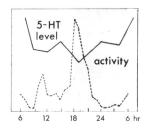

FIGURE 57. Circadian variations of locomotor activity and pars intercerebralis neurosecretory activity evaluated by the percentage of loaded neurosecretory cells, RNA and protein synthesis, and brain 5-HT level in *Acheta domesticus*. Dashed line: locomotor activity. (Modified from Cymborowski, 1970b,c; Cymborowski and Dutkowski, 1969, 1970a; Muszyńska-Pytel and Cymborowski, 1978a.)

roendocrine control occurs in the former (see Chapter 6), nervous control occurs in the latter. Removal of the corpora cardiaca–corpora allata complex and the pars intercerebralis has no effect on behavior. However, removal of the brain or severance of the circumesophageal connectives leads to the disappearance of the circadian rhythm and to increased activity, showing that rhythm control is exerted by the brain and transmitted nervously. Removal of the subesophageal ganglion or cutting the connectives behind this organ permanently immobilizes the insects. These results, similar to those obtained by Brady (1967a,b), Nishiitsutsuji-Uwo *et al.* (1967), and Nishiitsutsuji-Uwo and Pittendrigh (1968) on cockroaches, suggest that the subesophageal ganglion stimulates activity in normal insects while the brain inhibits it. The severance of the ventral nerve cord behind the subesophageal ganglion allows circadian activity to continue in the anterior part of the body, while the posterior part remains motionless (Dupont-Raabe, 1956b).

It seems that activity rhythm and activity level are linked processes governed by distinct mechanisms, and investigators do not always clearly distinguish between them. So in *Locusta*, the postprandial rest, which is not concerned with circadian rhythm, appears to have a neurohormonal control, for fasting insects injected with the hemolymph of recently fed insects or with corpora cardiaca homogenates reduce their locomotor activity (Bernays, 1980). In *Drosophila*, a neuroendocrine control was also demonstrated, but it is difficult to decide whether it concerns the circadian rhythm or the activity level. It has been shown that the brain of short-period mutant flies implanted into arrhythmic hosts induces bursts of activity in the latter (Handler and Konopka, 1979).

7.3. ECLOSION RHYTHM

Species such as the saturniid moths *Hyalophora cecropia* and *Antheraea pernyi* always emerge at a definite time, which varies for each species. Eclosion regulation is under brain control, and if the brain is removed, eclosion occurs at random. A brain implantation or the injection of brain extracts triggers the characteristic behavior that precedes eclosion, and this occurs within 1–2 hr, even in isolated abdomens. By making reciprocal implantations in both species, the eclosion rhythms obtained correspond to the donor and not to the host (Truman and Riddiford, 1970; Truman, 1971), but the characteristic behavioral programs of each species are retained. Hence the implantation of a *Hyalophora* brain in *Antheraea* induces the program of the latter, but at the time of *Hyalophora*.

The brain thus plays a humoral role. The active factor, termed eclosion hormone, is secreted in the pars intercerebralis (Truman, 1973b) and stored in the corpora cardiaca. It triggers eclosion behavior, which includes several successive steps in *Hyalophora*: a phase of rapid rotational abdominal movements, a rest period, a new phase of violent activity consisting of peristaltic movements, and finally eclosion. This is followed by spreading of the wings, caused by the pressure of the blood that fills them, which is pulsed by contraction of the abdominal longitudinal muscles. Once this process is complete, the degeneration of the abdominal longitudinal muscles begins under the control of eclosion hormone (Finlayson, 1956; Truman, 1973a), (Fig. 58). Al-

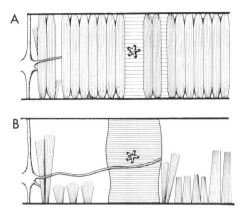

FIGURE 58. Posteclosion breakdown of muscles in the tobacco moth *Manduca sexta*. Internal view of the musculature of the right half of the fifth abdominal segment. (A) Moth at eclosion; (B) 3 days after eclosion. The right margin of the drawing is the dorsal midline; the portion of nerve cord on the left marks the ventral midline. Muscles extending to the heart and nerve cord have been omitted. (Modified from Truman, 1973a.)

though no degeneration takes place in isolated abdomens, it does take place if the hormone is injected.

Eclosion hormone acts on the central nervous system, where it triggers the initiation of the species-specific program. This was demonstrated in isolated abdomens, where the action potential of the nerves controlling the muscles involved in preecdysial contractions was recorded. The results differ widely according to whether the eclosion hormone is present or absent. If it is present, many similar bursts of spikes are first observed in the successive ganglia, but reverse from one side to the other, corresponding to rotational movements. In the next phase of peristaltic contractions, the impulses are identical on both sides, but are phase-shifted from back to front (Truman and Sokolove, 1972). When isolated central nervous systems are subjected to the eclosion hormone, they respond by the generation of a program of motor activity that mimics the one expected during the preeclosion and eclosion behaviors. The initiation of these behaviors is elicited by only a few minutes' exposure of the central nervous system to the hormone (Truman, 1978).

The tanning processes begin as soon as the young moth has spread its wings. The question arises whether bursicon, the tanning hormone, is different from eclosion hormone. By using different organs of the tobacco hornworm *Manduca sexta*, Truman tested eclosion hormone activity on *Antheraea* and bursicon activity on *Manduca* (1973c). These experiments revealed that eclosion hormone is located in the brain and in the corpora cardiaca, whereas bursicon is found in the abdominal ganglia and in the perisympathetic organs (Table XXX). Experiments in which different parts of the brain were removed indicate that eclosion hormone is located in the pars intercerebralis and not in the lateral protocerebral neurosecretory cells.

Eclosion hormone is not constantly present in the brain–corpora cardiaca complex. It appears in *Antheraea* at the onset of adult development, namely

TABLE XXX. Location of Bursicon and Eclosion Hormone in the Central Nervous System and Neurohemal Organs of Two Lepidopterans[a]

Hormone	Insect	Donor					
		Organ[b]					
		Brain	cc + ca	th gg	ab gg + po	ab gg	po
Eclosion hormone	*Manduca sexta*	++	++		0		
	Antheraea pernyi	++	+++	0		0	
Bursicon	*Manduca sexta*	0	0	0	+++	+	+++

[a]After Truman (1973b).
[b]Abbreviations used: ab gg, abdominal ganglia; cc + ca, corpora cardiaca + corpora allata; po, perisympathetic organs; th gg, thoracic ganglia.

on the 7th day of pupal life, increases progressively to reach a plateau on the 10th day in the brain and on the 13th day in the corpora cardiaca. It then diminishes considerably in the corpora cardiaca on the 19th day, namely, just before eclosion, while its level in the hemolymph rises sharply (Truman, 1973b). The time of appearance of eclosion hormone in the hemolymph of *Manduca* was determined by isolated wing bioassay (Reynolds, 1977). It was demonstrated that this hormone is released in the blood only at a restricted time that is determined by a circadian clock once development has been completed (Reynolds *et al.*, 1979).

Eclosion hormone is present in another lepidopteran, the silk moth *Bombyx mori* (Morohoshi and Fugo, 1977). However, it was not found in the cockroach *Leucophaea maderae* or the linden bug *Pyrrhocoris apterus*. The brain and the corpora cardiaca extracts of these species have no effect on *Antheraea* (Truman, 1973b). Nevertheless, a factor necessary for ecdysis was demonstrated in Odonata. In the median anterior part of the brain in the larva of the dragonfly *Aeshna cyanea*, two pairs of type A neurosecretory cells exist, which are heavily loaded at the time of ecdysis, and empty in permanent larvae, where molting is suppressed by removal of the molting glands. Selective electrocoagulation of these cells prevents ecdysis (Charlet and Schaller, 1976) (Fig. 59).

Like most other neurohormones, eclosion hormone acts through cyclic nucleotides. The injection of cAMP or cGMP into isolated abdomens of *Hyalophora* produces the same effects as eclosion hormone, and causes a significant rise in the cAMP content of the central nervous system, but cGMP is still two times more potent than cAMP (Truman *et al.*, 1976, 1979).

7.4. ENDOGENOUS NERVE ACTIVITY

Investigations into behavioral control have frequently demonstrated the role played by the neurohormones of the pars intercerebralis or corpora cardiaca. Studies of spontaneous nerve activity of the central nervous system also reveal the action of neurohormones in this process. Working on the cockroaches *Blaberus giganteus* and *Periplaneta americana*, Ozbas and Hodgson (1958) used corpora cardiaca extracts and obtained a drop in the endogenous nerve activity of the ventral nerve cord *in vitro* and a decrease in coordination and locomotor activity *in vivo* (Fig. 60).

Milburn *et al.* (1960) also demonstrated the activity of the corpora cardiaca of the cockroach *Periplaneta* on the phallic nerve, where the extracts trigger a volley of nervous rhythmic impulses.

Corpora cardiaca also act in stress situations. Prolonged immobilization of the insects, giving rise to a continuous reaction of struggling, or a series of mechanical or electrical stimulations causing hyperactivity induces the release

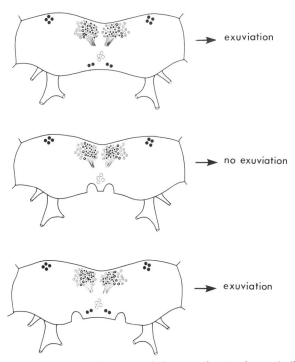

exuviation

no exuviation

exuviation

FIGURE 59. Neurosecretory cell electrocoagulation experiments demonstrating the location of the exuviation factor in *Aeshna cyanea* larvae. (Top) Normal insect. (Center) Electrocoagulation of paramedian neurosecretory cells. Exuviation is suppressed. (Bottom) Electrocoagulation of an area located near the protocerebral paramedian cells. Exuviation is maintained. (Modified from Charlet and Schaller, 1976.)

into the blood of a paralyzing substance that can be transmitted from one insect to another by injection or parabiosis (Beament, 1958). Hodgson and Geldiay (1959) have shown that stresses such as electrical stimulations or successive turning over of the insects induce the disappearance of neurosecretory materials in the corpora cardiaca, and also cancel the effect exerted by the extracts of these organs on the spontaneous nerve activity of the central nervous system. This suggests that the same substance may act on behavior as well as on endogenous nerve activity. It undoubtedly originates in the brain, because corpora cardiaca extracts lose their activity when the brain and corpora cardiaca have been disconnected (Ozbas and Hodgson, 1958).

The corpora cardiaca factor is thermostable, resistant to chymotrypsin, and different from adrenaline, noradrenaline, serotonin, or dopamine (Milburn and Roeder, 1962). According to Gersch and Richter (1963), Strejčková *et al.* (1964), and Birkenbeil (1971), the substance that stimulates the phallic nerve

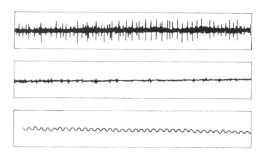

FIGURE 60. Endogenous nerve activity in the cockroach ventral nerve cord *in vitro*. (Top) Ten minutes after adding saline; (center) corpora cardiaca extracts; (bottom) time scale (50 oscillations/sec). (From Ozbas and Hodgson, 1958.)

in *Periplaneta* is factor D_1 of the central nervous system and the corpora cardiaca, which possesses many other physiological functions that are discussed later in this work.

7.5. EGG CARE

The dermapteran *Labidura riparia* has a very unusual behavior related to pre- and postoviposition. During the 10 days of embryonic development, the female takes care of its eggs, turning and licking them, does not eat, and acquires an aggressive behavior. Pars intercerebralis cauterization switches off this behavior, thus suggesting a neurohemal control (Caussanel *et al.*, 1978). Ultrastructural study of the aortic wall during the reproductive cycle shows that, whereas one peak of exocytotic profiles occurs at oviposition, a continuous exocytotic activity takes place during the egg-care period (Juberthie and Caussanel, 1980).

7.6. POLYMORPHISM AND PHASE

Polymorphism is a widespread phenomenon in insects with many determining factors and different aspects, the regulation of which is still poorly known.

The role of the corpora allata has been established in the triggering of green pigmentation, wing development, rate of maturation of the ovaries, certain behavioral characteristics of the solitary phase in locusts (Joly and Joly, 1953; Staal, 1961), the phase determinism of their offspring (Cassier and Papillon, 1968), and bee and termite cast determination (see Lüscher, 1976; Noirot, 1977; Lebrun, 1978). In aphids, the corpora allata also appear to be involved in morph determination: in response to crowding, nutrition, photoperiod, and temperature, the aphid gives rise to parthenogenetic females or

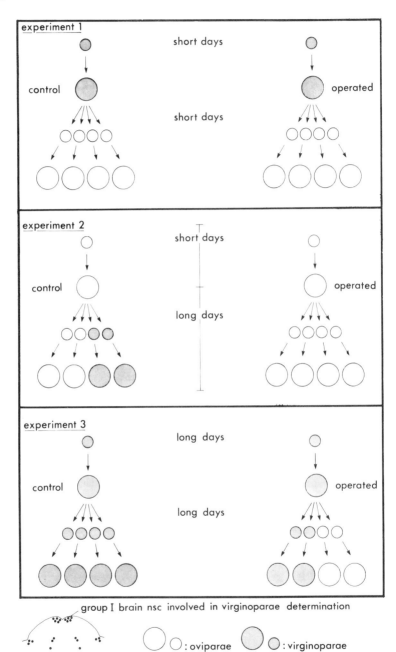

FIGURE 61. Regulation of polymorphism in the aphid *Megoura viciae. nsc, Neurosecretory, cells.* (From the data of Steel and Lees, 1977; conceived by the author.)

to sexuparae, to viviparae or to oviparae, to alatae or to apterae. Experiments with topically applied juvenile hormone extracts and analogues seemed to divert alate development toward the apterous condition, an effect similar to that obtained by long-day photoperiodic treatment (Lees, 1978; Hardie, 1980). Moreover, juvenile hormone acts upon the morph determination of the ped-ogenetic embryos developing in the abdomen of the treated larvae, which switch from oviparae to parthenogenetic viviparae (Lees, 1978).

The involvement of a neurohormone has been demonstrated in the vetch aphid *Megoura viciae* by microcautery of different groups of brain neu-rosecretory cells in insects reared in various conditions (Steel and Lees, 1977). The morph of the offspring can be distinguished in young larvae by the presence of growing eggs in oviparae and by chains of embryos located in the ovarioles in the virginoparae. As shown in Fig. 61, the operated insects behave normally when reared in short-day conditions; in long-day conditions, oviparae appear in the offspring, and the response to long days in insects previously maintained in short days, i.e., production of virginoparae instead of oviparae, is prevented by microcautery of median neurosecretory cells. Thus these neurosecretory cells appear to be the origin of a factor controlling virginoparae production in response to increased day length. Because they send one of their processes to the ventral nerve cord, their secretory products may be delivered directly to the abdomen and perhaps to the reproductive system itself. In fact, lesions made in the protocerebrum where the nccI emerge from the brain do not interfere with the photoperiodic response of the insect. It is not yet known if the brain action is a direct one or if it is exerted through the corpora allata.

7.7. CONCLUSIONS

The data available on behavioral regulation deal with rather dissimilar processes, lending plausibility to the hypothesis of a diversity of control mechanisms. Although discrepancies exist in a few cases, some general conclusions can nevertheless be drawn.

Neurohormones act in behavioral processes that involve such periodic phenomena as migratory flight, developmental stages such as eclosion or prepupal retraction in the fly, and possibly circadian rhythms.

The pars intercerebralis acts on flight as well as on eclosion, but its action in locomotor activity is not firmly established. The corpora cardiaca display a very clear-cut action on spontaneous electrical activity but appear to act on behavior through metabolic processes. In fact, they are indispensable to sustained flight but not to simple locomotor activity.

Aphid polymorphism is controlled by a cerebral neurohormone that favors production of virginoparous females.

8

Osmoregulation

Control of water metabolism in insects is achieved in a variety of ways depending on their mode of life, aquatic or terrestrial, and the nature of their diet. For some insects the elimination of water is a vital problem. For others there is a constant struggle against the lack of water; their small size, large surface area, and intense respiratory metabolism lead to considerable loss of water.

Significant water movements take place across the cuticle and the spiracles. These movements intensify during locomotor activity, which requires an increase in respiratory metabolism. As for water intake through food, it is important in herbivores, blood feeders, and sap suckers.

How do insects achieve their water regulation? They can avoid desiccation by behavior that causes them to seek shade and moisture. They also seem capable of regulating their transpiration and of adjusting their respiratory quotient, but the regulation of water metabolism is essentially achieved by sophisticated organs such as the Malpighian tubules and the rectum.

From the hemolymph, the Malpighian tubules absorb a liquid that flows through the rectum. They exhibit passive permeability to water and to substances of molecular weight below 5000, a large part of which is reabsorbed at the rectum. The latter displays anatomical arrangements that are designed for reabsorption, e.g., cryptonephridia and rectal papillae. Depending on the insect's physiological state, the rectum reabsorbs variable amounts of water; it also removes amino acids, sugars, fats, and certain ions, especially potassium. Hence the regulation of the water balance of the insect involves two organs that act in opposite ways, the Malpighian tubules and the rectum. Activation of the Malpighian tubules causes a loss of water, while activation of the rectum increases the insect's body water.

Investigations into water metabolism control began in 1956 with research dealing with the bee *Apis mellifica* and the larva of the beetle *Anisotarsus cupripennis*, which lives in the ground. In these two early studies, the complexity of the problem emerged with the divergent results obtained on the two species: diuretic function of the corpora cardiaca in the bee and antidiuretic function of the same organs in the beetle. Subsequent investigations dealt with blood feeders such as the blood-sucking bug *Rhodnius prolixus*, the mosquito *Anopheles freeborni*, and the tsetse fly *Glossina austeni*, all of which must lose weight rapidly after a meal in order to avoid losing their mobility and their capacity for flight; also investigated were species that display different life and feeding modes in relation to water metabolism, including cockroaches, locusts, stick insects, and butterflies.

8.1. TECHNIQUES

Different techniques have been employed for research on water metabolism. Experiments conducted *in vivo* mainly involved measurement of the ratio

$$\frac{\text{(initial fresh weight } - \text{ final fresh weight)}}{\text{dry weight}}$$

following manipulations designed to alter the osmotic pressure of the hemolymph, e.g., organ removals, ligatures, electrical stimulation, or distilled or saltwater injections. Tritiated water injections were sometimes made at the end of the experiment, the body water extracted by lyophilization, and the radioactivity measured by liquid scintillation to determine the amount of tritiated water retained by the insect (Proux and Buscarlet, 1976).

Sláma and Zdarek (1974), working on the linden bug *Pyrrhocoris apterus*, estimated the amount of water drunk and the amount excreted. The water drunk is sucked by the insect through a graduated capillary tube, enabling its measurement. The amount of water excreted is calculated from the area of the patches formed by urine drops on paper impregnated with bromophenol blue.

Mordue and Goldsworthy (1969) injected a solution of amaranth into locusts whose hemolymph was then punctured every 5 min and examined with a spectrophotometer. This procedure allowed an estimation of the speed of elimination of the dye by the Malpighian tubules.

The majority of the experiments were conducted *in vitro* and followed two basic methods. The first was Ramsay's method (1954) or one of its variants (Maddrell, 1972). This consists of measuring the amount of urine produced by the Malpighian tubules isolated in a drop of deproteinized serum placed in a small container filled with liquid paraffin (Fig. 62). The severed Malpighian tubules are ligated and cut at one end to allow the urine to escape. The urine produced is measured with an ocular micrometer. This technique was employed by Maddrell (1963) and Berridge (1966). Other investigators proceeded somewhat differently, isolating the Malpighian tubules

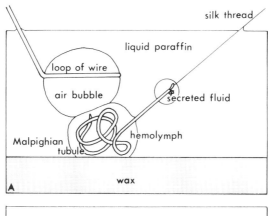

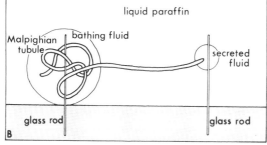

FIGURE 62. Experimental arrangement used to investigate fluid secretion by an isolated Malpighian tubule. (A, modified from Ramsay, 1954; B, modified from Maddrell, 1972.)

with a short portion of the digestive tract in a dish or glass capsule containing physiological saline. The gut is ligated on one side, while a capillary tube is inserted in the other. Diuretic variations are estimated by changes in the liquid level in the capillary tube (Mills, 1967; Cazal and Girardie, 1968).

The second method allows measurement of the urine secreted by the use of a dye that is retained by the cells of the Malpighian tubules. The latter, alone or combined with the midgut, are placed in an isotonic dye solution. They are rinsed after 30 min and extracted with an alcoholic solution of which the dye content is estimated colorimetrically (Altmann, 1956, 1958) or measured on the spectrophotometer, thus yielding the rate of fluid secretion (Mills and Nielsen, 1967).

The reabsorption activity of the rectum was evaluated by a method similar to that used for the Malpighian tubules. The rectum is ligated at one end and a graduated capillary tube inserted at the other. The movement of the liquid in the capillary serves to measure the reabsorption (Vietinghoff, 1966a; Cazal and Girardie, 1968). In other experiments (Mordue, 1969), the rectum operates inside out, reversing the direction of water flow. Immersed in a saline solution, its activity is revealed by its weight increase.

An *in vivo* method consists of injecting 14_C-labeled potassium urate into the insect. The concentration of radioactive substance present in the rectum, which denotes the secretory activity of the Malpighian tubules, is measured by a scintillation counter (Mordue, 1969).

Regulation of ion metabolism has not been extensively investigated. The methods for its study are atomic emission–absorption spectroscopy, transepithelial potential measurements, electron probe X-ray microanalysis, and the use of radioisotopes, ion-selective microelectrodes, and ionophores.

8.2. DIURETIC HORMONE

The first really thorough study of water metabolism control was achieved by Maddrell (1963, 1964a,b) working on the blood-sucking bug *Rhodnius prolixus*. This insect takes a single large blood meal, representing 10 times its body volume, at the onset of each larval instar and immediately excretes a large quantity of urine. A diuretic hormone is present in the blood of freshly fed insects (Fig. 63) but absent in fasting insects. By organ removals and extract injections, the source of this hormone was revealed. Later investigations (discussed below) confirmed the existence of a diuretic hormone in all the insects examined: cockroaches, stick insects, locusts, crickets, grasshoppers, bees, beetles, butterflies, flies, mosquitoes, and bugs.

8.2.1. Synthesis Sites

Investigations aiming to identify the sites of synthesis of diuretic hormone dealt chiefly with the brain, the corpora cardiaca, and the ganglia of the ventral nerve cord.

Histophysiological studies were conducted on different insects subjected to hydrating and dehydrating conditions. In most cases, significant storage of type A neurosecretory products was observed in the pars intercerebralis and corpora cardiaca of insects placed in a dry or dehydrating environment, or injected with saline, thus suggesting the existence of a diuretic factor (Tables XXXI and XXXII).

The presence of a diuretic hormone in the pars intercerebralis of the brain has long been suggested by observations made during investigations into other problems. After removal of the pars intercerebralis, many authors reported the occurrence of swelling related to high water retention (Table XXXIII). More recent investigations show that except for *Rhodnius*, diuretic hormone is present in the brain and in the corpora cardiaca of the various species examined. By the use of surgery and the regenerate method (see Sec-

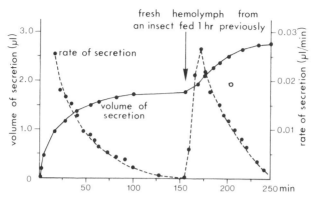

FIGURE 63. Demonstration of the presence of a diuretic hormone in the blood of a freshly fed *Rhodnius prolixus*. Adding hemolymph from an insect in diuresis strongly increases the rate of secretion of a resting set of isolated tubules. (Modified from Maddrell, 1963.)

tion 1.2.4), it was possible to confirm, in the desert locust *Schistocerca gregaria*, that the diuretic factor of the corpora cardiaca originates in the neurosecretory cells of the brain, and not in the glandular cells of the corpora cardiaca (Mordue and Goldsworthy, 1969).

The location of diuretic hormone in the brain was also investigated by methods other than cauterization of the pars intercerebralis. Pilcher (1970b) tested the activity of the different parts of the brain of the stick insect *Carausius morosus* and reported that the lateral and tritocerebral neurosecretory cells are devoid of activity, which is displayed only by the central region of the protocerebrum. Recent investigations succeeded in pinpointing the cells producing diuretic hormone. Not all the type A cells of the pars intercerebralis are involved, as previous studies appeared to indicate, but only four type A neurosecretory cells located slightly outside the pars intercerebralis in the migratory locust *Locusta migratoria* (Girardie, 1970). Heavily loaded in high-humidity conditions, these median subocellar cells are empty in low humidity. A sharp increase occurs in fresh weight and hemolymph volume after their electrocoagulation, whereas electrocoagulation of the pars intercerebralis has no effect on hemolymph volume and only gives rise to a temporary increase in fresh weight (Proux, 1979; Proux *et al.*, 1980). Reimplantation of the median subocellar cells in insects in which they have been destroyed restores a normal situation. Similar results were obtained by Charlet (1974) on the dragonfly *Aeshna cyanea*, which also has four type A median protocerebral neurosecretory cells. They occur in the release phase at the time of postecdysial swelling, and cauterization causes large swelling and a rise in hemolymph volume. In the cockroach *Leucophaea maderae*, similar cells were described by de Bessé (1978): their cauterization also induced water retention.

TABLE XXXI. Amount of Neurosecretory Material within Corpora Cardiaca and Type A Neurosecretory Cells of the Pars Intercerebralis in Various Hygrometric Conditions

Insect	Reference	Operative procedure	Neurosecretory material amount[a]			
			Dehydrating conditions		Hydrating conditions	
			pi	cc	pi	cc
Blaberus giganteus	Wall and Ralph (1962)	Dehydration or saline injection	+++			
Periplaneta americana	Nayar (1962)	Saline injection		+		
		Distilled water injection				+++
Carausius morosus	Pflugfelder (1937)	Dehydration		+++		
Schistocerca gregaria	Highnam *et al.* (1965)	Distilled water injection			+	
Melanoplus sanguinipes	Dogra and Ewen (1969)	Saline injection	+++			
		Distilled water injection			+	
Gryllodes sigillatus	Awasthi (1975)	Saline injection	+++	+++		
		Distilled water injection			+	+
Acheta domesticus	Geldiay and Edwards (1976)	Dryness	+++			
		Humidity			+	
Iphita limbata	Nayar (1957, 1960); Seshan (1968)	Salt water drinking	+++	+++		
		Dehydration	+++			
		Hydration			+	
Cenocorixa bifida	Jarial and Scudder (1971)	Dehydration	+			
		Hydration			+++	
Sphaerodema rusticum	Awasthi (1975)	Saline injection	+			

[a] +, Weak amount; +++, strong amount; cc, corpora cardiaca; pi, pars intercerebralis.

TABLE XXXII. Amount of Neurosecretory Material within Neurosecretory Cells of the Ventral Nerve Cord in Various Hygrometric Conditions

Insect	Reference	Neurosecretory cells studied	Neurosecretory material amount[a]	
			Dehydrating conditions	Hydrating conditions
Blaberus giganteus	Wall and Ralph (1962)	Thoracic A cells	+	
Leucophaea maderae	de Bessé (1965)	Thoracic and abdominal A cells	+	+ + +
Clitumnus extradentatus	Raabe (1965b)	Thoracic and abdominal A cells	+	+ + +
Schistocerca gregaria	Delphin (1965)	Abdominal A2 cells	+	+ + +

[a]+, Weak amount; + + +, strong amount.

TABLE XXXIII. Water Retention Caused by
Pars Intercerebralis Removal

Insect	Reference
Gryllus limaculatus	Girardie (1966a)
Schistocerca gregaria	Highnam *et al.* (1965)
Locusta migratoria	Cazal and Girardie (1968)
Melanoplus sanguinipes	Elliott and Gillott (1977)
Leptinotarsa decemlineata	de Wilde (1966)
Anisotarsus cupripennis	Nunez (1956)
Calliphora erythrocephala	Thomsen (1952)

The ventral nerve cord was investigated in many species: the cockroach *Periplaneta americana*, the stick insect *Carausius morosus*, the migratory locust *Locusta migratoria*, the cabbage white butterfly *Pieris brassicae*, the tsetse fly *Glossina austeni*, the phantom midge *Corethra plumicornis*, the mosquitoes *Anopheles freeborni* and *Aedes taeniorhynchus*, and the bug *Rhodnius prolixus* (Table XXXIV). These studies revealed the frequent presence of a diuretic factor in the ventral nerve cord ganglia.

Taking an overall look at the positive data obtained by one or more authors, it appears that diuretic hormone is distributed throughout the ventral nerve cord at least in small amounts in stick insects and locusts, but is present only in the subesophageal and thoracic ganglia of *Pieris*, and only in the abdominal ganglia of cockroaches, *Corethra,* and *Rhodnius*. Because systematic assays were not applied to all the ganglia, only partial data are available. However, they indicate the diuretic activity of the thoracic ganglia in mosquitoes, of the thoracoabdominal nerve mass in the tsetse fly, and the slight activity of the latter in the cotton stainer *Dysdercus fasciatus*.

Contradictory results are sometimes obtained with extracts of the same organ on a given species and on related species. This probably derives from changes in the hormonal content of the nerve ganglia due to variations in environmental conditions. The brain of *Periplaneta* displays diuretic activity when the insects are placed in 95% relative humidity, while antidiuretic activity is observed in dehydrated insects (Wall and Ralph, 1964; Wall, 1965) and disappears completely in hydration conditions (Wall, 1966). In *Locusta*, a change in activity was also observed in the extracts depending on the hygrometric conditions (Cazal and Girardie, 1968). Corpora cardiaca extracts show stronger diuretic activity in dehydrated insects, and increased antidiuretic activity in hydrated insects. In the ventral nerve cord, the diuretic factor is only clearly apparent in insects deprived of brain diuretic cells (Proux, 1978c, d). Moreover, the sensitivity of the effectors varies according to dehydration, as in the rectum of the cockroach (Steele and Tolman, 1980).

The source cells of the diuretic factor have only been investigated in

TABLE XXXIV. Location of the Diuretic Factor

Column groups: **Method** spans *In vivo* and *In vitro*. **Compared activity[a]** spans Brain, cc, ca, vnc (se gg, th gg, ab gg), and po.

Insect	Reference	Method		Compared activity[a]			vnc			
		In vivo	*In vitro*	Brain	cc	ca	se gg	th gg	ab gg	po
Periplaneta americana	Mills (1967)		Mt	0	0	+	0	0	+++	
	Wall (1966)		Mt	+++	0	+++			+++	0
Carausius morosus	Pilcher (1970b)		Mt	+++	+++	0	+++			
	Vietinghoff (1966a,b)[b]		r	+++			———	+	+++	
Locusta migratoria	Cazal and Girardie (1968)		Mt	+++						
	Proux (1978a,b)									
Schistocerca gregaria	Highnam et al. (1965)	pi−, ca−		+++	+++			+	+	
	Mordue (1969)	pi−, ca−	Mt, r	+++	+++					
Melanoplus sanguinipes	Elliott and Gillott (1977)	pi−		+++						
Gryllus bimaculatus	Girardie (1966a)	pi−		+++	+++	++				
Apis mellifica	Altmann (1956, 1958)		Mt, r			+++				
Anisotarsus cupripennis	Nunez (1956)	brain−, cc−, lig			——— +++ ———		0			
Leptinotarsa decemlineata	de Wilde (1966)	pi−		+++						
Pieris brassicae	Nicolson (1976)[c]		Mt	++	+++		+	+		
Danaus plexippus	Dores et al. (1979)	lig		+++	——— ++ ———			++	0	
Calliphora erythrocephala	Thomsen (1952)	pi−, Wr−		+++	+++					
	Schwartz and Reynolds (1979)		Mt	+				+++		

(Continued)

TABLE XXXIV. (Continued)

Insect	Reference	Method		Compared activity[a]						
		In vivo	In vitro	Brain	cc	ca	se gg	vnc th gg	vnc ab gg	vnc po
Glossina austeni	Gee (1975)		Mt	+++				+++		——
Corethra sp.	Gersch (1967, 1969)	elec stim						0	+++	+++
Aedes taeniorhynchus, larva	Maddrell and Phillips (1978)		Mt	+++				+++		
Anopheles freeborni, adult	Nijhout and Carrow (1978)		Mt	+++				+++		
Rhodnius prolixus	Maddrell (1963, 1964a)	gg m−, lig	Mt	+	0	0	0	0	+++	+++
Dysdercus fasciatus	Berridge (1966)	pi−, cc.ca−		+++	+	0	0	+		——
Pyrrhocoris apterus	Sláma (1964)			0	0	0				

[a]0, +, ++, +++, Zero, weak, medium, strong activity; ca−, cc−, gg m−, pi−, Wr−, corpora allata, corpora cardiaca, nerve ganglionic mass, pars intercerebralis, Weismann ring ablation; ab gg, abdominal ganglia; elec stim, electrical stimulation; lig, ligature; Mt, Malpighian tubules; po, perisympathetic organs; r, rectum; se gg, subesophageal ganglion; th gg, thoracic ganglia; vnc, ventral nerve cord.
[b]Chromatographed extracts.
[c]Boiled extracts.

Rhodnius. The posterior part of the thoracoabdominal nerve mass, which corresponds to a fused mass of all the abdominal ganglia, is the only active part (Maddrell, 1963). It contains three distinct types of neurosecretory cells: three A cells, six phloxinophilic B_1 cells, and five hematoxylinophilic and phloxinophilic B_2 cells on each side (Baudry, 1968). Their study, during feeding and diuresis, suggests that the B_1 cells are responsible for synthesizing the diuretic factor for they release their neurosecretory products in the minutes that follow the meal (Baudry, 1969). After investigating the diuretic activity of isolated neurons located in the same region, Berlind and Maddrell (1979) demonstrated that the active cells correspond to B_1 cells (Fig. 64).

8.2.2. Release Sites

Diuretic hormone can be released very rapidly, at least in blood feeders: in 3 min in *Rhodnius* (Maddrell, 1963) and in 1 min in *Anopheles* (Nijhout and Carrow, 1978). Its diffusion is accelerated in *Rhodnius* by strong peristaltic contractions of the midgut that begin immediately after the blood meal (Maddrell, 1964a).

The release sites can be identified by using Maddrell and Gee's method (1974), in which the presumed release areas are placed in potassium-rich solutions. In this method the neurosecretory endings undergo depolarization, causing hormone release in the incubation medium. This is not the case when the cell bodies are subjected to the same treatment (Fig. 65). The use of this method in *Rhodnius* and *Glossina* reveals peripheral release sites. Whereas the incubation medium of the nerve mass was inactive in both species, the proximal part of the first three pairs of abdominal nerves in *Rhodnius* and the

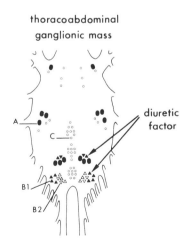

thoracoabdominal
ganglionic mass

diuretic
factor

FIGURE 64. Location of the neurosecretory cells involved in diuretic hormone secretion in *Rhodnius prolixus*. (Modified from Maddrell, 1963, and Baudry, 1968.)

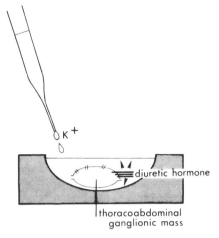

FIGURE 65. Stimulation of diuretic hormone release from neurohemal areas of *Rhodnius prolixus* abdominal nerves by increasing the potassium concentration in the incubating medium. (From data of Maddrell and Gee, 1974.)

peripheral abdominal nerves in *Glossina* were highly active. These nerves have characteristic ultrastructural features (Maddrell, 1966; Maddrell and Gee, 1974) and may be considered as distal perisympathetic organs. In the blowfly *Calliphora erythrocephala*, the release of diuretic hormone probably also takes place from the nerves, the extracts of which are more potent than those of nerve ganglia (Schwartz and Reynolds, 1979).

In other insects, nothing is known about the release sites of the diuretic factor produced in the ventral nerve cord. As for the diuretic factor of the pars intercerebralis, it appears to be released in the corpora cardiaca and to be transported by the nccI in *Schistocerca* (Goldsworthy *et al.*, 1972b).

8.2.3. Release Control

It would be interesting to identify the sensory receptors implicated in the release of diuretic hormone. It may be assumed that they differ according to the insect's diet. Few investigations have been devoted to this matter, and some of them, such as those using nerve severance, are always debatable because they fail to demonstrate whether interruption of the information is related to transport from the periphery to the brain or vice versa. In any event, water retention is induced by severance of the ventral nerve cord in many species: the beetles *Anisotarsus* (Nunez, 1956) and *Leptinotarsa* (de Wilde, 1966), the mosquito *Anopheles* (Nijhout and Carrow, 1978), and the locust *Locusta* (Proux, 1978b). Decapitation creates serious problems in *Anopheles* (Nijhout and Carrow, 1978) and *Glossina austeni* (Gee, 1975). This suggests that the information is integrated at the brain level. The nature and the source of the information are poorly known, but we do know that feeding triggers diuretic hormone release in locusts (Highnam *et al.*, 1966; Mordue,

1967) and the cotton stainer *Dysdercus fasciatus* (Berridge, 1966). Moreover, Gwadz (1969) reported the presence of tension receptors in the gut wall of *Aedes*, which regulate the volume of the blood meal and may also control diuresis. By various experiments performed on *Rhodnius*, Maddrell (1964b) determined that the source of the initial stimulus of diuresis was not the intake of food, the increase in internal pressure, or the osmotic change in the blood, but rather abdominal distension. In fact, diuresis is not modified by decapitation, by expulsion of the ingested meal, or by hypo- or hypertonic nutrients, and the introduction of air into the digestive tract fails to cause diuresis in fasting insects. If, however, a small gash is made in the abdominal wall, through which the midgut is teased out so that the meal ingested does not cause the abdomen to swell, the insects feed far more abundantly and do not display greater diuresis, thus suggesting that diuresis is controlled by abdominal distension.

8.2.4. Circulating Hormone Rate, Hormone Destruction

Diuretic hormone can often be revealed in the blood (*Periplaneta, Carausius, Glossina, Rhodnius, Dysdercus*). However, its rate varies with water conditions and in some species it is undetectable (*Pieris, Anopheles*), probably owing to its instability.

To provide an idea of the quantity of circulating hormone, attempts were made to measure the effective doses initiating fluid secretion. In *Rhodnius*, the third part of the ganglionic mass is sufficient to induce a strong secretion (Maddrell, 1964a) (Fig. 66). In *Schistocerca*, hormone release is likely to correspond to the quantity produced by a pair of corpora cardiaca during a 10-

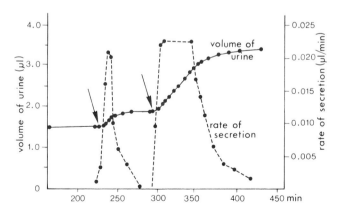

FIGURE 66. The response of a set of *Rhodnius prolixus* isolated tubules to successive doses of diuretic hormone. First arrow, 0.078 ganglionic mass/100 μl hemolymph; second arrow, 0.30. (Modified from Maddrell, 1964a.)

to 12-hr period in the male or 4–6 hr in the female (Mordue, 1969), and the active doses are about 0.04 pair of corpora cardiaca per 100 μl of hemolymph. Similar levels were estimated in *Carausius* (Pilcher, 1970b) and *Dysdercus* (Berridge, 1966).

The breakdown rate of the corpora cardiaca diuretic hormone of *Schistocerca* has been estimated to be 0.05–0.15 pair of corpora cardiaca per hour (Mordue, 1969). In *Rhodnius* (Maddrell, 1964a), diuretic hormone, which is abundant in the hemolymph after a meal, disappears in about 2 hr, i.e., at the end of diuresis. This may be due to arrested release or to destruction. The latter hypothesis is confirmed by the following experiment: two drops of active blood are kept under paraffin, one in the presence of a single set of Malpighian tubules, the second in the presence of four sets. In the second case, urine secretion stops far sooner, demonstrating that diuretic hormone is destroyed by the Malpighian tubules themselves, which remain capable of resuming their function with fresh extracts (in *Glossina*, Gee, 1975; in *Pieris*, Nicolson, 1976). Apart from its destruction by the Malpighian tubules, diuretic hormone can be destroyed by an enzyme of the central nervous system. Boiled extracts are more active and more stable than unboiled extracts, and if an unboiled extract is added to a boiled extract, the latter quickly loses its activity (Gee, 1975) (Fig. 67).

In *Dysdercus* (Berridge, 1966), *Carausius* (Pilcher, 1970b), and *Glossina* (Gee, 1975), the Malpighian tubules destroy diuretic hormone and can thus control its hemolymph titer. In *Calliphora*, however, diuretic hormone, while inactivated by hemolymph, is not inactivated by Malpighian tubules (Schwartz and Reynolds, 1979).

8.3. ANTIDIURETIC HORMONE

The existence of an antidiuretic hormone was demonstrated in cockroaches, stick insects, locusts, bees, and the larvae of the phantom midge *Corethra*, chiefly by *in vitro* experiments (Table XXXV). This hormone is

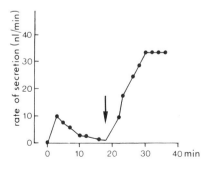

FIGURE 67. Fluid secretion rate of an isolated Malpighian tubule of *Glossina austeni* bathed by an unboiled homogenate of thoracic ganglia. Effect of replacing this with a boiled homogenate at the time indicated by the arrow. (Modified from Gee, 1975.)

TABLE XXXV. Location of the Antidiuretic Factor

| Insect | Reference | Method | | Compared activity[a] | | | vnc | | | |
		In vivo	In vitro	Brain	cc	ca	se gg	th gg	ab gg	po
Periplaneta americana	de Bessé and Cazal (1968)		Mt, r	+++	+++					+++
Blabera fusca	Wall and Ralph (1962); Wall (1965)		Mt, r	+++	+++	++	+++	++	+	
Crausius morosus	Vietinghoff (1966a,b)[b]		r	+++	+++	+++				
Clitumnus extradentatus	de Bessé and Cazal (1968)		Mt, r	+++	+++					+
Locusta migratoria	Cazal and Girardie (1968)	pi−	Mt, r	+++	+++					+++
	de Bessé and Cazal (1968)		Mt, r		+++					
Schistocerca gregaria	Mordue (1970)		r	+++	+++					
Apis mellifica	Altmann (1956, 1958)		Mt, r	+++	+++					
Corethra plumicornis	Gersch (1967, 1969)	elec stim						+++		

[a]+, ++, +++, Weak, medium, strong activity; ab gg, abdominal ganglion; ca, corpora allata; cc, corpora cardiaca; elec stim, electrical stimulation; Mt, Malpighian tubules; pi−, pars intercerebralis ablation; po, perisympathetic organs; r, rectum; se gg, subesophageal ganglion; th gg, thoracic ganglion; vnc, ventral nerve cord.
[b]Chromatographed extracts.

found in the brain at least in some species, and its presence in the ventral nerve cord was reported whenever investigated. It is also present in the corpora cardiaca and the perisympathetic organs. It is not possible to state exactly where in the brain it is produced.

Histophysiological studies revealed changes in the load of the neurosecretory cells of the ventral nerve cord after changes in water conditions. Thoracoabdominal type A cells of cockroaches, stick insects, and locusts are heavily loaded in insects subjected to hydrating conditions, and release their secretory material in dehydrating conditions (Table XXXII).

We know practically nothing about the stimuli that act in controlling antidiuretic hormone release. The release sites are probably the corpora cardiaca and the perisympathetic organs, both of which are very active in cockroaches, stick insects, and locusts (de Bessé and Cazal, 1968). In the stick insects, the load of the perisympathetic organs undergoes a significant decrease when the insects are kept in a dry environment (Raabe, unpublished results) (Fig. 68).

8.4. DIURETIC OR ANTIDIURETIC FUNCTION OF OTHER HORMONES

Do other hormones play a part in controlling water metabolism?

Some investigators have reported that corpora allata extracts display an activity comparable to that of corpora cardiaca. It may be due in some cases to contamination by neurosecretory products contained in the corpora car-

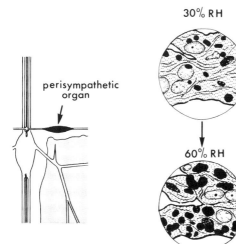

FIGURE 68. Increase of neurosecretory material stored in perisympathetic organ as a function of strongly increased environmental humidity in *Carausius morosus*. RH, relative humidity.

diaca. However, removal of the corpora allata induces swelling in the locust and the fly (Mordue, 1969; Day, 1943). It is difficult to state at present whether the effect is direct, or whether the lack of the corpora allata causes retention of cerebral neurohormones, as claimed by various authors (Thomsen and Lea, 1968; Highnam *et al.*, 1963).

Ecdysone may also be a factor in water metabolism. Ryerse (1978) showed that the functioning of the Malpighian tubules of the pupa of the skipper butterfly *Calpodes ethlius* normally stops just before pupariation, and continues, *in vitro* or *in vivo*, after an ecdysone injection. In the young tsetse fly, moreover, the addition of ecdysone or ecdysterone to the incubation medium of the Malpighian tubules containing a diuretic extract increases the excretion rate (Gee *et al.*, 1977).

Apart from controlling the activity of the Malpighian tubules and the rectum, other endocrine regulations of water metabolism may exist. The opening and closing rhythm of the spiracles appears to be controlled by the nervous system, but transpiration, which depends on the permeability of the cuticular wax layer, may be regulated by hormonal factors (Treherne and Willmer, 1975). In *Periplaneta,* in fact, decapitation induces a rapid loss of water, which may derive from transpiration, for it does not occur in the Malpighian tubules or in the spiracles. Brain and corpora cardiaca extracts sharply reduce this water loss, so that a neurohormone may effectively control transpiration.

8.5. PURIFICATION OF DIURETIC AND ANTIDIURETIC HORMONES

Attention has seldom been drawn to purification and characterization of diuretic and antidiuretic hormones, and the identity of the factors at work in insects is not really established (Table XXXVI). Interspecific activity has been reported in some cases: diuretic hormone of *Acheta* acts on the Malpighian tubules of *Periplaneta* (de Bessé and Cazal, 1968); diuretic hormone of *Periplaneta, Glossina,* and *Rhodnius* acts on the Malpighian tubules of *Schistocerca* (Bernstein and Mordue, 1978); and diuretic hormone of *Carausius* is active on *Rhodnius*, but diuretic hormone of *Rhodnius* is inactive on *Carausius* (Maddrell *et al.*, 1969).

As for the stability of hormone preparations, their fragility at ambient temperature has been noted in some cases *(Rhodnius)*, doubtless owing to the presence of proteolytic enzymes in the extracts. In most species, boiling of the preparations prolongs their life and fails to diminish activity, thus indicating the hormone's thermostability.

Diuretic and antidiuretic hormones are probably proteins or polypeptides like most neurohormones, because they are destroyed by proteolytic enzymes

TABLE XXXVI. Some Properties of Diuretic and Antidiuretic Hormones

	Reference	Thermostability	Solubility in methanol	Molecular weight	Peptidergic nature
Antidiuretic hormones					
Periplaneta americana	Goldbard *et al.* (1970)			8.000	
Apis mellifica	Altmann (1958)	+			
Diuretic hormones					
Periplaneta americana	Goldbard *et al.* (1970)			30,000	
Schistocerca gregaria	Bernstein and Mordue (1978)	+	+		+
Apis mellifica	Altmann (1958)	+			
Anopheles freeborni	Nijhout and Carrow (1978)	+			
Glossina austeni	Gee (1975)	+	+	1,200	+
Dysdercus fasciatus	Berridge (1966)	+			
Rhodnius prolixus	Maddrell *et al.* (1969)	+			
	Aston and White (1974)		0	60,000 and 2,000	+
Danaus plexippus	Dores *et al.* (1979)	+		3,000	+

(*Locusta, Rhodnius, Glossina*). Moreover, amylase and neuraminidase destroy diuretic hormone of *Glossina*, thus suggesting a sialoglucoproteic nature (Gee, 1975). These two enzymes are inactive in *Rhodnius* (Aston, 1979).

Attempts have been made to isolate the active factors in the cockroach, stick insect, tsetse fly, and bug. In *Periplaneta*, Goldbard et al., (1970) used the terminal abdominal ganglion as a source of active factors, and the rectum *in vitro* as effector. Extracts chromatographed on polyacrylamide gels P30 and P10 allowed separation of two factors, a diuretic factor of molecular weight above 30,000, and an antidiuretic factor of molecular weight from 8000 to 10,000.

The activity of factors C_1 and D_1, isolated from the ventral nerve cord of *Periplaneta* by Gersch and co-workers (see Chapters 2 and 6), was assayed *in vitro* on the rectum of *Carausius* (Vietinghoff, 1966b, 1967) and on the Malpighian tubules of the same species (Unger, 1967). It was found that both factors C_1 and D_1 are antidiuretic to the Malpighian tubules but diuretic to the rectum.

Diuretic and antidiuretic hormones of the corpora cardiaca have also been investigated in locusts (Mordue and Goldsworthy, 1969; Mordue et al., 1970). Assays were conducted *in vivo* by the method described above. According to these authors, the diuretic and antidiuretic factors may be identified with factors C_1 and D_1.

In *Rhodnius* (Aston and White, 1974), a search was made for diuretic hormone in extracts of the thoracoabdominal nerve mass, in the K^+-rich incubation medium, in the blood, and in extracts of the head and thorax. After chromatography on polyacrylamide gels P60, P30, and P2, two similar activity peaks appeared in the extracts of the head, thorax, and thoracoabdominal nerve mass. This involves a substance of molecular weight above 60,000 and another substance of molecular weight below 2000. The latter may be the active factor, because it is the only one present in the blood and in the K^+-rich incubation media of the nerve ganglion.

A continuation of these investigations by Aston (1979) and Hughes (1977, 1979) demonstrated the fragility of diuretic hormone and the difficulties encountered in its extraction, storage, and purification. Its release is not facilitated by ionophores or low temperatures. Ultrasonication of the organs yields inconsistent results, which are improved by subsequent mechanical homogenization. Good protection against enzymes is obtained by enzyme inhibitors, but significant losses of substances by adsorption were reported during purification. Like other proteins, diuretic hormone appears to be stabilized by the use of bovine serum albumin.

New separation tests by centrifugation, chromatography, and electrophoresis confirmed the existence of two peaks, the first of high, the second of low molecular weight. Centrifugation followed by extraction with ammonium sulfate yields the first factor, while chloroform/methanol delipidation followed by chromatography on Sephadex G-25 yields the low-molecular-weight active factor.

Investigations on the subject still have a long way to go, for the data gathered on the different species have never been subjected to comparative analysis. It is possible that several diuretic and antidiuretic factors exist, even within a single species. The experiments of Wall and Ralph (1962), which showed that the activity of thoracic ganglia and corpora cardiaca homogenates is reversed according to whether the solvent employed is ethanol or ethyl acetate, give support to this hypothesis.

It seems fairly well established that diuretic hormone is not identical to 5-HT (Maddrell et al., 1969; Aston and White, 1974). However, the latter stimulates the activity of the Malpighian tubules in many species and displays several similarities to diuretic hormone. Its inhibitors, tryptamine and tyramine, diminish the secretion of the Malpighian tubules of *Rhodnius*. Adrenaline was also reported to act as a stimulant in locusts, but, as for 5-HT, pharmacodynamic action is probably involved.

Antidiuretic hormone of vertebrates has no effect on *Rhodnius* (Maddrell, 1969), but a substance related to vasopressin was found in the subesophageal ganglion of the stick insect (Rémy et al., 1977, 1979), *Acheta* (Strambi et al., 1979), and locust (Rémy et al., 1979). It does not appear to exist in the pars intercerebralis or corpora cardiaca. Its concentration in the subesophageal ganglion of *Acheta* varies with humidity and is lower in insects raised in a moist environment (Strambi et al., 1978). It is located in two medioventral cells whose axons lead to the brain, where they are observed in the neurohemal area described by Geldiay and Edwards (1973). As above, further research will be necessary to determine whether physiologically active substances are involved there.

8.6. MODE OF ACTION OF DIURETIC HORMONE

Some investigators tested the action of cAMP on the Malpighian tubules. Diuresis was stimulated in all experimental cases. Aminophylline and other phosphodiesterase inhibitors display the same effect. Moreover, a very rapid rise in cAMP was reported in the Malpighian tubules after they were placed in contact with diuretic hormone (Aston, 1975). As in other regulations, cAMP thus plays the part of second messenger. It probably acts by stimulating the apical cationic pump of the Malpighian tubules. After stimulation by 5-HT, which mimics diuretic hormone effects in some species, a similar increase in cAMP and transepithelial potential was observed (Table XXXVII).

In the salivary glands, which also serve for the transport of water and molecules, the action of 5-HT, cAMP (Berridge and Patel, 1968), and calcium (Prince et al., 1972) has been reported, acting together to change the mem-

TABLE XXXVII. Effect of Various Substances on the Secretory Activity
of the Malpighian Tubules

Insect	Reference	Adrenaline	5-HT	cAMP
Periplaneta americana	Steele and Tolman (1980)			+
Carausius morosus	Maddrell *et al.* (1969)		+	
	Maddrell *et al.* (1971)			+
Locusta migratoria	Mordue (1972)	+	0	+
	Anstee *et al.* (1980)		0	+
Schistocerca gregaria	Maddrell and Klunsuwan (1973)		0	+
	Bernstein and Mordue (1977)	+		+
Pieris brassicae	Nicolson (1976)			+
Aedes taeniorhynchus	Maddrell and Phillips (1978)		+	+
Rhodnius prolixus	Maddrell *et al.* (1969)		+	
	Maddrell *et al.* (1971)			+

brane permeability and the activity of the cationic pumps (Berridge *et al.*, 1976).

It is probable that in this case, and also in many other tissues, ATP supplies the energy necessary for potassium transport, a membrane ATPase being specifically activated by Na^+–K^+ (Anstee and Bell, 1975; Peacock, 1976; Peacock *et al.*, 1976). Ouabain acts to inhibit Na^+–K^+ ATPase activity in some locusts (Anstee *et al.*, 1980), while in others no effect on the Malpighian tubules (*Rhodnius,* Maddrell, 1969; *Carausius*, Pilcher, 1970a; *Glossina*, Gee, 1976) has been recorded (see Anstee and Bowler, 1979).

In vivo and *in vitro* experiments performed on the cockroach *Periplaneta* show that injection of corpora cardiaca–corpora allata extracts into intact insects produces in the rectal tissue a decrease in the concentration of glycogen and an increase in phosphorylase activity, oxygen consumption, and water transport: *in vitro*, these effects are mimicked by dibutyryl cAMP, and require the presence of Na^+ (Steele and Tolman, 1980; Tolman and Steele, 1980a,b).

8.7. ION METABOLISM

Insects are able to regulate the osmotic and ionic concentration of their hemolymph. Many recent investigations have dealt with this matter, but endocrine regulation has only been demonstrated in two cases.

Ionic resorption in *Rhodnius* occurs in the lower part of the Malpighian tubules. It is stimulated by 5-HT (Maddrell and Phillips, 1975) and by a substance located in the thoracoabdominal nerve mass and the abdominal nerves. Diuretic hormone does not seem to be involved in this case, for the extracts of the neurosecretory cells that act on diuresis have no effect on ion absorption. The latter depends, at least in part, on the ion content of the hemolymph (Maddrell and Phillips, 1976). In the rectum of *Locusta*, cAMP and corpora cardiaca homogenate increase the short-circuit current and trans-epithelial electropotential difference, stimulating electrogenic transport of Cl^- (Spring *et al.*, 1978; Spring and Phillips, 1980).

8.8. CONCLUSIONS

Investigations into the regulation of water metabolism show that all insects possess a diuretic hormone, while only some of them also appear to possess an antidiuretic hormone (Table XXXVIII).

If one considers insect diet, it appears that two antagonistic hormones exist in most omnivores and herbivores, while only one hormone (diuretic hormone) is present in species that absorb large amounts of liquid, such as the sap suckers and blood feeders. It may be assumed that antidiuretic hormone is really useless in such species, but it is also highly possible that the presence of an antidiuretic factor has escaped the attention of investigators. In fact, if a single organ, such as the corpora cardiaca (Mordue, 1969) or the pars intercerebralis (Cazal and Girardie, 1968), or the entire central nervous system (Unger, 1957) contains two antagonistic factors, it is clear that, depending on the physiological state of the insect, one may prevail and the other go undetected. Furthermore, variations in the sensitivity of the Malpighian tubules and rectum, reported in relation to food, ovarian state, and water conditions, may have given rise to certain negative experimental results. Moreover, it is possible that the two effectors, the Malpighian tubules and the rectum, display different responses to neuroendocrine factors.

It should be noted that the Malpighian tubules are rarely innervated, making the existence of diuretic hormone necessary. The rectum, on the contrary, is provided with many neurosecretory endings, which can exert local control (see Chapter 1).

The synthesis sites of diuretic hormone and antidiuretic hormone are not accurately known in all species, because investigations have rarely dealt with all the nerve ganglia. However, it is probable that the brain and the ventral nerve cord ganglia produce both hormones in most cases.

The release sites of cerebral diuretic and antidiuretic hormones are located in the corpora cardiaca. As for the ventral nerve cord, the perisym-pathetic organs of cockroaches, stick insects, and locusts play a part in the release of antidiuretic hormone, while diuretic hormone of *Rhodnius* and *Glossina* is released in distal neurohemal organs.

The mode of action of diuretic hormone on the Malpighian tubules has been thoroughly examined. Like 5-HT, this hormone acts through cAMP,

TABLE XXXVIII. Occurrence of Diuretic and Antidiuretic Factors
in the Central Nervous System

Insect	Developmental stage	Diet	Diuretic factor	Anti-diuretic factor
Periplaneta	Larva, adult	Omnivorous	+	+
Carausius	Larva, adult	Phytophagous	+	+
Locusta	Larva, adult	Phytophagous	+	+
Schistocerca	Larva, adult	Phytophagous	+	+
Apis	Adult	Sap suckers	+	+
Anisotarsus	Larva		+	
Leptinotarsa	Adult	Phytophagous	+	
Pieris, Danaus	Adult	Sweet water suckers	+	
Glossina	Adult	Hematophagous	+	
Calliphora	Adult	Facultative blood feeders	+	
Corethra	Larva	Carnivorous	+	+
Aedes	Adult	Hematophagous	+	
Anopheles	Adult	Hematophagous	+	
Rhodnius	Larva, adult	Hematophagous	+	
Dysdercus	Larva, adult	Sap suckers	+	

which activates the cationic pump of the apical membrane of the Malpighian tubules.

Purification and characterization assays of diuretic and antidiuretic hormones were conducted with different organs and insect species, i.e., the central nervous system and the terminal ganglion of *Periplaneta,* the corpora cardiaca of the locust, and the thoracoabdominal nerve mass of the tsetse fly and the blood-sucking bug *Rhodnius*.

The active factors are thermostable polypeptides. Data concerning their molecular weights are contradictory. The antidiuretic factor of *Periplaneta* has a molecular weight of 8000–10,000, while the values obtained for diuretic hormones of *Periplaneta* and *Glossina* are respectively 30,000 and over 1200. In *Rhodnius*, diuretic factors with molecular weights of 60,000 and 2000 have been isolated.

Metabolism

All aspects of metabolism and metabolites—storage, release into the blood, and use—are of fundamental importance to all functions and particularly to those that require high energy consumption. Hence, it is important to be thoroughly familiar with metabolic mechanisms and their regulation, which sometimes differ from one species to another, in order to understand fully the control mechanisms of the major vital functions.

9.1. PROTEASES AND AMYLASES

The protease level in the intestine rises during feeding periods. In the linden bug *Pyrrhocoris apterus*, the adult female feeds at the beginning of each oogenetic cycle, and at that time the level of its intestinal proteases increases. On the other hand, in the adult male, feeding is relatively scarce, and the protease levels are consequently low, as in starved insects (Hrubešová and Sláma, 1967) (Fig. 69). The protease level in the intestine does not depend on the food itself, but on the quantity of proteins contained in the digestive tract in the cockroach *Leucophaea maderae* (Engelmann, 1969), and especially in blood feeders such as mosquitoes (Fisk, 1950; Briegel and Lea, 1975), tsetse flies (Gooding, 1974), and blood-sucking bugs (Persaud and Davey, 1971).

Different experiments have suggested that a neurohormone acts in controlling the secretion of intestinal proteases. Decapitation of the mealworm *Tenebrio molitor* 1 day before eclosion prevents the normal increase of proteases (Dadd, 1961). Head ligation also causes a decrease in proteases in the tsetse fly *Glossina morsitans* (Langley, 1967b). The injection of the blood of fed insects into starved insects elevates the proteases in *Tenebrio* (Dadd, 1961). In the latter and in the larva of the beetle *Morimus funereus*, maintenance at a temperature above normal induces a decrease in the enzymes of the digestive tract. The injection of neuroendocrine complex extracts restores a normal rate of proteolytic activity (*Morimus* larvae, Ivanović *et al.*, 1978;

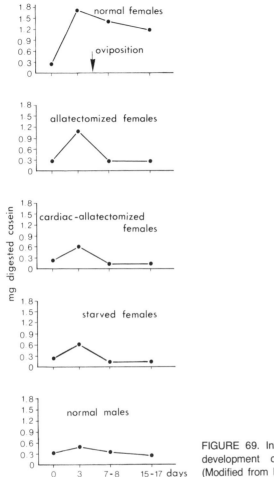

FIGURE 69. Intestinal protease activity during development of adult *Pyrrhocoris apterus*. (Modified from Hrubéšová and Sláma, 1967.)

adult *Tenebrio*, Janković-Hladni *et al.*, 1976). These extracts also act on the amylolytic activity of the *Tenebrio* larva (Ivanović *et al.*, 1978) and adult (Janković-Hladni *et al.*, 1978).

Histophysiological data also show that the neurosecretory cells of the pars intercerebralis exhibit modifications related to protein ingestion. In the blowfly *Calliphora erythrocephala* (Thomsen, 1965), the neurosecretory cells are heavily loaded in insects fed with carbohydrates, slightly loaded after protein feeding, and a significant release into the axons occurs in relation to the meat meal. In the virgin female of the North American grasshopper *Melanoplus sanguinipes* (Dogra and Gillott, 1971), the meal induces a sharp elevation in proteases that is also accompanied by a high release of neurosecretory materials. In *Morimus* larvae, temperature elevation causes the storage of neurosecretory materials in the pars intercerebralis and in the ventral neurosecretory cells of the subesophageal ganglion, with a concomitant decrease in the proteolytic and amylolytic enzymes of the digestive tract (Ivanović *et al.*, 1978).

Accurate experiments involving removal of the pars intercerebralis were conducted on many insects (*Calliphora*, Thomsen and Møller, 1959; the grasshopper *Gomphocerus rufus*, Loher, 1966; the mealworm *Tenebrio*, Mordue, 1967; the mosquito *Aedes aegypti*, Briegel and Lea, 1979; and the bug *Dysdercus cingulatus*, Muraleedharan and Prabhu, 1979). They cause a decrease in the midgut proteases. Thus in meat-fed *Calliphora*, the protease level in operated insects is similar to that of carbohydrate-fed flies. In *Tenebrio*, proteases are normally elevated at the onset of imaginal life but are inhibited when an operation is performed just before or after molting.

Removal of the corpora cardiaca–corpus allatum complex also causes a drop in proteases in the adult female of *Pyrrhocoris* (Hrubešová and Sláma, 1967), whereas allatectomy has a different effect. Therefore, it is probable that the active factor originates in the brain. This has been confirmed experimentally in *Tenebrio* by experiments involving severance of the nccI (Mordue, 1967).

Implantations and injections performed on operated insects restore normal situations: implantation of corpora cardiaca in *Pyrrhocoris* (Hrubešová and Sláma, 1967), implantation of corpora cardiaca–corpora allata complexes in *Calliphora* (Thomsen and Møller, 1963), brain extract injections into neckligated *Glossina morsitans* (Langley, 1967b), and implantation of the neurosecretory cells of the pars intercerebralis in parsectomized *Dysdercus fasciatus* (Muraleedharan and Prabhu, 1979).

A neurohormone originating in the pars intercerebralis and released into the corpora cardiaca may, therefore, ensure regulation of the secretory cells of the midgut. Although this concept of neurohormone control of protease secretion is shared by many authors, it is questioned by others who state that the effect is indirect and is exerted by means of food intake, which can be disturbed by the operation (*Sarcophaga argyrostoma*, Engelmann and Wilkens, 1969; *Leucophaea*, Engelmann, 1969). These authors postulate that the midgut proteases are regulated by a secretagogue mechanism, and that contact with the ingested proteins stimulates the secretory activity of the cells of the midgut wall. This view is advanced by Fisk (1950) for mosquitoes, Engelmann (1969) and Gordon (1968) for cockroaches, Engelmann and Wilkens (1969) and Foster (1972) for the flies *Sarcophaga bullata* and *Glossina austeni*, Persaud and Davey (1971) and Muraleedharan and Prabhu (1979) for the bugs *Rhodnius* and *Dysdercus*.

Furthermore, the gut wall appears to contain a stimulatory substance in the American cockroach *Periplaneta americana*, which may originate in neurosecretory axons or in the peripheral neurosecretory cells and the amount of which displays a circadian rhythm (Rounds, 1968).

A peptide resembling gastrin in its immunolgical reactions, molecular weight, and sensitivity to pronase was found in the brain of the tobacco hornworm *Manduca sexta* (Kramer *et al.*, 1977). However, it appears to be absent from the digestive tract and the hemolymph, thus making its function in digestion debatable.

It may be concluded that midgut proteases and amylases are affected by one or more neurohormones of the pars intercerebralis, but not necessarily by direct action.

9.2. PROTEINS

There is considerable evidence that protein syntheses are of considerable importance for the physiology of the insect during its development, at metamorphosis, and in the adult during reproduction (Fig. 70). Hence the study of their regulation is of great interest, but raises a number of difficulties. The species differ from each other especially in the variation of the hemolymph protein concentration during development and reproduction; some of them store proteins and vitellogenin in the hemolymph, while others store these substances in the fat body. The extent of the fat body reserves also depends

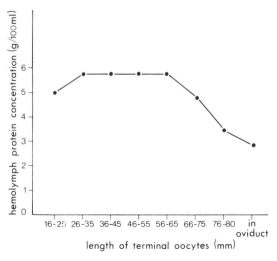

FIGURE 70. Mean hemolymph protein concentration of *Schistocerca gregaria* females during oocyte growth. (Modified from Hill, 1962.)

on the type of food ingested by the insect; cockroaches have large reserves, while locusts do not. The protein level in the hemolymph, like that in the fat body, is not constant during development and reproduction. In *Leucophaea*, for example, the blood protein concentration is different during the first and second reproductive cycles (Engelmann and Penney, 1966). Good prior knowledge of the species is therefore essential for any investigation; and it is indispensable to operate on animals of known age.

Different methods are employed in studying the regulation of protein metabolism. The investigations carried out often deal with changes in total proteins and not with changes in individual proteins, thus yielding results that are not comparable. Moreover, protein separation is carried out by different techniques—chromatography, electrophoresis, radioimmunoassay—which do not yield the same accuracy. At least some hormonal effects on protein metabolism may be indirect. Thus, cautery of the pars intercerebralis causes a significant increase in the blood volume, which is reflected in the blood protein concentration.

Hormone activities are also estimated by various methods: *in vivo* in relation to the hemolymph protein concentration, or *in vitro* in relation to the protein syntheses in the fat body taken from normal or operated insects and incubated in the presence of a labeled precursor and sometimes endocrine glands. The following experimental protocols are encountered:

1. Operated insect: measurement of hemolymph protein concentration.
2. Operated insect: fat body sampling and measurement of protein syntheses *in vitro*.
3. Normal insect: fat body sampling and measurement of protein syntheses *in vitro* in the presence of different organs.

The incubation media employed for *in vitro* investigations are very different, and this may have an effect on the results obtained, as noted by Osborne *et al*. (1968). The high amino acid concentration of the hemolymph requires their presence in the incubation liquid.

Thus, in many respects the study of protein metabolism is very difficult; and the diversity in the methods used explains the many controversial data.

During development of the insect and at metamorphosis, protein syntheses are indispensable to growth. Variations in the protein level of the hemolymph and the fat body have been effectively observed during the larval instars and at metamorphosis, when the hemolymph proteins are taken up by the fat body, where they accumulate as in the skipper butterfly *Calpodes ethlius* (Locke and Collins, 1968) and the cabbage white butterfly *Pieris brassicae* (Chippendale and Kilby, 1969).

Variations in the hemolymph protein concentration also occur in the adult female during the gonadotropic cycle. They consist essentially in the synthesis of specific female proteins, the vitellogenins, which are incorporated into the vitellus of which they make up the bulk (75–90%). Vitellogenins are proteins, sometimes glycoproteins, often lipoproteins. They rarely exist in the male, are indispensable to vitellogenesis, and accumulate in the hemolymph of castrated females. They are synthesized mainly in the fat body, but also in other tissues such as the midgut, the follicle cells (*Rhodnius*, Vanderberg, 1963), and the basement membrane of the ovary (*Leucophaea*, Wyss-Huber and Lüscher, 1972).

Many investigations have dealt with the function of the corpora allata in the protein metabolism of the adult female. They generally point out that vitellogenin synthesis is controlled by the corpora allata. Additional accumulated evidence shows that the protein secretions of the colleterial glands of the female cockroach also depend on the corpora allata (Zalokar, 1968; Shaaya and Sekeris, 1970). The action of juvenile hormone on protein metabolism is also revealed in many other ways: after removal of the corpora allata, a decrease in total urates is observed (Thomas and Nation, 1966b), as well as an increase in hemolymph amino acids (Coles, 1965) and modifications of the transaminases (Steele, 1976).

In the adult female of the mosquito *Aedes aegypti*, ovarian ecdysone stimulates the production of vitellogenin by the fat body, thus assuming the role played by the corpora allata in most insects (Fallon *et al.*, 1974). In some other species such as the cockroach *Gromphadorhina* (Bar-Zev and Kaulenas, 1975), ecdysone appears to exert a comparable action by increasing RNA and fat body protein syntheses in the adult female.

The frontal ganglion also exerts an effect on protein syntheses, but in a different manner. In the adult *Locusta*, its removal is followed by a significant decline in protein syntheses of the incubated fat body (Bignell, 1974). In the larva of the same species, removal of the frontal ganglion also decreases RNA and protein syntheses of the fat body and the midgut (Gillott, 1964; Clarke and Gillott, 1967a,b) and gives rise to cytological and enzymatic modifications (elevation in succinic dehydrogenase) that indicate a decrease in protein syntheses (Clarke and Anstee, 1971a,b). The effects caused by removal of the frontal ganglion are not necessarily neuroendocrine, because this organ performs a very important function in the regulation of food intake and in the transmission to the brain of nerve impulses related to feeding.

The neurosecretory cells of the pars intercerebralis appear to contribute to protein regulation in the adult female, where, as we have seen, they stimulate protease secretion. Thus, by favoring digestion of the proteins, they are likely to play a role in their metabolism. A number of investigations support the view that these cells may also have a more direct function.

The earliest major study on this matter was made by Hill (1962) on the adult female of the locust *Schistocerca gregaria* (Fig. 71, Table XXXIX). Simultaneously investigating the neurosecretory cells, vitellogenesis, and the hemolymph protein concentration in insects subjected to various conditions,

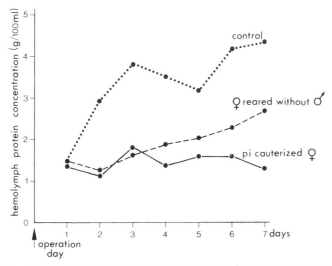

FIGURE 71. Mean hemolymph protein concentrations in *Schistocerca gregaria* females after cautery of the median neurosecretory cells; in operated control females; and in females of similar age as controls and reared without mature males. pi, Pars intercerebralis. (Modified from Hill, 1962.)

he reached the conclusion that the gonadotropic action of the pars intercerebralis is exerted by means of protein metabolism. In fact, when the blood protein concentration is low, the neurosecretory cells of the pars intercerebralis are heavily loaded, whereas the release of neurosecretory material is associated with rising or high blood protein concentration. Furthermore, removal of the pars intercerebralis suppresses the elevation of hemolymph proteins, and the implantation of corpora cardiaca in females before oviposition, that is, when protein concentration is low, causes the latter to increase in

TABLE XXXIX. Storage of Neurosecretory Material, Blood Protein Concentration, and Vitellogenesis in *Schistocerca gregaria* Females[a]

Female rearing condition or surgical state	Type A nsm stored in pi[b]	Blood protein concentration	Oocyte state
Females reared with males	Release, then storage	Increasing, then decreasing	Vitellogenesis
Females reared without males	Important storage	Low	No vitellogenesis
Females deprived of pi		Low	No vitellogenesis

[a]After results of Hill (1962).
[b]nsm, Neurosecretory material; pi, pars intercerebralis.

4 hr. Neurohormone action is not exerted by means of the ovary, because the increase of blood proteins precedes vitellogenesis. Moreover, castrated females display a high concentration of blood proteins (Hill, 1963).

These results have been confirmed *in vitro* on fat bodies of female locusts injected with pars intercerebralis or corpora cardiaca homogenates; 4 hr after the injection, RNA and protein syntheses are sharply increased (Osborne *et al.*, 1968). Furthermore, the *in vivo* incorporation of [^{14}C]glycine or [^{3}H]leucine in the fat body proteins is reduced by cautery of the pars intercerebralis, as occurs with removal of the corpora allata (Hill, 1963; Hill and Izatt, 1974). The presence of these latter organs is necessary but not sufficient for vitellogenin synthesis, because locusts that have undergone cautery of the pars intercerebralis followed by an application of juvenile hormone do not synthesize vitellogenin (Engelmann *et al.*, 1971).

In the locust, the pars intercerebralis thus appears to act side by side with the corpora allata in protein syntheses in the adult female. This hypothesis has been confirmed in other species.

In the grasshopper *Gomphocerus,* cautery of the pars intercerebralis causes a depression in blood protein concentration, which does not take place after allatectomy (Loher, 1966). In the grasshopper *Melanoplus*, removal of the pars intercerebralis, like that of the corpora allata, brings about a drop in blood protein concentration (Elliott and Gillott, 1977), both operations exerting their action via the fat body (Elliott and Gillott, 1978).

In *Tenebrio*, the total concentration of hemolymph proteins in the female depends on mating and age. Low at eclosion, it is very high in 8-day-old mated females. Cautery of the pars intercerebralis inhibits the normal increase in protein concentration if carried out on the pupa, and causes a decrease in hemolymph concentration similar to that brought about by removal of the corpora allata if it is carried out on the young female (Mordue, 1965c).

In the Colorado beetle *Leptinotarsa decemlineata* (de Loof and de Wilde, 1970), the maintenance of the insect in short-day photoperiod causes genital diapause. The hemolymph proteins of diapausing females are different from vitellogenin, which is only present in long-day females. Juvenile hormone injection has no effect in diapausing females, but corpora cardiaca or brain implantation induces the appearance of vitellogenin, thus demonstrating the action of a neurohormone. To determine whether its role is direct, comparative experiments are conducted on diapausing females deprived of their corpora cardiaca–corpora allata complex. After corpora cardiaca implantations or corpora cardiaca implantations combined with juvenile hormone injections, vitellogenin appears only in the latter case, so that its synthesis requires the presence of both juvenile hormone and a neurohormone (Table XL).

Many investigations into the regulation of protein syntheses were conducted on different species of cockroaches, and different methods were employed. Standard removal of corpora allata and pars intercerebralis (Engelmann and Penney, 1966) in *Leucophaea*, followed after 3–4 weeks by

TABLE XL. The Role of the Pars Intercerebralis and Corpora Allata in the Regulation of Protein Synthesis[a]

Insect	Reference	Pars inter-cerebralis	Corpora allata
Periplaneta americana	Menon (1965); Thomas and Nation (1966a); Bell (1969)		⊕
Leucophaea maderae	Engelmann and Penney (1966); Engelmann (1969); Scheurer and Lüscher (1968); Brookes (1969); Engelmann *et al.* (1971); Koeppe and Ofengand (1976)		⊕
	Scheurer (1969a,b)	+	⊕
Nauphoeta cinerea	Lüscher (1968a); Lanzrein (1974); Barth and Sroka (1975)		⊕
Melanoplus sanguinipes	Weed-Pfeiffer (1945)		+
	Elliott and Gillott (1977, 1978)	+	+
Locusta migratoria	Minks (1965); McCaffery (1976)		+
	Bentz *et al.* (1970); Goltzené-Bentz *et al.* (1972)		⊕
Schistocerca gregaria	Hill (1962, 1963)	+	+
	Engelmann *et al.* (1971)		⊕
Gomphocerus rufus	Loher (1966)	+	0
Tenebrio molitor	Mordue (1965c)	+	+
	Laverdure (1970)		⊕
Leptinotarsa decemlineata	de Loof and de Wilde (1970)	+	⊕
Manduca sexta	Sroka and Gilbert (1971)		+
Antheraea polyphemus	Blumenfeld and Schneiderman (1968)		⊕
Danaus plexippus	Pan and Wyatt (1971)		⊕
Sarcophaga bullata	Wilkens (1968, 1969)	0	⊕
Phormia regina	Mjeni and Morrison (1976)		⊕
Oncopeltus fasciatus	Schreiner (1977)		+
Rhodnius prolixus	Coles (1964, 1965); Pratt and Davey (1972); Baehr (1974)		⊕
Triatoma protracta	Mundall and Engelmann (1977)	0	⊕

[a] ⊕, Vitellogenin synthesis; +, total protein synthesis; 0, no effect.

hemolymph protein assay, shows that the corpora allata are indispensable to vitellogenin synthesis; cautery of the pars intercerebralis has heterogeneous effects, disturbing the formation of vitellogenin and vitellogenesis in about half of the cases and allowing the normal mechanisms to subsist in the other half. The complementary experiment of juvenile hormone application to isolated abdomens (Engelmann, 1969) confirms the function performed by corpora allata in the induction of vitellogenin synthesis in this insect.

Decapitation, followed by implantation of various organs and accompanied by injections of [^{14}C] alanine, was carried out by Lüscher (1968a) on *Nauphoeta cinerea*, a species that offers the advantage of continuing normal oogenesis in fasting conditions. One hour after the injection of the labeled amino acid, the proteins synthesized in the fat body were assayed. Proteo-synthesis, which is reduced to one-third in decapitated females, is restored by corpora allata implantation but not by that of the corpora cardiaca, thus confirming Engelmann's results obtained by a different method.

On the other hand, *in vitro* investigations on *Leucophaea* yielded contra-dictory results. When the fat body is incubated in the presence of corpora cardiaca or brain extract, proteosynthesis, which is estimated by the incor-poration of labeled alanine, is stimulated; but when the corpora allata are used in the same manner, no effect is obtained (Wyss-Huber and Lüscher, 1966; Lüscher *et al.*, 1971). If, however, instead of adding glandular organs to the incubation medium, the insect is operated upon and the fat body removed and studied *in vitro*, it appears that removal of the corpora allata results in a significant decline in proteosynthesis.

The contradictory results and interpretations concerning protein metabo-lism regulation in cockroaches were resolved by Scheurer (1969a,b) in a very thorough investigation on *Leucophaea*. It was shown that both the pars intercerebralis and the corpora allata play a part in the regulation of protein metabolism, but that each organ controls the synthesis of different proteins. The experiments were conducted on both normal fasting females and decapi-tated females implanted with brain, corpora cardiaca, or corpora allata. The hemolymph protein concentration was then analyzed by electrophoresis on acrylamide gel. Examination of the variations in total protein concentration revealed a weak effect of the corpora allata and a stronger effect of the corpora cardiaca or the brain (150 and 200% increases). Considering what happens with respect to the different proteins, the results differ widely. The control females possess 11 hemolymph proteins, 6 of them in significant quantities, also found in the ovary. Their variations are synchronous and correspond to variations in total blood proteins. They increase during vitellogenesis and fall to a minimum at ovulation. In insects implanted with corpora allata, only one protein displays a significant increase—protein G, which is vitellogenin. In insects implanted with brain or corpora cardiaca, protein G undergoes no variation, but proteins F, H and L increase sharply, the brain being slightly more active than the corpora cardiaca. Finally, a fifth protein, protein D, responds only to brain implants. These experiments thus demonstrate the parallel role of the two organs and help to understand the contradictory results obtained on the species mentioned earlier (Table XLI).

The control of protein metabolism by neurohormones has been confirmed by the study of urates (protein degradation products). Bodenstein (1953b) investigated the urate cells of the fat body in *Periplaneta* after removal of the corpora cardiaca and corpora allata on the one hand, and of the corpora allata alone on the other. The urates persist in the latter case, whereas in the former

TABLE XLI. Regulation of
Vitellogenin Synthesis in
Leptinotarsa decemlineata[a]

Short-day females: Organ[b]		
Removed	Implanted or injected	Vitellogenin synthesis[c]
	JH	0
	cc	+
	Brain	+
cc–ca	cc	0
cc–ca	cc + JH	+
ca	JH	+

[a]After data of de Loof and de Wilde (1970).
[b]ca, Corpora allata; cc, corpora cardiaca; JH, juvenile hormone.
[c]0, No effect; +, vitellogenin synthesis.

they disappear in 2 weeks but reappear with corpora cardiaca implantation. The transaminases of the fat body, the flight muscles, and the Malpighian tubules decrease significantly in locusts after cautery of the pars intercerebralis and reappear if corpora cardiaca extracts are injected into operated insects.

A recent investigation provided highly interesting clarifications concerning the role of the corpora cardiaca. In the migratory locust *Locusta migratoria*, the neurohemal lobe stimulates protein synthesis *in vivo* and *in vitro*, while the glandular lobe exerts an inhibitory action. The thermostable factor involved, which is destroyed by the proteases, may be identical to the adipokinetic hormone, and this would explain its presence in species in which no adipokinetic action has been detected (Carlisle and Loughton, 1979).

The neurosecretory cells of the ventral nerve cord may also take part in controlling protein metabolism, but little data are available on this question. Dutkowski and Cymborowski (1973), working on the house cricket *Acheta domesticus*, showed that subesophageal ganglion implantations depress RNA syntheses in the neurosecretory cells of the brain and in the follicular epithelium of the ovary, but not in the fat body. In the stick insect *Carausius morosus*, perisympathetic organ extracts induce a premature decrease in nonvitellogenic hemolymph protein in the freshly molted adult female. However, they have no effect on the hemolymph proteins of the laying female (Raabe and Demarti-Lachaise, 1973) (Fig. 72).

9.3. CARBOHYDRATES

In insects, the most readily accessible and most often used energy material, especially for flight, is trehalose—the most abundant carbohydrate transported by the hemolymph. Glucose is present only in small quantities in

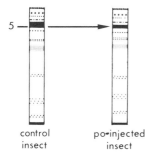

FIGURE 72. Effect of perisympathetic organ extracts on the blood protein level in the stick insect *Carausius morosus*. Injection of a perisympathetic organ (po) extract into a young adult causes a decrease in fraction 5, characteristic of this stage. (Left) control insect; (right) injected insect. (Modified from Raabe and Demarti-Lachaise, 1973.)

the blood although it is used by the muscle, where trehalose is hydrolyzed to glucose under the action of a trehalase. Trehalose is stored in the fat body and also in other tissues in the form of glycogen. The latter can be of direct or indirect alimentary origin; in the latter case, it originates in proteins and lipids.

In some species, the fuel used for flight is not trehalose but another carbohydrate, such as glucose or fructose. This has been demonstrated recently in the honeybee (Brandt and Huber, 1979). Furthermore, the source of flight metabolites may be the muscle and not the fat body (Downer and Parker, 1979).

Hemolymph trehalose concentration varies during postembryonic development, as reported by many investigators (*Carausius*, Dutrieu and Gourdoux, 1967; *Locusta*, Hill and Goldsworthy, 1968; *Bombyx*, Dutrieu, 1962; *Pieris*, Dutrieu, 1963; *Calliphora*, Normann, 1975). It increases during muscular exercise and in such circumstances as stress (Divakar and Němec, 1973; Downer, 1979) or injury (Wyatt, 1961). It also undergoes regular variations during the circadian cycle in *Acheta* (Nowosielski and Patton, 1963) and *Periplaneta* (Gersch, 1976).

During insect growth, histogenetic periods require a larger energy supply, which is obtained from carbohydrates and especially from glucose. Hence it is not surprising that operations that inhibit development have an effect on sugar metabolism. In the silkworm, ligation of the pupa behind the molting gland arrests development and causes an increase in glycogen comparable to that of the diapausing insect. Implantation of the molting gland restores both development and the normal glycogen level (Ito and Horie, 1957). Decapitation has the same effect as ligation; brain implantation, like that of the molting gland, causes development to resume and diminishes the glycogen reserves (Kobayashi, 1957). If ecdysone and labeled glucose are injected simultaneously into decerebrated diapausing pupae of *Samia cynthia* and *Bombyx mori*, unlike what is observed in control insects, incorporation is very low in the fat body glycogen but high in the hemolymph trehalose, the latter increasing considerably in quantity (Kobayashi and Kimura, 1967; Kobayashi *et al.*, 1967). The direct or indirect role of ecdysone in carbohydrate metabolism is disputed, and could better be resolved by investigations *in vitro*. Carbohydrate metabolism is also influenced by the corpora allata, removal of which inhibits the deposition of ovarian glycogen (Engels and Bier, 1967; Schreiner, 1977) and causes glycogen accumulation in the fat body in many species: stick insects (L'Hélias, 1953), flies (Orr, 1964; Liu, 1974; Thomsen, 1952; Butterworth and Bodenstein, 1969), and bugs (Janda and Sláma, 1965). The blood trehalose concentration is also modified by allatectomy in the adult female of *Musca domestica,* but this appears to be due to the absence of ovarian development brought about by this operation (Liu, 1973), because in the male of another fly *(Phormia regina)*, removal of the corpora allata does not modify the hemolymph trehalose level (Chen and Friedman, 1977b). The action of the corpora allata on carbohydrates may be indirect, as in the case of *Musca,* and exerted through the ovary or through the protein synthesis that involves the action of carbohydrate energy. The possibility also exists that the action of the corpora allata may be direct, and represents one aspect of reproductive behavior in the adult insect.

9.3.1. Diapause Hormone

As previously mentioned (Chapter 3), the offspring of some strains of the silkworm *Bombyx mori* exhibit diapause. The existence of a diapause hormone was demonstrated, originating in the subesophageal ganglion (Hasegawa, 1957). A relationship exists between diapause and carbohydrate content of the ovary; and ovaries that produce diapause eggs are far richer in glycogen. Diapause hormone exerts its action at the time of vitellogenesis and acts on ovarian trehalase (Yamashita and Hasegawa, 1967; Shimada and Yamashita, 1979) (Fig. 73). Removal of the subesophageal ganglion causes it to decrease in the ovary, while the other enzymes active in glycogen synthesis are unchanged.

The action of diapause hormone on the ovary depresses the fat body glycogen and hemolymph trehalose (Hasegawa and Yamashita, 1965). It also affects the lipids, which become more abundant in the ovary and less so in the fat body.

9.3.2. Hypertrehalosemic Hormone

Steele (1961, 1963) first noted the existence of a hyperglycemic neurohormone in *Periplaneta*. Corpora cardiaca extracts injected in this insect give rise to a sharp increase in hemolymph trehalose, which is elevated by 50% in 30 min and reaches a peak in 5 hr (150%). The extracts exhibit very strong activity, maximum reactions being obtained with a dose corresponding to 1/10 pair of corpora cardiaca, and even a dose of 1/500 pair inducing a clear reaction.

Simultaneous with this increase in hemolymph trehalose, hyperglycemic

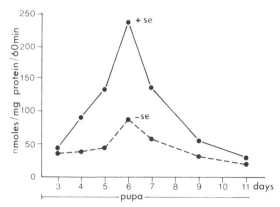

FIGURE 73. Effect of subesophageal ganglion on the trehalase activity in pupal ovaries during adult development in *Bombyx mori*. + se, Control insect; − se, insect whose subesophageal ganglion was removed on the day of pupation. (Modified from Yamashita and Hasegawa, 1967.)

hormone (here termed hypertrehalosemic hormone) causes a drop in fat body glycogen. It appears to have no effect on the glycogen of either the digestive tract or the muscles, but acts on the glycogen of the perineurium of the ventral nerve cord, in which it activates the formation of trehalose as in the fat body. This process, which is slower, appears to occur only in extreme cases (Hart and Steele, 1973).

The hormone acts by activating a fat body phosphorylase in the cockroach (Steele, 1963; Wiens and Gilbert, 1967) and the locust (Goldsworthy, 1970), in both of which it behaves like the vertebrate hormones, glucagon and catecholamines.

The trehalose is used by the muscle after its hydrolysis to glucose under the action of a trehalase, which may exist in an inactive form. Candy's experiments (1974) on *Schistocerca* point out that this enzymatic activation is independent of neurohormone factors.

The effect of corpora cardiaca extract on hemolymph trehalose concentration, demonstrated by Steele in *Periplaneta*, was confirmed in many *in vivo* and *in vitro* investigations (*Leucophaea*: Wiens and Gilbert, 1965; *Periplaneta*: Ralph, 1962a; Bowers and Friedman, 1963; Ralph and McCarthy, 1964; Natalizi and Frontali, 1966; *Carausius*: Dutrieu and Gourdoux, 1967; *Locusta*: Chalaye, 1969; Goldsworthy, 1969; *Tenebrio*: Dutrieu and Gourdoux, 1974; *Phormia*: Friedman, 1967; *Calliphora*: Normann and Duve, 1969; Vejbjerg and Normann, 1974) (Fig. 74).

Hypertrehalosemic activity of the corpora cardiaca cannot be demonstrated at all times (Table XLII). In the fly *Phormia* (Friedman, 1967), corpora cardiaca extracts only display an effect *in vivo* in starved flies,

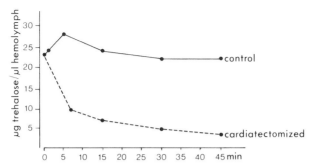

FIGURE 74. Concentration of hemolymph trehalose in resting flies and after various periods of sustained flight in *Calliphora erythrocephala*. (Modified from Vejbjerg and Normann, 1974.)

because in this insect, which is a high carbohydrate consumer, trehalose can be formed directly from glucose, and is only synthesized from glycogen if glucose is lacking. Because the hormone acts through glycogen, it is inactive in glucose-rich, fed insects. Locusts also display irregular reactions. In different periods of their life, they are unresponsive to the hypertrehalosemic hormone of corpora cardiaca. While a clear response is observed in 6-day-old adult males, no reaction is shown to injections of corpora cardiaca extracts in younger, mature or unfed males (Goldsworthy, 1969). The ineffectiveness of hypertrehalosemic hormone in young and unfed adults is probably due to an inadequate quantity of stored glycogen, and in the mature male to the saturation of the phosphorylase system, or to modifications in the phosphorylase content of the fat body. The latter appears to increase from the onset of eclosion to the sixth day, and then remains at a high level. The absence of a response in the unfed insect has also been observed in the cockroach *Periplaneta* (Ralph and McCarthy, 1964).

The irregular responses of the locust and its low sensitivity as compared to that of the cockroach *Periplaneta* have led investigators to use the latter to test the hypertrehalosemic activity of the corpora cardiaca of the locust.

TABLE XLII. Hemolymph Blood Trehalose Concentration following an Injection of Corpora Cardiaca Extract[a,b]

Adult *Phormia regina*		Adult male *Periplaneta americana*		Adult male *Locusta migratoria*		
Fed	Fasting	Fed	Fasting	Just molted	6 days old	Fasting
0	++	++	0	0	++	0

[a]From data of Friedman (1967), Ralph and McCarthy (1964), and Goldsworthy (1969).
[b]0, No effect; ++, increase of blood trehalose concentration.

9.3.2.1. Source of Hypertrehalosemic Hormone

The source of hypertrehalosemic hormone was located in locusts by separately testing on *Periplaneta* the glandular and the neurohemal parts of the corpora cardiaca of *Locusta* (Mordue and Goldsworthy, 1969). They are both active, but the glandular part is far more active than the neurohemal part (Table XLIII). Because the presence of a hypertrehalosemic factor in the neurohemal lobe of the corpora cardiaca may result from contamination during separation of the two lobes, additional experiments were conducted by Highnam and Goldsworthy (1973). These authors sectioned the nccI and nccII, making it possible to obtain entirely neurohemal corpora cardiaca regenerates, after a certain time interval, from the sectioned extremity of the nerves and in particular that of the nccI. These regenerates display hypertrehalosemic activity, thus suggesting the existence of two factors—one synthesized in the glandular lobe and one originating in the brain.

According to Cazal (1971), who carried out selective removal of the glandular lobe in *Locusta*, the latter contains a hypertrehalosemic–hyperglycemic factor, whereas the neurohemal lobe stores a hypertrehalosemic–hypoglycemic factor.

The search for hypertrehalosemic activity in the brain revealed the latter to be slightly hypertrehalosemic in *Periplaneta* (Ralph, 1962a; Bowers and Friedman, 1963; Ralph and McCarthy, 1964), but inactive in locusts (Highnam and Goldsworthy, 1973) and in *Carausius* (Gäde, 1979b). These results are surprising, because the neurohemal part of the corpora cardiaca is active and corpora cardiaca regenerates are active too. This probably means that two antagonistic factors, which mask each other, may exist in the pars intercerebralis. Some recent data support this view, for a hypotrehalosemic neurohormone has been demonstrated in the pars intercerebralis (see Section 9.3.3). As for the ventral nerve cord, negative results were obtained by Steele (1961) and also by Gersch (1974b) who, working on *Periplaneta*, conducted electrical stimulation experiments on different nerve ganglia of which the incubation medium was then injected into other insects. The prothoracic ganglion and the sixth abdominal ganglion were totally inactive, whereas the subesophageal ganglion showed slight activity. In *Carausius* (Gäde, 1979b) and *Manduca* (Ziegler, 1979), which have very active corpora cardiaca, no activity was detected in the ventral nerve cord ganglia.

9.3.2.2. Interspecific Activity

The hypertrehalosemic factor of the corpora cardiaca of *Locusta* and *Phormia* is very active on *Periplaneta* (Friedman, 1967; Chalaye, 1969), and the corpora cardiaca factor of *Carausius* acts on *Tenebrio* (Dutrieu and Gourdoux, 1967). The latter is hypertrehalosemic in *Periplaneta* and increases

TABLE XLIII. Location of the Hypertrehalosemic Hormone

Insect	Reference	Hypertrehalosemic effect[a]				
			Corpora cardiaca			Ventral nerve cord
		Brain	Glandular lobe	Neurohemal lobe	Corpora allata	
Periplaneta americana	Steele (1961, 1963)	0				0
	Bowers and Friedman (1963)	++	++		++	
	Ralph and McCarthy (1964)	+	++		+	
Periplaneta americana and *Blaberus discoidalis*	Wiens and Gilbert (1965)		++			
Carausius morosus	Dutrieu and Gourdoux (1967)	(+)	+++		(+)	
	Gäde (1979b)	0	+++		0	0
Locusta migratoria	Mordue and Goldsworthy (1969)	0	+	++	0	
Locusta migratoria and *Schistocerca gregaria*	Lafon-Cazal and Roussel (1971)		++	++		
Tenebrio molitor	Dutrieu and Gourdoux (1974)			++		
Phormia regina	Friedman (1967)		++			
Calliphora erythrocephala	Normann and Duve (1969)		+++			

[a] 0, No effect; (+), very weak effect; +, weak effect; ++, medium effect; +++, strong effect.

the hemolymph lipid concentration in the locust; curiously, it exerts neither of these two effects on the stick insect (Gäde, 1979b) (Fig. 75). Extracts of *Schistocerca* are active on *Pyrrhocoris* (Divakar and Němec, 1973). Thus, the substance present in these different species may be the same.

9.3.2.3. Control of Hypertrehalosemic Hormone Release

The release of a hypertrehalosemic factor from the corpora cardiaca is controlled by the brain, as seen by electrical stimulation exerted on this organ in *Phormia* (Normann and Duve, 1969). In *Periplaneta*, brain stimulation is transmitted to the corpora cardiaca by the nccII and involves adrenergic mechanisms; it initiates the release of the hormone (Gersch *et al.*, 1970);

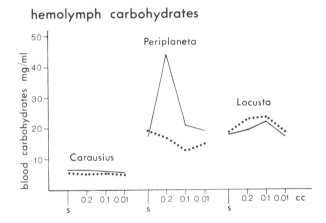

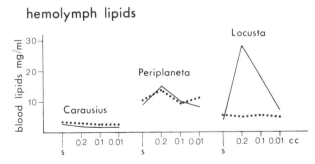

FIGURE 75. Concentration of total hemolymph carbohydrates (top) and lipids (bottom) in control insects and corpora cardiaca-injected insects. Corpora cardiaca material was obtained from adult specimens of *Carausius morosus* and the extracts injected into *Carausius, Periplaneta,* and *Locusta.* s, Saline. (Modified from Gäde, 1979b.)

nccII severance, on the contrary, causes a considerable decrease in hemolymph trehalose (Gersch, 1974a). The circadian variations in blood trehalose displayed by this species are under central nervous system control, and removal of the brain, subesophageal ganglion, or mesothoracic ganglion causes the blood trehalose concentration to remain at its lowest level (Gersch, 1976).

The hemolymph trehalose concentration appears to exert a feedback regulation on the fat body, at least in some cases. In the giant silk moth *Hyalophora cecropia* (Murphy and Wyatt, 1965), the presence of large quantities of trehalose in the fat body incubation medium increases the incorporation of [14C]glucose by the fat body glycogen, to the detriment of trehalose, and in *Phormia*, the quantities of trehalose synthesized *in vitro* depend on the amount of trehalose in the incubation medium (Friedman, 1967). In locusts, however, these facts were not substantiated, for dilution of the hemolymph by saline injection does not change the hemolymph trehalose concentration (Goldsworthy, 1969).

9.3.3. Hypotrehalosemic Hormone

Removal of the corpora cardiaca and the neurosecretory cells of the pars intercerebralis has served to identify a neurohormone that is antagonistic to hypertrehalosemic hormone, i.e., has hypotrehalosemic effects. It has been identified essentially in dipterans, but also in one lepidopteran. In *Musca,* removal of the corpora cardiaca–corpora allata complex causes a progressive elevation of hemolymph trehalose, which continues for at least 4 days (Liu, 1973). In *Calliphora* (Normann, 1975; Duve, 1978), removal of the pars intercerebralis or corpora cardiaca, or disconnection of the corpora cardiaca, causes a sharp increase in hemolymph trehalose within 20 hr, rising from 27 g/liter before the operation to 63 and 50 g/liter afterwards, while the injection of neurosecretory cell extract restores the normal hemolymph trehalose concentration in 2 hr (Normann, 1975). The existence of a hypotrehalosemic brain factor was confirmed by Chen and Friedman (1977a) working on another fly, *Phormia,* in which removal of the corpora cardiaca–corpora allata complex causes 150% hypertrehalosemia in 2 days, which is not due to the absence of the corpora allata, for their removal has no effect. In the adult tobacco hornworm *Manduca sexta,* corpora cardiaca–corpora allata extracts not only induce hypotrehalosemia, but also cause a significant drop in fat body glycogen, which may correspond to the action of a glycogenolytic factor (see Section 9.3.4). Partial purification indicates that these effects are due to two different peptides (Tager *et al.,* 1976). The mechanism of action of hypotrehalosemic hormone has been investigated in *Phormia* (Chen and Friedman, 1977b). It appears that the hormone does not act on fat body glycogen synthetase, but that it induces a decrease in UDP glucose:glucose phosphate glucosyltransferase activity.

9.3.4. Glycogenolytic Hormone

Results obtained in some insects suggest that a glycogenolytic hormone may also occur. In fact, pars intercerebralis removal causes glycogen accumulation in the fat body in *Calliphora* (Thomsen, 1952), *Aedes* (Van Handel and Lea, 1965), and *Locusta* (Goldsworthy, 1971). Implantation of the pars intercerebralis into parsectomized *Aedes* restores a normal glycogen level in the fat body (Lea and Van Handel, 1970), while injection of corpora cardiaca extracts has no effect on operated locusts (Goldsworthy, 1971). Thus, a glycogenolytic hormone distinct from hypertrehalosemic hormone appears to exist and to be synthesized in the neurosecretory cells of the pars intercerebralis. It is interesting to note that its effects are similar to those of the corpora allata and the molting gland.

9.3.5. Neurohormones and Blood Glucose Concentration

The factors involved in sugar regulation appear to be numerous (Table XLIV). Apart from their action on hemolymph trehalose, the corpora cardiaca may act on blood glucose concentration, although this action is disputed (Table XLV). Some investigators report that the corpora cardiaca do not act on hemolymph glucose (*Periplaneta*, Steele, 1961; McCarthy and Ralph, 1962). However, Bowers and Friedman (1963), using very pure substrates and enzymes, have shown that corpora cardiaca extracts cause a rise in hemolymph glucose in the South American cockroach *Blaberus discoidalis*. Lafon-Cazal and Roussel (1971) obtained the same effect on hemolymph trehalose and

TABLE XLIV. Factors Involved in Carbohydrate Metabolism Regulation

Factor	Organ[a]	Insects
Hypertrehalosemic hormone	cc	*Periplaneta, Blaberus, Carausius*
(1 or 2 factors)	cc (nh)	*Tenebrio, Phormia, Calliphora, Locusta*
	cc (gl)	*Locusta*
Hypotrehalosemic hormone	pi cc	*Calliphora, Phormia, Manduca*
Hyperglucosemic hormone	cc	*Blaberus*
	cc (gl)	*Locusta, Schistocerca*
	pi	*Calliphora*
Hypoglucosemic hormone	cc (nh)	*Locusta, Schistocerca*
Glycogenolytic hormone	pi cc	*Locusta, Aedes, Manduca*
Octopamine	cns	*Periplaneta*

[a]cc, Corpora cardiaca; cns, central nervous system; gl, corpora cardiaca, glandular lobe; nh, corpora cardiaca, neurohemal lobe; pi, pars intercerebralis.

TABLE XLV. Effect of Corpora Cardiaca and Corpora Allata on Glucose and Trehalose Blood Levels[a]

Insect	Reference	Pars intercerebralis		Corpora cardiaca		Corpora allata	
		G	T	G	T	G	T
Periplaneta americana	Steele (1961)			0		0	
	McCarthy and Ralph (1962); Ralph (1962a)			0		–	
Blaberus discoidalis	Bowers and Friedman (1963)			++		0	
Locusta migratoria and *Schistocerca gregaria*	Lafon-Cazal and Roussel (1971)			nh –	++	++	
				gl ++	++		
Calliphora erythrocephala	Duve (1978)	–	–				

[a]0, No effect; ++, increase; –, decrease; G, glucose blood level; gl, corpora cardiaca glandular lobe; nh, corpora cardiaca neurohemal lobe; T, trehalose blood level.

hemolymph glucose concentration with the glandular part of the corpora cardiaca of *Locusta* and *Schistocerca*, but the opposite effect with the neurohemal part, which proves to be hypertrehalosemic and hypoglucosemic. Hypoglucosemic (hypoglycemic) activity has also been identified in the pars intercerebralis of *Calliphora* (Duve, 1978). It is associated with the hypotrehalosemic activity, and may derive from a unique substance resembling insulin, for after purification, the active factor of the pars intercerebralis cross-reacts with antibodies to bovine insulin. Its precise origin was determined by the use of anti-insulin antiserum. Six to eight insulin-immunoreactive cells were identified among the 24–26 median type A neurosecretory cells present in *Calliphora vomitoria* (Duve and Thorpe, 1979).

The corpora allata may also influence hemolymph glucose levels, but the data are fragmentary and contradictory. Lafon-Cazal and Roussel (1971) postulate that the corpora allata are stimulatory in *Locusta,* but according to McCarthy and Ralph (1962) they are inhibitory in the cockroach *Periplaneta.* This subject requires more detailed investigation.

9.3.6. Amines

When cockroaches are handled, they react by glycogenolysis and rapid trehalose release, which also occurs in neck-ligated insects (Matthews and Downer, 1974; Hanaoka and Takahashi, 1976). Thus, hypertrehalosemic hormone cannot be involved. The action of a neurotransmitter of the central nervous system has been considered, and the activity of different neurotransmitters and their inhibitors was tested *in vivo* on the hemolymph trehalose concentration in the adult male of *Periplaneta* (Downer, 1979) (Table XLVI). The results obtained reveal a moderate effect of dopamine, tyramine, and normetanephrine, and a strong effect of octopamine. This suggests that the octopamine of the central nervous system (Robertson and Steele, 1974), which acts on its phosphorylase content, may play a part in controlling hemolymph trehalose concentration. In locusts, octopamine is present in the blood and its level increases rapidly during flight, especially during the period of utilization of trehalose as fuel for flight (Goosey and Candy, 1980).

Serotonin has been identified in the corpora cardiaca of *Periplaneta* (Gersch *et al.*, 1961; Colhoun, 1963). It elevates hemolymph trehalose (Mordue and Goldsworthy, 1969), but its concentration in the corpora cardiaca is too low for it to be physiologically responsible for the effect of these organs, as illustrated by the experiments of Downer where it proved to be inactive.

In addition to octopamine and serotonin, adrenaline was reported recently to be active in the bee, in which it elevates hemolymph trehalose and hemolymph glucose levels (Bounias and Pachéco, 1979a).

TABLE XLVI. Blood Trehalosemic Level following
Injection of Neurotransmitters and Their Inhibitors[a]

Substance used	Trehalose (μg/μl)	Substance used	Trehalose (μg/μl)
Controls	20	Acetylcholine	20
Corpora cardiaca	**31**	α-Aminobutyric acid	19
Octopamine	**36**	*Normetanephrine*	27
Dopamine	27	Isoproterenol[b]	21
Tyramine	27	Chlorpromazine[c]	20.5
Synephrine	20	Dichloroisoproterenol[d]	20.5
Glutamic acid	23	Propranolol[d]	20
Aspartic acid	20	Nialamide[e]	21
Histamine	21	Iproniazid phosphate[e]	21
Serotonin	19		

[a]After Downer (1979).
[b]Receptor activator.
[c]Dopamine inhibitor.
[d]β-Adrenergic blocking agent.
[e]Monoamine oxidase inhibitor.

9.4. LIPIDS

Lipids are extremely important in the energy metabolism of insects, especially during flight. While flight consumes carbohydrates in dipterans and amino acids such as proline in the tsetse fly and Colorado beetle, many other insects—such as locusts, aphids, and lepidopterans—consume lipids, which they sometimes convert from alimentary carbohydrates.

In locusts, the first 20 min of flight, which is particularly rapid, uses carbohydrates as an energy source; but these are then supplanted by lipids, especially diglycerides, of which the hemolymph level rises by a factor of 5 during flight (Beenakkers, 1965). The use of lipids in migratory flights is preferable to that of carbohydrates, because they have a higher energy value, furnish metabolic water, and occupy less storage space.

Lipids also perform an important function in the adult female, in which triglycerides represent one of the major constituents of the vitellus (Sroka and Barth, 1975).

Lipids are stored in the fat body mainly in the form of triglycerides, but the fat body also contains small amounts of free fatty acids and mono- and diglycerides.

The circulating form of lipids is represented by the diglycerides released by the fat body. In some cases, the fat body also releases triglycerides (Cook and Eddington, 1967), which are hydrolyzed in the blood by a lipase (Downer and Steele, 1973). Diglycerides bind with hemolymph proteins or lipoproteins

(Siakotos, 1960, in *Periplaneta*; Chino and Gilbert, 1964, in *Hyalophora cecropia* and *Melanoplus*; Mayer and Candy, 1967, in *Schistocerca;* Thomas and Gilbert, 1969, in *Hyalophora gloveri;* Mwangi and Goldsworthy, 1977b, in *Locusta*). These lipid-transport substances are synthesized in the fat body (Thomas, 1972, in *Hyalophora cecropia*; Peled and Tietz, 1973, in *Locusta*).

The hemolymph lipid concentration varies during the life of the insect, in the adult female as a function of reproduction and, in general, in accordance with activity, especially flight.

The corpora allata perform an important function in lipid metabolism. Allatectomy causes lipid overload of the fat body, as reported by various investigators in *Melanoplus* (Weed-Pfeiffer, 1945), *Periplaneta* (Bodenstein, 1953b), locusts (Odhiambo, 1966; Minks, 1967; Strong, 1968; Walker and Bailey, 1970; Hill and Izatt, 1974), *Leptinotarsa* (de Loof and Lagasse, 1970), *Spodoptera* (El-Ibrashy and Boctor, 1970), *Drosophila* (Vogt, 1949), *Calliphora* (Thomsen, 1952), and *Phormia* (Orr, 1964). Reciprocally, supernumerary corpora allata implantation decreases the lipids of the fat body.

It may be supposed that the effect of the corpora allata on the fat body results, in the adult female, from their action on vitellogenesis. However, ovariectomy has no effect on the lipid content of the fat body in *Phormia* (Orr, 1964), *Schistocerca* (Hill and Izatt, 1974), or *Locusta* (Lauverjat, 1977), and the male displays the same reactions as the female to allatectomy. If one considers the part played by the corpora allata in protein synthesis, and the fact that proteins are generally bonded to lipids, it is not surprising to note that operations that cause a decrease in protein synthesis and release are accompanied by lipid storage. Reciprocally, because hemolymph lipids bind to proteins, the absence of the latter may induce their storage. Some effects occuring in lipid metabolism that are due to removal of the corpora allata thus appear to be indirect, while on the whole their action is probably direct, as demonstrated *in vitro* (Gilbert, 1967). Corpora allata do not appear to exert any inhibition on lipid release (Walker and Bailey, 1971a; Hill and Izatt, 1974), and may act, according to Walker and Bailey (1971b), as on other tissues, by controlling growth and differentiation. However, this view is not shared by all investigators, and it is proposed that the corpora allata may inhibit lipid syntheses as shown in experiments by Vroman *et al.,* (1965) conducted *in vivo* on the cockroach *Periplaneta,* and those of Gilbert (1967) and Hill and Izatt (1974) carried out *in vitro* on the fat body of *Leucophaea* and *Schistocerca*.

9.4.1. Adipokinetic Hormone

A hormone exerting an effect contrary to that of the corpora allata has been identified in the corpora cardiaca. In *Schistocerca* and *Locusta*, it causes the release of fat body diglycerides during flight, and is accompanied by an increase in their hemolymph titer (Beenakkers, 1969; Mayer and Candy, 1969; Goldsworthy *et al.*, 1972a). This hormone has been termed adipokinetic hormone (Mayer and Candy, 1969) (Fig. 76).

Adipokinetic hormone is present in large quantities in the hemolymph, during and after flight. If the hemolymph of a locust that has just flown is injected into *Schistocerca* at rest, the diglyceride content of its hemolymph increases. The similarity between the hemolymph factor and the corpora cardiaca factor is demonstrated by chromatography (Mayer and Candy, 1969).

The reactivity of insects to adipokinetic hormone may vary. In *Locusta*, the nymph reacts very slightly and the adult only displays typical reactions between day 5 and day 35 (Mwangi and Goldsworthy, 1977a) (Fig. 76). These reduced responses do not derive from a deficiency of adipokinetic hormone in the corpora cardiaca, for corpora cardiaca extracts of locusts of different ages show the same activity (Gäde and Beenakkers, 1977).

Adipokinetic hormone acts directly on the fat body, causing diglyceride release *in vitro* (Mayer and Candy, 1969). It appears not only to induce conversion of the triglycerides and the fatty acids of fat body to diglycerides, but also may induce the formation of diglycerides in the hemolymph from the fatty acids contained in the latter (Beenakkers, 1969). Moreover, during flight, it causes the formation of diglyceride-carrier lipoproteins (Van Der Horst *et al.*, 1979).

Investigations on locusts, in which the flight metabolites change during flight, provide some evidence concerning the probable action of adipokinetic hormone on the muscle itself. In the adult male of *Schistocerca*, the hormone inhibits the use of glucides and stimulates the use of lipids by the muscle (Robinson and Goldsworthy, 1974). The injection of diglycerides before flight

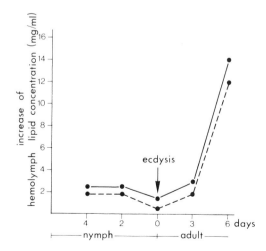

FIGURE 76. Adipokinetic response of locusts during development to corpus cardiacum extracts (0.01 pair of corpora cardiaca) from insects of the same age (dashed line) or from 12-day-old adults (solid line). (Modified from Gäde and Beenakkers, 1977.)

diminishes the use of carbohydrates, and this effect is considerably enhanced by simultaneously injecting an extract of the glandular lobes of the corpora cardiaca. Adipokinetic hormone thus acts on the muscle (Robinson and Goldsworthy, 1976), enhancing the penetration of diglycerides.

9.4.1.1. Source of Adipokinetic Hormone

The investigations of Goldsworthy *et al.*, (1972a,b, 1973) clearly showed that in locusts, adipokinetic hormone is absent from the pars intercerebralis, present in very small amounts in the neurohemal part of the corpora cardiaca, and very abundant in the glandular lobe of these organs, even after sectioning of the nccI (Fig. 77). The quantity released during a 30-min flight corresponds to 0.0025 glandular lobe (Goldsworthy *et al.*, 1972b), and it causes a 300-fold increase in hemolymph lipids.

Ultrastructural study of the glandular part of the corpora cardiaca of *Locusta* was carried out comparatively in insects at rest and in insects that had flown for 5 and 60 min. After prolonged flight, the secretory activity of the Golgi apparatus increases significantly together with cellular exocytotic activity (Rademakers and Beenakkers, 1977).

In other species, however, such as the mosquito *Aedes taeniorhynchus* (Van Handel and Lea, 1965) and the termite queen *Odontotermes assmuthi*, the pars intercerebralis appears to control lipid metabolism (Kallapur and Basalingappa, 1977). In the monarch butterfly *Danaus plexippus,* adipokinetic activity is present not only in the brain and in the corpora cardiaca–corpora allata complex, but also in the thoracic ganglia; on the contrary, the abdomi-

corpora cardiaca regenerates

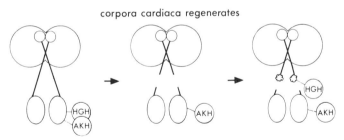

FIGURE 77. Demonstration of the site of production of metabolic neurohormones by the corpora cardiaca regeneration technique. The hormonal activity of the corpora cardiaca is compared in intact locusts and in locusts with the nccl sectioned. The hyperglycemic hormone (HGH) produced in the pars intercerebralis disappears from the corpora cardiaca after nccl section, whereas the adipokinetic hormone (AKH) produced in the glandular cells of the corpora cardiaca remains. Corpus cardiacum regenerates exhibit DH and HGH activity. (From data of Goldsworthy *et al.*, 1972b, and Highnam and Goldsworthy, 1973.)

nal ganglia are devoid of activity (Dallmann and Herman, 1978). In stick insects, corpora allata and ventral nerve cord ganglia are inactive, whereas corpora cardiaca are very potent (Gäde, 1980).

9.4.1.2. Regulation of Adipokinetic Hormone Release

Adipokinetic hormone is used by the locust for flight after an initial period of high-speed flight that uses trehalose as fuel. The question thus arises whether the hemolymph trehalose level is the determining factor for adipokinetic hormone release. Trehalose injection during flight does not modify hormone release (Houben and Beenakkers, 1974), and the level remains very low as shown by accurate measurements taken after extraction and purification of the hemolymph (Cheeseman et al., 1976). On the other hand, the hemolymph lipid concentration exerts a negative feedback on adipokinetic hormone release (Cheeseman and Goldsworthy, 1979). In locusts, although adipokinetic hormone is synthesized in the corpora cardiaca, its release requires intact nerve connections between the brain and the corpora cardiaca (Goldsworthy et al., 1972a). The cells exerting a major role appear to be aminergic lateral protocerebral cells, whose activity may be controlled by acetylcholine (Samaranayaka, 1976, 1977).

9.4.1.3. Interspecific Activity

The adipokinetic factor appears to exert interspecific activity (Table XLVII). Corpora cardiaca extracts of *Schistocerca* act on *Tenebrio* and vice versa (Goldsworthy et al., 1972a). Corpora cardiaca extracts of the stick insect *Carausius* are also active on the locust, but exert no activity on the stick insect itself (Gäde, 1979b). Surprisingly enough, in the cockroach *Periplaneta,* the corpora cardiaca exert a diametrically opposite effect, that of decreasing the hemolymph glycerides and increasing those of the fat body *in*

TABLE XLVII. Interspecific Effect of Corpora Cardiaca
Extracts on Blood Lipid Concentration[a]

Corpora cardiaca donor	Effector[b]		
	Periplaneta	*Locusta*	*Carausius*
Periplaneta americana	−	+	
Locusta migratoria	−	++	
Carausius morosus		+	0

[a]After data of Downer and Steele (1969, 1972), Downer (1972), Goldsworthy et al. (1972a), and Gäde (1979b).
[b]0, No effect; +, increase; ++, strong increase; −, decrease.

vivo as *in vitro* (Downer and Steele, 1969, 1972; Goldsworthy *et al.*, 1972a), and this insect reacts in the same manner to corpora cardiaca extracts of *Locusta*. Nevertheless, its own corpora cardiaca exert an adipokinetic effect in *Locusta* (Downer, 1972). Hence it may be inferred that the two species contain the same substance, or two closely related substances to which the effector reacts differently. This may be explained by the role played by lipids in the two species. Whereas they represent a metabolite of prime importance in the locust, they are rather a reserve material in the cockroach, which essentially burns glucides. In *Danaus*, which also uses diglycerides during migratory flight, an adipokinetic factor was demonstrated. Like *Schistocerca*, *Danaus* reacts to *Tenebrio* extracts, but unlike *Schistocerca*, it does not react to *Periplaneta* or bee extracts; therefore it probably possesses a different adipokinetic hormone (Dallmann and Herman, 1978).

9.4.2. Hypolipemic Factor

The occurrence of a hypolipemic factor located in the storage lobe of the corpora cardiaca has been demonstrated in locusts (Orchard and Loughton, 1980). The storage lobes were isolated in saline and after stimulating the nccI or nccII, the bathing saline was injected into locusts whose hemolymph lipids were estimated after 45 min. Stimulation of the nccI failed to cause release of the hypolipemic factor, whereas stimulation of the nccII did cause its release.

9.5. RESPIRATORY METABOLISM

During the life of the insect, respiratory metabolism undergoes significant modifications, corresponding to changes in energy expenditures. Processes such as morphogenesis, vitellogenesis, locomotor activity, and metabolic processes require large energy inputs and lead accordingly to increases in respiratory metabolism. Diapause and fasting, on the other hand, cause a decrease in respiration (Fourche, 1969). Thus it is obvious that all the hormones and neurohormones that act on the foregoing processes have an influence on respiratory metabolism.

Ecdysone, which provokes RNA and protein syntheses, brings about changes in respiratory metabolism that are observed at the end of diapause, during the intermolt period, and in the pupal instar. The corpora allata, which also take part in protein syntheses, stimulate the respiratory metabolism of larvae (*Galleria*, Sehnal and Sláma, 1966), and of the adult female, in connection with reproductive processes (*Calliphora*, Thomsen, 1949; *Leptinotarsa*, de Wilde and Stegwee, 1958; *Leucophaea*, Sägesser, 1960; *Locusta*, Roussel, 1963; *Pyrrhocoris*, Sláma, 1964). Their removal decreases respiratory metabolism, and their implantation elevates it.

The question arises as to the mode of action of the corpora allata. According to Altmann (1960), working on the bee and the locust, and according to Sláma (1964, 1965), working on the bug *Pyrrhocoris*, their action is indirect and related to the quantity of metabolically active tissues. The assumption that the corpora allata have a direct action on respiratory metabolism has nevertheless been supported by several investigators (Thomsen, 1949; de Wilde and Stegwee, 1958; Sägesser, 1960; Lüscher, 1968a), for in the adult female, removal of these organs has the same effect in normal insects and in castrated insects (*Calliphora, Leptinotarsa, Leucophaea*).

The respiratory modifications of the fat body cells *in vitro* under the action of corpora allata substantiate this point of view. They have been observed in *Leucophaea* (Lüscher, 1965; Müller and Engelmann, 1968) and *Nauphoeta* (Lüscher, 1968a), and the study of total homogenates of *Leptinotarsa* (de Wilde and Stegwee, 1958) yields the same results.

Experiments have also been conducted with isolated mitochondria. When the latter are obtained from total homogenates of *Plodia* larvae, a stimulatory activity of the corpora allata is observed (Firstenberg and Silhacek, 1973). In *Locusta*, the mitochondria of the thoracic muscles are stimulated by corpora allata homogenates (Clarke and Baldwin, 1960; Minks, 1967), while those of the fat body are also stimulated in the adult but inhibited in fifth-instar larvae (Clarke and Baldwin, 1960). These results, however, are contradicted by those of Minks (1967) on the same insect.

The various neurohormones controlling metabolism, reproduction, and activity exert a definite influence on respiratory metabolism. In *Pyrrhocoris*, respiratory metabolism of the adult female declines sharply after removal of the corpora cardiaca (Sláma, 1964). This also applies to adults of *Schistocerca* and *Locusta* who have gone beyond the somatic growth period (Lafon-Cazal and Roussel, 1971) and to the adult male of *Blaberus discoidalis* (Keeley and Friedman, 1967).

As for the corpora cardiaca, the question arises whether the action is direct or indirect. Sláma (1964) postulated that the effects of removal of the corpora cardiaca are due to digestive disturbances, engendered by the absence of intestinal proteases. However, in the adult female of *Leucophaea* (Lüscher, 1965; Lüscher and Leuthold, 1965; Müller and Engelmann, 1968) and the adult male of *Leucophaea* (Wiens and Gilbert, 1965) and of *Blaberus* (Keeley and Friedman, 1967), the effects of corpora cardiaca on respiration are also obtained *in vitro* on pieces of the fat body. The action of corpora cardiaca may thus be exerted directly on the fat body. However, it does not necessarily act on respiratory metabolism, for, as we have seen, the carbohydrate, lipid, and protein metabolisms of the fat body are regulated by neurohormones, which may bring about the respiratory modifications observed.

A search for the source of the active factor was made by corpora cardiaca disconnection experiments in *Leucophaea* and in *Blaberus*. While the activity of these organs persists in the adult female of *Leucophaea* for several months (Engelmann and Müller, 1966; Müller and Engelmann, 1968), it disappears in the adult male of *Blaberus* some 30 days after the operation (Keeley and Van Waddill, 1971) Hence a positive conclusion cannot be drawn from these results. Other experiments involving removal of the pars inter-cerebralis show a decrease in oxygen consumption in adult females of both *Pyrrhocoris* (Sláma, 1964) and *Oncopeltus* (Conradi-Larsen, 1970). Moreover, extracts of the pars intercerebralis added to fat body pieces increase their respiratory activity in *Leucophaea* (Lüscher, 1965). The brain and corpora cardiaca are also active on homogenates of the fat body and the muscles of *Periplaneta, Leucophaea,* and *Blaberus* (Ralph and Matta, 1965). It appears, therefore, that the active factors are likely to originate at least in part in the pars intercerebralis. They may include hypertrehalosemic hormone from the pars intercerebralis and also adipokinetic hormone originating in the cells of the corpora cardiaca.

By what pathway do neurohormones act on respiratory metabolism? Examination of the respiration of fat body mitochondria shows a significant decline in *Blaberus* deprived of corpora cardiaca for 30 days, apparently due to the loss of electron-transport capacity of the mitochondria (Keeley and Friedman, 1969; Keeley, 1971). However, corpora cardiaca extracts do not act on mitochondrial respiration if added to the incubation medium, but increase it if injected daily into the insect (Keeley and Van Waddill, 1971), thus suggesting that the active factor does not act directly on the mitochondria.

9.6. PURIFICATION OF METABOLIC NEUROHORMONES

9.6.1. Hypertrehalosemic Hormone

The separation by paper chromatography of the factors present in the corpora cardiaca of the cockroach *Periplaneta* led Brown (1965) to isolate many factors that are active on the contractility of the heart and the hindgut, and two distinct factors active on hemolymph trehalose concentration.

The investigations of Natalizi and Frontali (1966) and Natalizi *et al.* (1970) were also designed to isolate the heart-accelerating and hyper-trehalosemic factors in *Periplaneta*. Like Brown, these authors reported that the active factors are destroyed by proteolytic enzymes and are different from serotonin. By using heads of *Periplaneta* and *Apis* and corpora cardiaca of *Periplaneta,* and by practicing chromatography on different Sephadex gels and on Cellex CM, they found, in addition to substances acting on the heartbeat and on spontaneous nerve activity, two peaks exhibiting hypertrehalosemic activity.

Mordue and Goldsworthy (1969) separated four fractions in the corpora cardiaca of *Locusta* and *Schistocerca*. Two of them increased hemolymph trehalose concentration in *Periplaneta,* the third one, located in the neurohemal lobe, was slightly active, while the fourth, originating in the glandular lobe, displayed far greater activity. Both fractions also acted on heartbeat and diuresis.

Thorough purification of corpora cardiaca hormones was carried out by Stone *et al.,* (1976), Jones *et al.,* (1977), and Holwerda *et al.* (1977a,b). These investigations succeeded in isolating the adipokinetic factor of the locust; the factor was also shown to have hypertrehalosemic activity (Holwerda *et al.,* 1977a; Jones *et al.,* 1977).

In the cockroach, the hypertrehalosemic factor is distinct from the adipokinetic factor, and both hormones are different from the adipokinetic–hypertrehalosemic factor of the locust (Holwerda *et al.,* 1977b). This observation agrees with the results of Downer and Steele (1972), who purified the hypertrehalosemic factor of *Periplaneta* and showed it to be devoid of adipokinetic activity. It also explains the differences in the reactions of the cockroach and the locust to the same extract.

9.6.2. Adipokinetic Hormone

Adipokinetic hormone (AKH) as been investigated by many authors. The earliest, Chino and Gilbert (1965), showed that the hemolymph factor causing lipid mobilization could be a protein because it is thermolabile and non-dialysable. The peptidic nature of adipokinetic hormone was confirmed in *Schistocerca* by Mayer and Candy (1969). These investigators pointed out that this heat-stable substance is inactivated by proteolytic enzymes.

Stone *et al.* (1976) and Cheeseman *et al.* (1977) identified the structure of adipokinetic hormone by using the lipid mobilization capacity of the adult male *Locusta* as a bioassay. One AKH unit was defined as the quantity of hormone causing the release of 1 μg of lipid per microliter of hemolymph in 1 hr (Fig. 78).

The starting material for isolation of the hormone consisted of 3000 glandular lobes of the corpora cardiaca of *Schistocerca* and *Locusta*, representing a mass of 750μg. The dissected material, placed in methanol, was subjected to ultrasonication, centrifugation, and then taken up by distilled water. This was followed by chromatography in a glass bead column eluted with distilled water. The residue, dissolved in methanol, was subjected to thin-layer chromatography on silica gel. The chromatogram was developed in isopropanol–water–acetic acid, which separated out an ultraviolet-absorbing area of R_f 0.67 containing adipokinetic hormone.

The amino acid composition of the hormone was determined by acidic

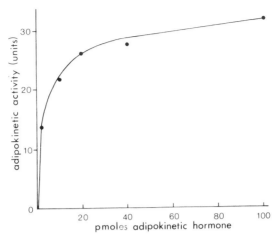

FIGURE 78. Dose–response curve for adipokinetic hormone, which elevates hemolymph lipid levels in *Locusta migratoria*. A unit of adipokinetic activity is defined as the amount of hormone that causes an increase of 1 μmole of lipid per microliter of hemolymph in 1 hr. (Modified from Jones *et al.*, 1977.)

hydrolysis, followed by mass spectrometry, and supplemented by Edman's method and enzymatic digestion by thermolysin.

The hormone proved to be a decapeptide with molecular weight of 1158, devoid of amine and free carboxyl groups. Its formula, given by Stone, is:

PCA-Leu-Asn-Phe-Thr-Pro-Asn-Trp-Gly-Thr-NH$_2$

In subsequent investigations, Cheeseman *et al.*, (1977), using different methods, further clarified the formula of adipokinetic hormone, which they showed to be:

Glu-Leu-Asn-Phe-Thr-Pro-Asn-Trp-Gly-Thr-NH$_2$

They estimated that 188 pmoles of adipokinetic hormone is present in one pair of corpora cardiaca, the half-life of the hormone being 19–24 min. Adipokinetic hormone was synthesized in 1977 by Broomfield and Hardy. Subsequently, 18 compounds comparable to adipokinetic hormone were synthesized in order to identify the structural characteristics of the fat body receptors (Stone *et al.*, 1978).

Holwerda *et al.* (1977a,b), whose works have already been mentioned, also attempted to purify the adipokinetic and hypertrehalosemic hormones of the locust and cockroach. Their purification methods are different from those of the other investigators. The organs, whole corpora cardiaca of *Locusta*, stored in deionized water at $-20°C$, were subjected to six successive freezings and thawings, then raised to boiling for 5 min, and centrifuged. They were then chromatographed on Biogel P-2 and Sephadex G-25, electrofocused, ultrafiltered, chromatographed on paper; amino acid analysis was performed with a Biocal analyzer after acidic hydrolysis. These investigations showed that, as already mentioned, the adipokinetic factor of the locust is different from that of the cockroach, and exhibits dual activity—adipokinetic and hypertrehalosemic. It appears to be an octapeptide containing the amino acids Asn, Glu, Gly, Leu, Phe, Pro, Ser, and Thr.

Recently, using corpora cardiaca of the locust *Schistocerca*, Carlsen *et al.* (1979) separated in addition to adipokinetic hormone another peptide with both adipokinetic activity in the locust and red pigment-concentrating activity in the shrimp *Leander adspersus*. The amino acid composition of this peptide is: Asp, Thr, Ser, Glu, Gly, Leu, Phe, and Trp. The new peptide was designated AKH II, AKH becoming I.

The factor contained in the corpora cardiaca of *Carausius* that causes an increase in hemolymph lipids in the locust has some properties close to those of the adipokinetic hormone of locusts; it is heat-stable and retains its activity after incubation with trypsin, carboxypeptidase A, and leucine aminopeptidase. However, activity is abolished when incubated with thermolysin and α-chymotrypsin (Gäde, 1980).

9.6.3. Adipokinetic Hormone and Crustacean Chromactivating Hormone

As we have seen in chapter 6, many investigators have demonstrated the presence, in the corpora cardiaca of insects, of a substance causing contraction of the red chromatophores of crustaceans. The crustacean red pigment concentrating hormone (RPCH) was identified recently, and was shown to be an octapeptide with a molecular structure resembling that of AKH I, thus explaining the activity of the corpora cardiaca of insects upon crustacean chromatophores.

Both insect hormones, AKH I and AKH II, induce contraction of the red chromatophores of prawns and shrimps (Mordue and Stone, 1976, 1977; Carlsen *et al.*, 1979), and RPCH provokes the release of fat body diglycerides in insects (Mordue and Stone, 1976, 1977). However, insect hormones are more potent in insects than in crustaceans, and vice versa, which is not surprising because their molecular structures are slightly different:

RPCH: Glu-Leu-Asn-Phe-Ser-Pro-Gly-Trp-NH_2
AKH I: pGlu-Leu-Asn-Phe-Thr-Pro-Asn-Trp-Gly-Thr-NH_2
AKH II: Asp-Thr-Ser-Glu-Gly-Leu-Phe-Trp-NH_2

9.7. METABOLIC HORMONES OF VERTEBRATES AND INSECTS

The similarity between carbohydrate and lipid metabolisms in insects and vertebrates has suggested the possibility of a resemblance or identity between their metabolic hormones.

With respect to carbohydrate metabolism, substances exhibiting the immunological reactions of insulin and glucagon were identified in the corpora cardiaca–corpora allata complex of *Manduca* larvae (Tager *et al.*, 1976).

Axons with immunoreactive anti-insulin staining have been located within the corpora cardiaca storage lobe of locusts (Orchard and Loughton, 1980) and purified head extracts of the blowfly *Calliphora vomitoria*. These extracts display a strong effect upon blood trehalose and glucose concentration, cross-react with bovine insulin antibodies, and display insulinlike biological activity on the isolated rat fat cells (Duve *et al.*, 1979).

Total protein extracts of *Drosophila* exhibit an insulinlike effect in the mouse, causing a decrease in blood sugar concentration similar to that caused by insulin (Meneses and Ortiz, 1975). Insulin injection has variable effects on insects, and sometimes no effect at all, as in the silkworm (Wenig and Joachim, 1936) and the blowfly *Calliphora* (Normann, 1975). It has a clear effect in other cases, decreasing hemolymph trehalose in *Vespa* larvae (Ishay *et al.*, 1976), *Manduca* (Tager *et al.*, 1976), and *Apis* (Bounias and Pachéco, 1979b).

Glucagon has been investigated less. In *Manduca*, it decreases the fat body glycogen (Tager *et al.*, 1976), similar to the effect demonstrated on the mosquito *Aedes* (Lea and Van Handel, 1970). Moreover, substances with the immunological reactivity of glucagon were demonstrated in the head of the adult indian meal moth *Plodia* and in the corpora cardiaca–corpora allata complex of adult *Periplaneta* (Tager *et al.*, 1976; Kramer, 1978) (Fig. 79).

These insulinlike and glucagonlike peptides were also demonstrated in hemolymph from larvae and nymphs of *Manduca* (Kramer *et al.*, 1980), which is a nice proof of their physiological role.

The best known insect hormone with respect to carbohydrate regulation is hypertrehalosemic hormone, which activates a phosphorylase of the fat body. By its action, it appears to be comparable to epinephrine. Few investigations have hitherto been devoted to the action of epinephrine on carbohydrate metabolism in insects, and the only data available derive from an ultrastructural study demonstrating a decrease of fat body glycogen under the action of epinephrine, in *Hyalophora* and in *Leucophaea* (Bhakthan and Gilbert, 1968). The effects of noradrenaline, dopamine, and dopa were investigated in adult honeybees. Noradrenaline injection causes an increase in glucose and trehalose blood concentration, whereas dopamine and dopa ap-

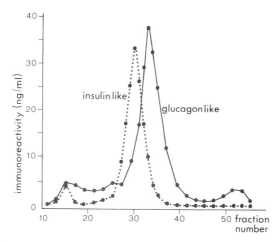

FIGURE 79. Profiles of insulinlike and glucagonlike immunoreactivity obtained after gel filtration of an extract of corpora cardiaca–corpora allata complexes from adult *Manduca sexta*. (Modified from Tager *et al.*, 1976.)

pear to require a conversion into an active form to have effect (Bounias, 1980).

Although these investigations are only in the early stages, it should be noted that the three vertebrate hormones appear to occur in insects, while, apart from a hypertrehalosemic factor, insects probably possess a hypotrehalosemic factor and a glycogenolytic factor.

Vertebrate hormones have also been tested on the lipid metabolism of insects by using *in vitro* the prelabeled fat body of pupae and adults of *Hyalophora,* adult males of *Periplaneta* and other cockroaches (Bhakthan and Gilbert, 1968), and also larvae of two lepidopterans, *Danaus plexippus* and *Agrius cingulata,* of which only *Danaus* displayed clear reactions (Chang, 1974a). The results are as follows.

ACTH has no effect on *Hyalophora* and *Periplaneta,* but growth hormone, gonadotropic hormone, and thyroxine increase the release of diglycerides and free fatty acids in *Hyalophora.*

Epinephrine also stimulates lipid release, but the lipids released in the cockroach, as in *Hyalophora*, are free fatty acids and not diglycerides, as normally occurs in the insect. Hence in this case, epinephrine exhibits the same effects as in vertebrates, but the effects are abnormal for the insect (Bhakthan and Gilbert, 1968). In *Danaus*, it appears to cause the release of tri- and diglycerides (Chang, 1974a).

Insulin, on the contrary, blocks lipid release in *Hyalophora* (Bhakthan and Gilbert, 1968), while in *Danaus* (Chang, 1974a) it stimulates the release of di- and triglycerides.

The action of glucagon has only been tested in *Danaus* (Chang, 1974a). Its effect appears to be similar to the effect caused in this species by the other vertebrate hormones, thus casting doubt on the validity of these experiments.

9.8. MODE OF ACTION OF METABOLIC HORMONES

The mode of action of metabolic hormones is not always clear. Evidence is available to show that diapause hormone of *Bombyx,* which raises the glycogen level of the ovary, acts through an ovarian trehalase (Yamashita and Hasegawa, 1967). Hypertrehalosemic hormone activates a fat body phosphorylase, giving rise to the formation of trehalose from the stored glycogen (Steele, 1963; Wiens and Gilbert, 1967). Its activity, however, may involve other enzymes as shown by Chen and Friedman (1977b). Adipokinetic hormone brings about the formation of diglycerides from triglycerides and free fatty acids, and also acts on the muscle, where it enhances the penetration of diglycerides (Mayer and Candy, 1969; Robinson and Goldsworthy, 1976). It also appears to perform a very different function—that of stimulating the formation of lipoproteins (diglyceride carriers) (Van Der Horst and Beenakkers, 1979).

Apart from the final effects just described, the metabolic hormones, like all peptidic hormones, used a second messenger, cAMP.

The function of cAMP in the action of hypertrehalosemic and adipokinetic hormones has been demonstrated by measurement of the cAMP of the fat body subjected to the action of these hormones, and by a comparison of the hemolymph trehalose and lipid concentrations induced by the hormones, by nucleotides, and by phosphodiesterase inhibitors.

Corpora cardiaca extracts elevate the fat body cAMP level in *Periplaneta* (Hanaoka and Takahashi, 1977; Gäde, 1977) and *Locusta* (Gäde and Holwerda, 1976; Gäde, 1979a). An increase in adenylate-cyclase is also observed in *Periplaneta* (Hanaoka and Takahashi, 1977). The increase in cAMP caused by the injection of corpora cardiaca extracts is related to their hypertrehalosemic effect in the cockroach, and adipokinetic effect in the locust. cAMP is elevated not only in experimental conditions, but also in natural conditions, as in the locust after flight (Gäde and Holwerda, 1976).

Added to the fat body incubation medium, cAMP brings about a hypertrehalosemic reaction in *Periplaneta,* resulting in the stimulation of the phosphorylase activity of the fat body (Steele, 1964; Hart and Steele, 1973). It also brings about an adipokinetic effect in the adult male of *Periplaneta* (Bhakthan and Gilbert, 1968) and in *Danaus* larvae (Chang, 1974a), with high fatty acid release in the former, and triglyceride release in the latter.

Dibutyric cAMP mimics the hypertrehalosemic effect in *Periplaneta* (Hanaoka and Takahashi, 1977) and the adipokinetic effect in *Schistocerca* (Spencer and Candy, 1976) and *Locusta* (Gäde and Holwerda, 1976). This is also true of caffeine and theophylline in *Periplaneta* (Steele, 1964; Hanaoka and Takahashi, 1977) and of theophylline in *Schistocerca* (Spencer and Candy, 1976).

9.9. CONCLUSIONS

The regulation of insect metabolism is a very complex and relatively unknown field, although the only insect hormone hitherto synthesized, adipokinetic hormone, is actually a metabolic hormone.

One of the difficulties of the matter arises from the great diversity of insects, which use different metabolites for the same function. Another difficulty derives from the fact that hormones endowed with the same activities may have different chemical structures and vice versa. Some rather enigmatic aspects have been pointed out: the existence in some insects of periods of unresponsiveness to a given hormone; the presence, in one species, of a substance that is active in another species, but inactive in the species posessing it, probably due to lack of a receptor.

Ecdysone and, above all, juvenile hormone play a part in metabolic regulation but to varying degrees. Neurohormones perform important functions in the regulation of all metabolism (Fig. 80). They stimulate the production of intestinal proteases and of proteins by the fat body. The hypertrehalosemic hormone active in carbohydrate metabolism is well known, but a hypotrehalosemic hormone and a glycogenolytic hormone also exist, at least in some insects (flies, mosquitoes, and butterflies). Some data are available on hemolymph glucose concentration control, but they are still very incomplete. On the other hand, abundant information is available concerning the diapause hormone of the silkworm, which brings about an accumulation of glycogen in the ovary. As for lipid metabolism, the hormone causing mobilization of fat body diglycerides in locusts has been isolated, purified, and synthesized. It is a decapeptide with a formula comparable to that of the chromactivating factor in crustaceans.

Vertebrate and insect hormones have been compared by reciprocal injection experiments and immunology. These experiments suggest the existence of possible analogies. It is interesting to note that crustacean hyperglycemic hormone, which has been synthesized, is a very large peptide quite different from the adipokinetic hormone of locusts.

The source of adipokinetic hormone in locusts is located in the glandular lobe of the corpora cardiaca. The hypertrehalosemic hormone appears to be synthesized either in the glandular lobe of the corpora cardiaca or in the pars

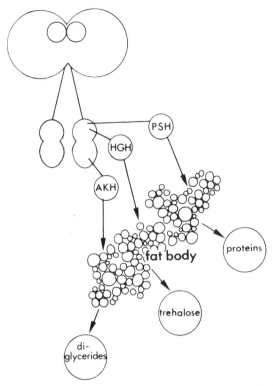

FIGURE 80. Schematic representation of the metabolic control exerted by the corpora cardiaca on the fat body.

intercerebralis, but the evidence is not conclusive. Apart from the diapause hormone, the other neurohormones originate in the neurosecretory cells of the pars intercerebralis. Because practically no investigations have been carried out concerning the ventral nerve cord, it is unknown whether metabolic hormones are synthesized there. It is also unknown whether a similarity or an identity exists between certain metabolic neurohormones. The adipokinetic hormone of the locust appears to be hypertrehalosemic, but two distinct factors exist in the cockroach, an adipokinetic hormone and a hyper-trehalosemic hormone. A recent work also showed that the adipokinetic hormone of locusts exerts an inhibitory effect on protein syntheses.

Neurohormones and Cuticle

Molting in insects is followed by many physiological processes: the sclerotization of the cuticle or tanning, which entails its hardening and darkening, water loss followed by secretion of the endocuticle, and sometimes its melanization. In the adult whose wings need to remain soft in order to be able to unfold, tanning is preceded by a change in the mechanical properties of the cuticle, called plasticization.

The regulation of these different processes has been investigated in flies, cockroaches, locusts, and beetles. The neurohormone bursicon was discovered in adult flies (Fraenkel and Hsiao, 1965) and confirmed in other species.

Studies of the control of puparation in flies served to point out the existence of two new factors, anterior retraction factor and puparium tanning factor.

10.1. TANNING

10.1.1. Flies

When the fly escapes from the puparial case, it is small in size, its cuticle is soft and pale, and its wings are wrinkled. It immediately begins to pump air, expands to its final size, and unfolds its wings. This takes about 20 min; then begins the tanning process, which is complete in 1–2 hr and imparts to the fly its final appearance.

Experimental investigations into the regulation of tanning in flies were conducted simultaneously by Cottrell on the blowfly *Calliphora erythrocephala* (1962a–c) and by Fraenkel and Hsiao (1962, 1963, 1965) on the flesh fly *Sarcophaga bullata* and other flies.

10.1.1.1. Chronology of Secretion

If a ligature is placed at the neck of a fly within the first 5 min after emergence, tanning is arrested. After this interval, bursicon secretion occurs despite ligation (Fraenkel and Hsiao, 1962, 1963, 1965). Hence, secretion of the tanning factor is limited to a very short period.

The injection of hemolymph can cause tanning to resume in a ligated fly (Fig. 81); the hemolymph, which was inactive on eclosion, becomes active 2–3 min later. In *Sarcophaga,* its activity then rises very rapidly, and after 30 min it reaches a level such that the blood of a single fly suffices to tan 100 flies (Fraenkel and Hsiao, 1962, 1963, 1965). One hour after emergence, hemolymph activity begins to gradually diminish.

10.1.1.2. Source of the Tanning Factor and Release Site

Tanning is triggered by a neurohormone that is already present in the extract of the brain and the thoracoabdominal ganglionic mass before emergence, and until 1 day later. The subesophageal ganglion seems to be devoid of activity (Cottrell, 1962b; Fraenkel and Hsiao, 1963).

The active portion of the brain is the pars intercerebralis. In the thoracoabdominal ganglionic mass, activity appears to be restricted to the abdominal region. This may result from the fact that the neurosecretory cells that produce bursicon are only present in this region, but it is also possible that the high activity of the abdominal region (six times that of the brain) is due to the fact that it contains the perisympathetic organs.

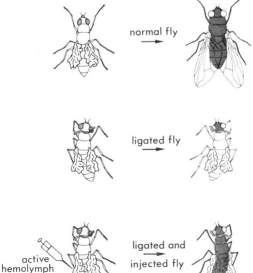

normal fly

ligated fly

active hemolymph

ligated and injected fly

FIGURE 81. Schematic representation of the effect of ligating newly emerged flies with and without injections of active hemolymph.

10.1.1.3. Release Stimulation

When the fly escapes from the puparium and climbs through the sand to reach the open air, the abdominal sensory receptors are stimulated. The nerve impulses are transported to the brain, which gives the signal to release bursicon. In fact, if the nervous connections between the brain and the ventral nerve cord are interrupted by severance or ligating at the neck, tanning does not take place, although bursicon is present throughout the central nervous system, even before emergence.

10.1.2. The Cockroaches *Periplaneta americana* and *Leucophaea maderae*

A humorally acting tanning factor was also demonstrated in two cockroaches by Mills (1965) and by Srivastava and Hopkins (1975).

10.1.2.1. Chronology of Secretion

The secretion chronology was studied *in vitro* and also *in vivo* by injecting insects ligated at the neck. These investigations showed that in *Periplaneta*, bursicon is released into the blood 20 min after molting; it reaches a peak in 90 min and declines after 2 hr. Five hours after the molt, the hemolymph can still cause slight tanning (Mills, 1966). In *Leucophaea* (Srivastava and Hopkins, 1975), the chronology is about the same, but bursicon appears earlier in the hemolymph, prior to ecdysis, and also disappears sooner (3 hr after molting).

10.1.2.2. Location of the Tanning Factor

The location of the tanning factor in cockroaches is widely debated. Different methods were employed for these investigations. In *Leucophaea* (Srivastava and Hopkins, 1975), bursicon was determined in organ extracts assayed on *Sarcophaga*. In *Periplaneta*, the study was carried out *in vitro* on pieces of the abdominal integument incubated with tissue homogenates of the same insect (Mills, 1965). According to Srivastava and Hopkins, bursicon is present throughout the central nervous system, in the corpora cardiaca, and even in the rectum. According to Mills, the corpora cardiaca and the ventral nerve cord ganglia exhibit moderate activity, whereas the brain and the terminal abdominal ganglion are highly active.

10.1.2.3. Release Site

Severance of the ventral nerve cord at the circumesophageal connective, between either the thoracic ganglia or the abdominal ganglia, considerably reduces the release of the tanning factor; a normal release occurs, however, if electrical stimulations are applied behind the section. Bursicon can thus be released from the ventral nerve ganglia. A leading role was ascribed to the terminal abdominal ganglion by Mills, but it now appears that the release of bursicon takes place in the last three pairs of perisympathetic organs. Concluding from ligature experiments, Srivastava and Hopkins (1975) consider that release occurs from the thoracic rather than the abdominal ganglia. These contradictions are difficult to reconcile. One explanation may be proposed: that different ganglia act at different times.

10.1.2.4. Release Stimulation

By placing ligatures at different times during ecdysis, Srivastava and Hopkins (1975) showed that the stimuli inducing bursicon release originate in the thorax and are related to the ecdysial split of the cuticle. If the latter is removed or sealed so as to be unbreakable, bursicon is not released into the hemolymph. Hence in the cockroach as in *Sarcophaga,* bursicon release occurs in response to peripheral stimulations.

10.1.3. The Migratory Locust *Locusta migratoria*

Molting in *Locusta* lasts rather a long time. Ten to fifteen minutes before ecdysis, the insect hangs head down, makes abdominal pumping movements, settles on its hind and middle legs, and begins longitudinal contractions of the abdomen that grow more intense and cause gradual breakage of the cuticle. Ecdysis is followed by a 10-min rest period after which the insect begins to spread its wings. Molting is completed in 60–90 min after the start of ecdysis, by which time tanning is already well advanced.

10.1.3.1. Source of the Tanning Factor and Chronology

Tanning investigations in *Locusta* were carried out by bioassays in which just-emerged *Phormia regina* blowflies were used (Vincent, 1972). Bursicon release into the hemolymph occurs in *Locusta* 6–10 min after the onset of ecdysis and is complete 60–80 min later. Extracts of the different nerve ganglia, which are inactive in the adult apart from the molting period, were tested at various developmental stages: at the start and the end of the fifth instar, and in the adult 5 hr after the onset of ecdysis. These determinations demonstrated that bursicon is present throughout the central nervous system.

10.1.3.2. Release Site

Because the entire nervous system contains bursicon, it remains to be determined whether this substance is released at all levels. Neck-ligated molting adults do not tan in the cephalic region, and this may indicate the absence of release from the brain and from the corpora cardiaca. However, as in all experiments of this type, the results may be explained by a lack of oxygen.

Nor is bursicon released from the thoracic ganglia, for its content in the latter does not diminish after molting and, in addition, an abdominal ligature inhibits thoracic tanning. Thus it appears that bursicon is released from the abdominal ganglia, which are slightly loaded at the time of molting. Vincent considers that the perisympathetic organs do not participate in bursicon release because of their inactivity in *Phormia* at the time of assay. However, this does not furnish definite proof. As demonstrated in *Tenebrio*, the activity of the perisympathetic organs changes with time.

10.1.4. The Mealworm *Tenebrio molitor*

Sclerotization in *Tenebrio* is far more time consuming than in the other species investigated and only part of it is dependent on bursicon, i.e., the tanning of the thoracic and the abdominal sternites (Delachambre, 1971).

10.1.4.1. Chronology of Secretion and Location of the Tanning Factor

Hemolymph activity during pupal life is estimated by injecting the blood of insects of different ages into young ligated adults (Delachambre, 1971; Grillot *et al.*, 1976). The hemolymph becomes slightly active on day 7; its activity then increases very rapidly, reaches a peak at the time of molting, and then remains fairly high for 2 days.

Extracts of the different ventral nerve cord ganglia and neurohemal organs (Fig. 82), taken throughout pupal life, were injected into 1-day-old pupae, which were ligated at the mesothorax and usually decapitated. The effects were assessed on 2-day-old adults (Grillot *et al.*, 1971b; Delachambre *et al.*, 1972; Provansal *et al.*, 1974; Grillot *et al.*, 1976) and revealed the presence of the active factor in all the ganglia of the central nervous system, in the corpora cardiaca, and in the perisympathetic organs. It was also shown that the activity of the different organs varies with time: nil in the prepupa, it reaches a peak between day 0 and day 5 depending on the organ, and disappears on day 7 when the hemolymph becomes active. The cephalic ganglia show early and durable activity (day 0 to day 5), while the thoracic and abdominal ganglia exhibit more delayed and transient activity (day 3). The anterior perisympathetic organs load on day 5, while the ganglia discharge. They release their contents rapidly, and their extracts are inactive from day 7. The posterior perisympathetic organs load later, on day 7, and appear to release more slowly. Thus, bursicon appears to be secreted throughout the ventral nerve cord, but the hypothesis of a transport from the head is also worth considering as in several similar cases.

If, during pupal life, the insects are ligated at different parts of the body

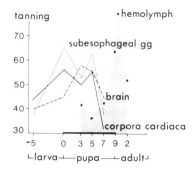

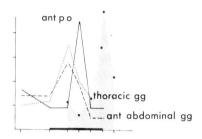

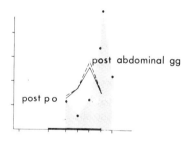

FIGURE 82. The tanning activity of corpora cardiaca, perisympathetic organs, and various parts of the central nervous system in *Tenebrio molitor* during the pupal stage and the beginning of the adult stage. ant, Anterior; gg, ganglion; po, perisympathetic organ; post, posterior. (Modified from Grillot *et al.,* 1976.)

(neck, thorax, in front of and in the posterior part of the abdomen), tanning is inhibited in the posterior portion until a critical period is reached, after which it proceeds normally despite the ligature. The critical period is approximately the same for cervical, thoracic, and anterior abdominal ligatures (between day 6 and day 7), but posterior abdominal ligatures show a later critical period (between day 7 and day 8). If one compares (Table XLVIII) the critical periods of ligatures to the times at which the different ganglia load and discharge, it may be observed that the critical period precedes the time of bursicon release but occurs significantly later than the time of ganglia loading. Hence it appears that the ligature interrupts a nerve release command rather than a

TABLE XLVIII. Chronology of Bursicon Storage and Release
in Tenebrio Molitor[a]

Organ assayed	Bursicon storage and release		Ligature critical period	
	Storage (days)	Release (days)	Ligature site	Critical period (days)
Brain	3–5	7		
Corpora cardiaca	0–5	7		
Subesophageal ganglion	0–5	7		
			Cervical ligature	6–7
Thoracic ganglia	3	7		
			Thoracic ligature	6–7
Anterior abdominal ganglia	3	7		
Anterior abdominal perisympathetic organs	5	7		
			Abdominal ligature	7–8
Posterior abdominal ganglia	7	9		
Posterior abdominal perisympathetic organs	7	9		

[a]After data of Delachambre (1971) and Grillot et al. (1976).

transport of active factor. This supports the first hypothesis postulated above, that bursicon is produced throughout the central nervous system.

10.1.4.2. Release site

The anterior perisympathetic organs have a very short loading period, which follows that of the ganglia. Their release, which can be observed with the electron microscope, just precedes the time when the hemolymph becomes active, thus suggesting that bursicon is released in these organs. (Fig. 83).

10.1.5. Lepidoptera

Truman (1973c), working on the tobacco hornworm *Manduca sexta*, tested the activity of eclosion hormone on tanning. He concluded that eclosion hormone and bursicon are two distinct factors, the former located in the brain and the corpora cardiaca, and the latter in the abdominal ganglia and the perisympathetic organs (see Chapter 7). Measurements of the titers of eclosion hormone and bursicon indicate that eclosion hormone is liberated into the blood during a 20-min period, 2½ hr before eclosion, and that bursicon release begins 2 min after eclosion and continues during a 10-min period. The target tissues of both hormones are responsive only during the last day of development (Reynolds et al., 1979).

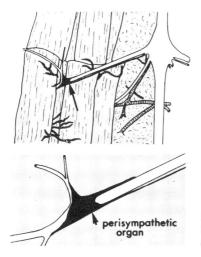

FIGURE 83. Location of perisympathetic organs in *Tenebrio molitor* pupa. (Modified from Grillot *et al.,* 1976.)

perisympathetic organ

10.1.6. Aphids

Selective electrocoagulation of identified neurosecretory cells in the brain of the aphid *Megoura viciae* demonstrated the source of a tanning factor to be two neurosecretory cells with two processes, one transporting neurosecretory material to the corpora cardiaca, the second along the ventral nerve cord (Steel, 1978b).

10.2. PLASTICIZATION

Tanning is preceded by an expansion stage that includes spreading of the wings. This occurs due to a change in the mechanical properties of the cuticle, which becomes soft and extensible. This plasticization is accompanied by hydration.

Plasticization in flies, which occurs during air pumping by the ptilinum (Cottrell, 1962d), is controlled by a neurohormone. Sephadex column fractionations of active blood seem to indicate that bursicon is involved (Reynolds, 1976), but its action is probably preceded by that of eclosion hormone as in *Manduca*. In *Manduca,* the wing cuticle, inextensible 3–4 hr before eclosion, then becomes extensible, thus enabling wing expansion. Once unfolded, the wing is tanned and finally becomes rigid. Following a number of purifications, it appears that two factors are involved, acting in succession: the first, identical to eclosion hormone, is located in the brain and corpora cardiaca, while the second, identical to bursicon, originates in the abdominal ganglia (Reynolds, 1977).

Plasticization in the blood-sucking bug *Rhodnius prolixus* takes place after molting and also after the blood meal which considerably inflates the abdomen. It is induced by a change in the intracuticular pH, which causes hydration of the cuticle (Reynolds, 1975). The active factor in this case is not bursicon; its action is mimicked by serotonin (Reynolds, 1974), as often occurs for neurohormones.

10.3. MELANIZATION

It is generally considered that sclerotization and melanization are two distinct processes, the former capable of occurring in the absence of the latter. In Chapter 6, we discussed the role of the ventral nerve cord ganglia in melanization, particularly in Lepidoptera. None of the investigations cited attempted to compare the melanization factor and bursicon. In two studies, however, which we shall now discuss, the data appear to demonstrate that bursicon intervenes in melanization.

In the first-instar larvae of the locust *Schistocerca gregaria* (Padgham, 1976a,b), this view is supported by the results of reciprocal hemolymph injections between *Schistocerca* and *Calliphora,* and ligatures that simultaneously influence both melanization and sclerotization. In *Sarcophaga,* ligatures and hemolymph injections yield the same results, as a function of time, for melanization and tanning, suggesting the possibility that bursicon acts in melanization processes. This hypothesis is substantiated by the melanizing activity of purified bursicon fractions. As for its mode of action, the inhibiting effect of α-MDH suggests that bursicon acts through DOPA decarboxylase (Fogal and Fraenkel, 1969).

10.4 ENDOCUTICLE SECRETION

Another bursicon-controlled process might be the secretion of the endo-cuticle, which takes place after tanning. This problem was investigated by two groups, the conclusions of which differ. Like tanning, endocuticle growth is inhibited in *Sarcophaga bullata* by early cervical ligatures and is restored by injections of active hemolymph or partly purified bursicon (Fig. 84) (Fogal and Fraenkel, 1969). However, different results were obtained by Schlein (1972a) in *Sarcophaga falculata.*

Ligatures and parabioses confirmed the action of a humoral factor causing endocuticle growth in *Sarcophaga*, but if one refers to the chronology of the active factor in the hemolymph (24 hr before eclosion) and to its source in the neurosecretory cells of the ocellar nerve, it appears that bursicon is not involved (Schlein, 1972b).

In *Locusta,* ligating the posterior part of the abdomen at the time of

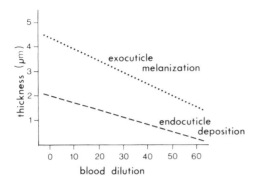

FIGURE 84. Effect of diluting active blood from just-molted *Sarcophaga bullata* on exocuticle melanization and endocuticle deposition. (Modified from Fogal and Fraenkel, 1969.)

molting diminishes endocuticle growth, as observed in the posterior femur (Vincent, 1971). Thus, because bursicon is thought to be released from the terminal ganglia, the inhibition resulting from the ligature suggests that bursicon is responsible for endocuticle growth.

Another process that may be controlled by bursicon is resilin deposition in adult insect tendons. In *Calliphora*, it depends on the neurosecretory cells of the pars intercerebralis (Sabaratnam, 1974).

10.5. DISAGGREGATION OF CELL FRAGMENTS OF THE WINGS

In the blowfly *Lucilia cuprina*, the hemolymph 1 hr after eclosion shows a large quantity of cell fragments that result from cytolysis of the wing epidermis. This transitory process disappears in 1 hr. Like tanning, it is suppressed by an early cervical ligature (after 5 min), whereas a later ligature (after 15 min) allows the cell debris to remain in the blood, while tanning proceeds. Hemolymph injection has different effects depending on the age of the fly. The appearance of the tanning factor is synchronous with that of the fragment disaggregation factor (totally inactive 2 min after emergence, the hemolymph becomes very active after 10 min). Hence it seems plausible that the factor acting on the wing (FDH) is identical to bursicon. Like bursicon, FDH is destroyed by proteolytic enzymes and, after filtration on Sephadex G-50 or separation by ammonium sulfate, the activities of both factors merge (Seligman and Doy, 1973). Furthermore, like bursicon, FDH acts via cAMP (Seligman and Doy, 1972) (see below).

10.6. DIURESIS LINKED TO TANNING

Strong diuresis occurs in insects during tanning. Is this caused by bursicon? Mills and Whitehead (1970) infer from their experiments on cockroaches that bursicon and a previously isolated diuretic hormone are probably identical (Goldbard *et al.,* 1970). By evaluating the diuretic activity of the hemolymph on isolated rectum preparations, they pointed out a similar variation in diuretic hormone and in bursicon. Like bursicon, purified diuretic hormone stimulates the incorporation of labeled tyrosine in the hemocytes.

10.7. PUPARIUM FORMATION

When a fly larva is about to pupate, it assumes a barrel shape by retracting the first three segments and by contracting; simultaneously, its cuticle plasticizes. It thus forms a puparium, which turns brown and hardens rapidly.

Endocrine control of pupariation was demonstrated in 1935 by Fraenkel. Subsequently, the investigations of Karlson and co-workers revealed the role of ecdysone in puparium formation; and the pupariation of the posterior portion of a ligated *Calliphora* larva was used for a long time for ecdysone bioassay (Karlson and Shaaya, 1964). Nevertheless, the work of Zdarek and Fraenkel shows that two other proteinic hormones intervene in pupariation. In fact, if *Calliphora* or *Sarcophaga* larvae are ligated a few hours before pupariation, the anterior part pupariates before the hind part, and the pupariation of the hind part, devoid of nerve ganglia, can be accelerated by the injection of hemolymph or nervous tissue homogenate. The active factors are present in the brain, in the ring gland, and also in the fused thoracoabdominal ganglionic mass. In the brain they are located in the pars intercerebralis; in the ring gland they are restricted to the corpus cardiacum (Zdarek and Fraenkel, 1969). Outside the central nervous system, active factors are also stored in peripheral sites, probably distal perisympathetic organs, from which they are released following an ecdysone injection (Fraenkel *et al.,* 1972).

If pupariation is regulated by neurohormones, what is the function of ecdysone? This hormone is important and initiates the whole process, as shown by the following. When the central nervous system is implanted in the hind part of the body that has not yet reached the critical period for ecdysone, pupariation does not occur. Conversely, it occurs when ligated insects are simultaneously injected with ecdysterone and nervous system extracts (Zdarek and Fraenkel, 1969). Pupariation can also be induced prematurely by injection of ecdysterone alone (Zdarek and Fraenkel, 1971), which implies that the latter stimulates the release of the neurohormones, or acts together with them in the pupariation processes.

Injections of hemolymph or nervous system extracts accelerate not only tanning but also anterior retraction; however, a comparison of hemolymph activity to that of nervous system extracts with respect to the two processes suggests that two distinct factors are involved (Fraenkel *et al.,* 1972). This hypothesis is also supported by an experiment involving the injection of α-

MDH (DOPA decarboxylase inhibitor), which allows retraction to subsist but prevents tanning. These two neurohormones have been termed respectively anterior retraction factor (ARF) and puparium tanning factor (PTF).

We shall see below that ARF and PTF differ in their molecular weights and in other properties, and that both are distinct from bursicon. Furthermore, hemolymph injection into a young fly, rich in bursicon, has no effect on larva pupariation, and the blood of a larva containing PTF fails to cause tanning in a young fly ligated at the neck (Zdarek and Fraenkel, 1969).

Several authors have suggested a pupariation-inhibiting factor in the higher Diptera, but its source and existence are still debated. In the tsetse fly *Glossina morsitans,* the housefly *Musca domestica* (Langley, 1967a), and other Diptera (Whitehead, 1974), this factor appears to originate in the brain, for a ligature placed behind this organ prevents pupariation of the fore part of the body but allows it to proceed in the hind part. In *Calliphora,* however (Price, 1970), the origin of the inhibitory factor appears to be different. In fact, if two ligatures are placed, the first behind the sixth segment and the second in front of the ninth segment, one observes an anterior pupariation, a posterior pupariation, and the absence of pupariation between the ligatures. This led Price to the conclusion that the inhibitory factor is located between the sixth and the ninth segments; as this part of the body does not contain the ventral ganglionic mass, it may be assumed that the inhibitory factor is contained in neurohemal organs.

Ratnasari and Fraenkel (1974b) concluded from various experiments conducted on *Sarcophaga* that the inhibition of tanning does not result from the presence of an inhibitory factor, but from deficient oxygenation of portions of the body separated from the spiracles. This opinion is shared by Chang (1974b), for in the tachinid fly *Lespesia archippivora,* administration of oxygen causes tanning of the fore part, while obstruction of the posterior spiracles prevents tanning in the posterior part of the body.

10.8. INTERSPECIFIC ACTIVITY

10.8.1. Bursicon

The interspecific activity of bursicon was assayed with the blood and organ homogenates of different species.

The blood of *Calliphora, Sarcophaga, Lucilia,* and *Phormia* shows strong interspecific activity (Cottrell, 1962b; Fraenkel and Hsiao, 1962). Tanning in *Sarcophaga* and *Calliphora* may be induced by the blood of several insects that have just molted or are at the point of doing so: *Periplaneta* nymphs and adults, the wax moth *Galleria mellonella,* the mealworm *Tenebrio molitor,* the milkweed bug *Oncopeltus fasciatus* (Fraenkel and Hsiao,

1965), and the desert locust *Schistocerca gregaria* (Cottrell, 1962b). The silkworm *Bombyx mori* appears to completely lack bursicon, both in the adult and in the larva or pupa (Fraenkel and Hsiao, 1965); however, the blood of the housefly *Musca domestica* is active on the larva of the cabbage white butterfly *Pieris brassicae* (Post, 1972).

Bursicon location has also been compared in the central nervous system and the corpora cardiaca of several species. In newly molted *Periplaneta* larvae and adults, it is present throughout the central nervous system and in the corpora cardiaca (Fraenkel and Hsiao, 1965). The brain, corpora cardiaca, and ventral nerve cord of stick insects and locusts also contain bursicon (Vincent, 1972).

A comparison of fly and cockroach bursicon by means of electrophoresis (Fraenkel and Hsiao, 1965) revealed the same location, thus suggesting that the origin of both factors may be identical.

10.8.2. Puparium Tanning Factor

PTF was investigated in the cockroach *Periplaneta* and the bug *Pyrrhocoris apterus*. In *Periplaneta*, the corpora cardiaca are active but the brain is not. In *Pyrrhocoris*, both organs are active (Zdarek and Fraenkel, 1969).

10.8.3. Bursicon and Other Neurohormones

Bursicon, which induces tanning in flies, has been compared to other neurohormones, brain hormone, gonadotropic hormone, and chromactivating factor, by assaying active fly blood for the presence of these different hormones. Brain hormone was assayed on diapausing pupae of *Bombyx* and *Philosamia*, gonadotropic hormone on flies incapable of developing their eggs owing to the removal of their cerebral neurosecretory cells, and chromactivating factor on the stick insect *Carausius*, which was prevented from spontaneous circadian color change by the removal of the brain. In none of these three cases did active hemolymph reestablish growth, vitellogenesis, or color change. Hence it appears that distinct factors are involved (Fraenkel and Hsiao, 1965). Bursicon is also distinct from eclosion hormone, ARF, and PTF.

10.9. NEUROHORMONE PURIFICATION

10.9.1. Bursicon

The properties of bursicon were investigated by Cottrell (1962b) on *Calliphora*, by Fraenkel and Hsiao (1963, 1965) and Fraenkel *et al.*, (1966) on *Sarcophaga,* and by Mills and Lake (1966) and Mills and Nielsen (1967) on *Periplaneta.*

Cottrell considers bursicon to be thermostable, resistant to boiling for 10 min, to cold of $-15°C$, and to desiccation at $120°C$. According to Fraenkel and Hsiao (1965), heating destroys a large part of the hemolymph activity, but not that of nerve tissue extracts. Both point out that bursicon is nondialysable and that it is inactivated by alcohol, acetone, and trichloroacetic acid, precipitated by ammonium sulfate, and destroyed by trypsin and pronase.

Electrophoretic separation of bursicon was carried out with *Sarcophaga* hemolymph and various nerve tissues of flies and cockroaches (pars intercerebralis and abdominal nervous system of *Sarcophaga,* corpora cardiaca and terminal abdominal ganglion of *Periplaneta*). The results obtained show the existence of two active fractions. The first is present in the hemolymph and the abdominal nerve ganglia of the fly, and in the corpora cardiaca and the terminal abdominal ganglion of *Periplaneta,* while it is absent from the pars intercerebralis of *Sarcophaga.* The second active fraction is found in the pars intercerebralis of *Sarcophaga* and in the corpora cardiaca of *Periplaneta* (Fraenkel *et al.,* 1966).

Purification, using whole cockroaches, was carried out by differential centrifugation and successive concentration by passage through various polyacrylamide and Sephadex gels, followed by a DEAE-cellulose column, and resulted in a 1700-fold purification (Mills and Lake, 1966). Bursicon is a protein possessing cationic properties. Its molecular weight is 38,000–40,000 as indicated by Fraenkel *et al.* (1966) and Mills and Nielsen (1967).

10.9.2. ARF and PTF

We have seen that ARF and PTF, which regulate fly pupariation, are two different substances, both distinct from bursicon. They were purified by Sivasubramanian *et al.* (1974) (precipitation by ammonium sulfate, ultracentrifugation, ultrafiltration, gel filtration, acrylamide gel electrophoresis, electroelution, polyacrylamide SDS gel electrophoresis) (Fig. 85). Two protein factors were isolated, both of which are destroyed by trypsin and pronase. They are nondialysable and are denatured by alcohol, acetone, and ammonium sulfate. ARF was purified 170-fold. Its molecular weight is 180,000, and it appears to contain two subunits of 90,000 in the blood. PTF was purified 200-fold. Its molecular weight is about 312,000. In the blood it appears to contain several subunits of 26,000 and in the central nervous system, four subunits of 80,000. ARF is thermolabile; PTF is thermostable in the central nervous system, but less stable in the hemolymph. Such a difference in heat-related behavior has also been observed in bursicon.

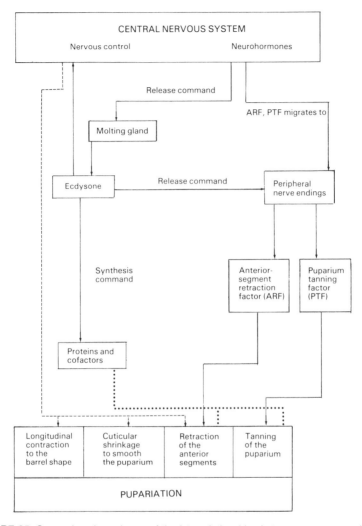

FIGURE 85. Comprehensive scheme of the interrelationships between neuromuscular and neuroendocrine events occurring during puparium formation in the fleshfly *Sarcophaga bullata*. (Modified from Sivasubramanian *et al.*, 1974.)

10.10. MODE OF ACTION OF TANNING HORMONES

The most frequent tanning in insects is quinonic tanning, involving *o*-quinones, which bond with the cuticular proteins through the latter's free amine groups and then combine with each other. Another type of fairly widespread tanning is β-sclerotization, wherein the protein quinone bond occurs via the β carbon. In both cases, the precursor is an o-diphenol, N-acetyldopamine.

The source of N-acetyldopamine in tyrosine metabolism was revealed by Sekeris and Karlson (1962). Karlson and Sekeris (1962a,b) pointed out its prominent role in sclerotization and demonstrated the role of ecdysone in the synthesis of DOPA decarboxylase. The metabolism of tyrosine generally includes the following steps:

1. Hydroxylation of tyrosine to DOPA by a tyrosine hydroxylase.
2. Decarboxylation of DOPA to dopamine by DOPA decarboxylase.
3. Acetylation of dopamine to N-acetyldopamine by N-acetyltransferase.

Sometimes, however, the metabolic path is somewhat different, i.e., the conversion of tyrosine to tyramine, the latter then being converted to DOPA, N-acetyldopamine, and N-acetyltyramine (Mills *et al.*, 1967).

As for the mode of action of bursicon, it appears that it acts neither on the exocuticle and hemocyte phenyloxidases, which reach a maximum concentration before its release (Mills *et al.*, 1968), nor on the synthesis of N-acetyldopamine and N-acetyltyramine (Mills *et al.*, 1967). It has been shown by Seligman *et al.* (1969) and Delachambre (1976) that it activates the formation of tyrosine hydroxylase, thus intervening at the very start of the metabolic chain in *Sarcophaga* and in *Tenebrio*. According to Mills *et al.* (1967), Whitehead (1969), Mills and Whitehead (1970), and Post (1972), the role of bursicon is different in *Periplaneta,* where it appears to increase the permeability of the hemocytes to tyrosine, which enters into contact with tyrosine decarboxylase contained in the hemocytes, is converted to tyramine and then to dopamine without passage through DOPA. The hemocytes transporting dopamine aggregate to the epidermis to provide it with sclerotization substrates.

The action of cAMP in sclerotization control was pointed out in many species. Its injection or the injection of dibutyric cAMP into ligated insects induces tanning in *Calliphora* (von Knorre *et al.*, 1972) and in *Periplaneta* (Vandenberg and Mills, 1974). The incorporation of labeled tyrosine in the cuticle and in the hemocytes was also pointed out in *Periplaneta* (Vandenberg and Mills, 1974, 1975). Adenyl cyclase was localized in the membranes of the hemocytes, which produce cAMP when incubated with [^{14}C]-ATP. Hence bursicon activates the membrane adenyl cyclase of the hemocytes to produce cAMP, thus increasing the permeability of the hemocyte membrane as it increases that of the Malpighian tubules (Maddrell *et al.*, 1971) and that of the salivary glands (Berridge, 1970).

In *Tenebrio,* the action of cAMP in tanning processes was demonstrated by radioimmunological determination of epidermal cAMP during development and after the injection of bursicon-rich hemolymph (Delachambre *et al.*, 1979a,b). Epidermal cAMP increases in the pupa simultaneously with blood bursicon, but it only increases in ligated insects after the injection of bursicon.

PTF also causes cAMP to act in *Sarcophaga bullata*, but this is not the case for ARF, whose action is not mimicked by any nucleotide (Fraenkel *et al.*, 1977; Seligman *et al.*, 1977). The injection of cAMP induces early tanning, accompanied by the incorporation of tyrosine in the cuticle (injection of cAMP and labeled tyrosine). Confirmation of the role played by cAMP is provided by an experiment involving the simultaneous injection of theophylline and blood diluted to the limit of its activity; tanning is thus achieved due to the action of theophylline, a phosphodiesterase inhibitor that prevents the destruction of cAMP.

Experiments were also conducted using inhibitors of RNA synthesis (BrdUrd, actinomycin D) and protein synthesis (puromycin and cycloheximide), alone or combined with cAMP or neurohormones. The results obtained suggest that PTF acts by accelerating the transcription of an mRNA necessary for tanning, while cAMP activates a process depending on the product of the gene derived by transcription of the gene activated by PTF.

10.11. CONCLUSIONS

Neurohumoral control of tanning has been demonstrated in many insects: *Sarcophaga, Calliphora, Periplaneta, Locusta, Tenebrio,* and *Manduca.* The active factor appearing in the blood at emergence is called bursicon. It has an interspecific activity and has been identified in seven orders.

Bursicon secretion and release are short-term processes that take place at molting, and are triggered by nerve impulses from the brain.

Bursicon is distributed throughout the central nervous system in *Locusta, Periplaneta,* and *Tenebrio;* in *Sarcophaga,* it is located only in the abdominal ganglionic mass and in the pars intercerebralis. The corpora cardiaca and the perisympathetic organs contain bursicon in most of the species investigated (*Periplaneta, Carausius, Schistocerca, Locusta, Tenebrio,* and *Manduca*) and are probably involved in its release.

Apart from tanning, bursicon plays a role in the control of other molting-related processes, which mainly involve the cuticle, plasticization, melanization, endocuticle secretion, disaggregation of wing cell fragments, and tanning-related diuresis. The data, however, are not firmly established in all cases. Bursicon intervenes after eclosion hormone, which triggers muscular contraction, eclosion, plasticization, wing spreading, and muscle breakdown (Fig. 86).

The specific tanning of the puparium of the higher Diptera is not

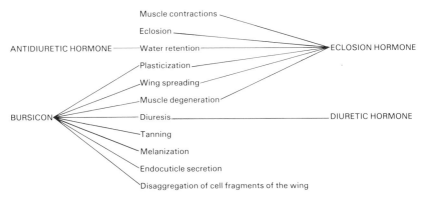

FIGURE 86. Processes involved in eclosion and their neurohormonal regulation.

controlled by bursicon but by a different factor, PTF, and puparium formation by retraction of the anterior segments depends on another neurohormone, ARF. These two substances are found throughout the central nervous system; their release seems to take place in abdominal neurohemal endings under the influence of ecdysone.

The comparison of bursicon with other neurohormones shows that bursicon is distinct from brain hormone, cerebral gonadotropic factor, chromactivating factor, eclosion hormone, ARF, and PTF. It may be identical to the diuretic factor in cockroaches, but this remains to be substantiated.

Bursicon, ARF, and PTF are nondialysable proteins. Bursicon has a molecular weight of 38,000–40,000. ARF and PTF have molecular weights of 180,000 and 300,000 respectively.

The mode of action of bursicon has been investigated in *Sarcophaga*, *Tenebrio*, and *Periplaneta*. In the two former species, it seems to induce the formation of tyrosine hydroxylase. In *Periplaneta*, it has been suggested that it modifies hemocyte permeability to tyrosine. In both cases, cAMP appears to play the part of a second messenger.

Concluding Remarks

General knowledge concerning insect neurohormones has advanced considerably in the past 20 years and numerous neurohormones have been discovered (Fig. 87).

The use of new methods, including autoradiography, electron microscopy, immunocytology, and axon impregnation, has shed considerable light on neurohormone synthesis and release sites, i.e., the neurosecretory cells, their pathways, and the neurohemal organs.

Experimental investigations have been largely developed thanks to the use of new techniques. Electrocoagulation has allowed selective destruction of individual neurosecretory cells within the nerve ganglia. The design of accurate bioassays (*in vitro* bioassays, as well as ecdysone and juvenile hormone titer determinations and others) has increased our experimental possibilities. Finally, the chemical purification of neurohormones has created new prospects.

The most striking facts that emerge from the research devoted to insect neurohormones are the neurohormones' great number and their intervention in all the major physiological regulations. Neurohormones control time-consuming phenomena such as growth and reproduction; they also control the various aspects of metabolism and rapid processes such as diuresis, contraction of the visceral muscles, ecdysis, pigment movements, and tanning. The functioning of the endocrine glands, the corpora allata, and the molting gland also depend on neurohormones (Fig. 88).

A quick comparison with vertebrates (Figs. 89 and 90) reveals that insects have about as many hormones as vertebrates. However, the distribution between hormones and neurohormones is different, and insects possess many more neurohormones than hormones. In most cases, the latter exert direct control on the effectors, while in vertebrates most of the controls involve two intermediaries—the adenohypophysis, which responds to the hypothalamic releasing hormones, and an endocrine gland controlled by the adenohypophysis.

One may raise the question whether all the effects attributed to particular neurohormones are due to distinct factors or whether several effects depend in some cases on a unique neurohormone. Investigations have been conducted to

```
1920
  |                              Growth ...................................... 1922
1929

1930
  |
1939

1940                           Growth ...................................... 1940
  |                            Molting gland stimulation ...................... 1946
  |                            Muscle contraction ........................... 1948
1949                           Physiological color change ..................... 1949

1950                           Diapause ..................................... 1951
  |                            Vitellogenesis ................................ 1954
  |                            Activity rhythm ............................... 1956
  |                            Morphological color change .................... 1956
  |                            Spontaneous nerve activity .................... 1958
1959                           Protease secretion ........................... 1959

1960                           Hypertrehalosemia ........................... 1961
  |                            Water retention .............................. 1962
  |                            Tanning ..................................... 1962
  |                            Protein synthesis ............................ 1962
  |                            Diuresis .................................... 1963
  |                            Glycogenolysis .............................. 1965
  |                            Corpora allata stimulation ..................... 1966
  |                            Oviposition .................................. 1968
  |                            Oocyte protection ........................... 1968
  |                            Anterior retraction ........................... 1969
  |                            Pupariation ................................. 1969
  |                            Adipokinesis ................................ 1969
1969                           Sexual differentiation ......................... 1969

1970                           Eclosion .................................... 1970
  |                            Sexual behavior ............................. 1971
  |                            Hypotrehalosemia ........................... 1973
  |                            Accessory gland activity ...................... 1974
  |                            Ovulation ................................... 1976
  |                            Juvenile hormone esterase production........... 1976
  |                            Gonial mitoses .............................. 1977
  |                            Polymorphism of offspring ..................... 1977
1979                           Ovarian ecdysone production .................. 1979

1980                           Hypolipemia ................................. 1980
```

FIGURE 87. Progressive discovery of neurohormones involved in control of physiological processes.

answer this question, but the data are often contradictory owing to the difficulty in separating the factors, for definite results can only be obtained with a high degree of purification. It seems, however, that some neurohormones act upon several effectors whereas others are quite distinct substances. It has been sometimes observed that certain neurohormones are found in insects in which they do not display the expected activity, suggesting that they regulate another function. Bursicon, the tanning factor, regulates many physiological processes distinct from tanning but is still distinct from several other neurohormones (e.g., brain hormone, gonadotropic hormone). Diuretic hormone acts upon

Malpighian tubules and also upon the rectum. Adipokinetic hormone triggers diglyceride release from the fat body and also diglyceride penetration in the muscle.

The existence of antagonistic factors has been observed in the regulation of various processes. They are often located in the same organ, rendering their identification particularly difficult. Another difficulty may originate from the fact that the substances present in the ganglia, in the neurohemal organs, and in the blood are not the same. Finally, the chemical nature of neurohormones may differ, even in closely related insects. So discrepancies appear in the data. They also result from the fact that all insects do not possess the same regulatory processes, for from a physiological standpoint they are sometimes very remote from each other.

The neurosecretory cells of insects are located in many sites, in the brain and in the different ventral nerve cord ganglia. The source of all neurohormones is not accurately known at the present time. Because most investigations have dealt with the pars intercerebralis of the brain, there is evidence that many neurohormones originate there, but it is generally unknown whether they are also synthesized in other neurosecretory parts of the brain and in the ventral nerve cord ganglia.

The presence of some neurohormones has been localized in the pars intercerebralis, but without an exact identification of the source cells. Two of these are eclosion hormone and bursicon. Greater accuracy has been achieved in other cases, at least for some species (Fig. 91). Thus it is known that the brain hormone is synthesized in one identified pair of lateral neurosecretory cells in the lepidopteran *Manduca sexta*. It is also known that four protocerebral anterior neurosecretory cells regulate exuviation in dragonflies and that tritocerebral neurosecretory cells control physiological color changes in stick insects. Diuresis appears to depend on four other anterior protocerebral neurosecretory cells in locusts and cockroaches, whereas in *Rhodnius* it is stimulated by abdominal neurosecretory cells. The functioning of the corpora allata depends on pars intercerebralis or protocerebral neurosecretory cells, or both. Several metabolic hormones originate in the pars intercerebralis, but adipokinetic hormone is produced in the glandular cells of the corpora cardiaca.

Data concerning the ventral nerve cord are not as advanced. The source of the embryonic diapause factor in *Bombyx* has been identified in two medioventral cells of the subesophageal ganglion, which may control the activity rhythm and which are active in the production of the ommochrome synthesis factor in *Carausius*. Many other neurohormones, such as the myotropic, egg-laying, chromactivating, and diuretic and antidiuretic factors, are present in the ventral nerve cord, either in all the ganglia or only in the thoracic (melanic pigment genesis factor) or abdominal region. An especially pronounced activity has been reported in the last abdominal ganglion in cock-

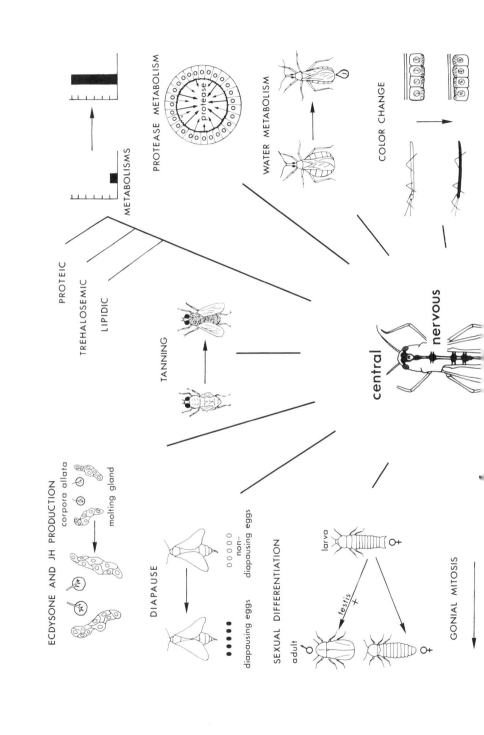

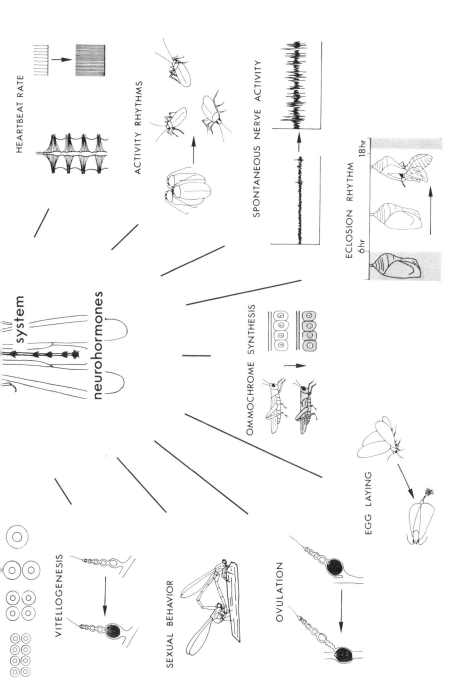

FIGURE 88. Schematic representation of the various regulations exerted by neurohormones in insects.

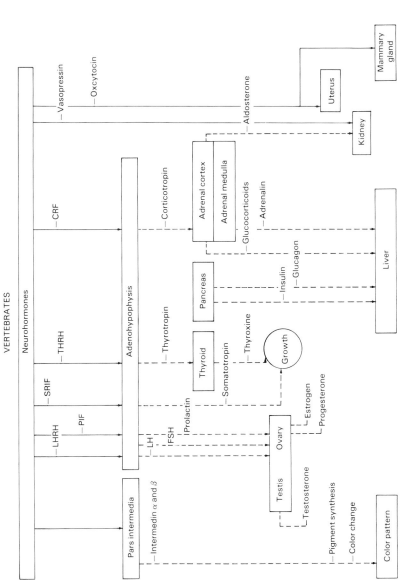

FIGURE 89. Neurohormonal and glandular regulations in vertebrates.

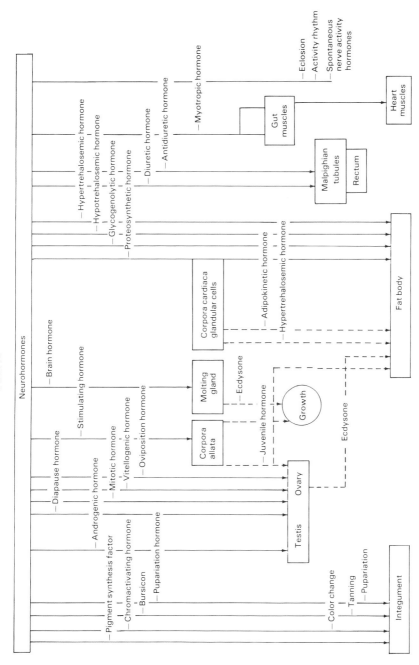

FIGURE 90. Neurohormonal and glandular regulations in insects.

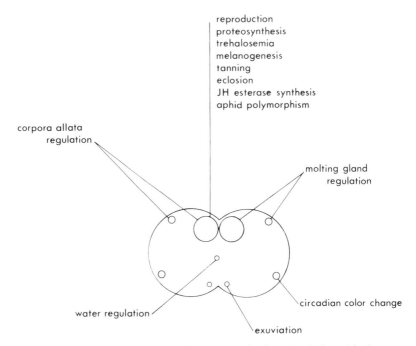

FIGURE 91. Location of neurohormone production sites in insect brain.

roaches. It appears that a greater specialization occurs in evolved insects, Diptera and Heteroptera, in which the thoracic and the abdominal ganglia are fused into a single mass. Hence bursicon and diuretic hormone are present throughout the central nervous system in Polyneoptera, whereas bursicon is restricted to the brain and to the abdominal nerve region in flies and diuretic hormone is confined to the abdominal region in *Rhodnius*.

Neurohormone release sites appear to be more variable than had been anticipated. Central sites exist (the corpora cardiaca and the perisympathetic organs), but in addition to them extensive peripheral sites lie along certain nerves and in the vicinity of the organs. They are involved in the release of diuretic hormone in *Glossina* and *Rhodnius*. Some neurohormones liberated in the effectors act as modulators or neurotransmitters. This seems to be the case for myotropic factors. Nevertheless, it has been shown that neurohormones, whose action can be mimicked by amines, particularly 5-HT, are distinct from the latter.

The neurosecretory axons also penetrate into the nerve ganglia, where they form synaptoid sites with neurons, as has been pointed out in vertebrates in the extrahypothalamic regions. Thus, some neurohormones appear to play a nerve modulator role in insects, as they do in vertebrates.

The action of neurohormones nearly always implies the intervention of cAMP, which functions as second messenger. The effects of neurohormones are mimicked by cAMP, and in numerous cases an important increase in the cAMP content is observed after hormonal stimulation.

Several research teams have attempted to purify neurohormones. Technological progress has facilitated investigations, but a major obstacle remains, that of the amount of material required to successfully carry out the investigations (2,000,000 heads of the silkworm were recently used for the purification of diapause hormone). All neurohormones are proteins or polypeptides. Their molecular weights range from 500–700 to 40,000. Two neurohormones have been identified and synthesized—adipokinetic hormone, a decapeptide, and proctolin, a myotropic pentapeptide.

Speculating on future progress and future trends in insect neurohormone research, it seems that purification of neurohormones will allow identification of the source cells of each hormone and its release sites, separation of antagonistic and distinct substances, and determination of hormone titer in hemolymph and in tissues. Physiological understanding will be achieved by knowledge of catabolic enzymes, carrier proteins, and receptors.

.

Addendum

A.1. SYNTHESIS, STORAGE, AND RELEASE OF NEUROHORMONES

A.1.1. Neurophysin and Neurohormones

A neurosecretory protein was isolated from the brain and the corpora cardiaca of *Locusta migratoria*. Histoimmunochemical studies showed its presence in the granules of the type A neurosecretory cells of the pars intercerebralis and in the storage lobe of the corpora cardiaca. Recently, this neurosecretory protein was found to be released from the corpora cardiaca storage lobe along with a diuretic hormone by high-potassium saline and electrical stimulation of the nccI. It therefore behaves like pituitary neurophysin (Friedel *et al.*, 1981; Orchard *et al.*, 1981).

A.1.2. Immunoreactive Compounds

Much important work has recently been accomplished concerning the occurrence in insects of vertebrate immunoreactive products. Thus in the brain of *Locusta migratoria* two pairs of metenkephalinlike neurons were identified. These cells, located in the lateral parts of the protocerebrum, send intracerebral collaterals along the nccI and nccII (Rémy and Dubois, 1981). In the brain of the blowfly, *Calliphora erythrocephala*, gastrin/cholecystokinin-immunoreactive neurons were detected (Duve and Thorpe, 1981). In the larva of *Eristalis aenus*, numerous immunoreactive cells were reported in the brain. They consisted of neurons found to be immunoreactive to glucagon, insulin, somatostatin, secretin, gastrin, substance P, enkephalin, endorphin, and pancreatic polypeptide (El-Salhy *et al.*, 1980; El-Salhy, 1981).

In a different field, antibodies raised to brain neurosecretory material appeared to inhibit both allatotropic activity and ovarian development in *Locusta migratoria* (Rembold *et al.*, 1980).

This Addendum includes papers published between November 1980 and November 1981.

269

A.1.3. Brain Neurosecretory Cells

The migration of neurosecretory cell groups already described in several insect orders during development was also reported in the primitive dipteran of the Tipulidae family. The axon pathways of neurosecretory cells were described (Panov, 1980).

A.1.4. Neurosecretory Cells of the Sympathetic Nervous System

The occurrence of type A and B neurosecretory cells was demonstrated in the frontal ganglion of *Achoea janata* and *Philosamia ricini* larvae (Awasthi, 1980).

A.1.5. Peripheral Neurosecretory Cells

Electron microscopy of the recently discovered abdominal neurosecretory organs of *Rhodnius prolixus* revealed that they contain peripheral neurosecretory neurons that release their products into the blood. Axons of central origin innervating these organs sometimes contained small aminelike granules, which did not seem to be released (Kuster and Davey, 1981).

A.1.6. Release Sites of Brain Neurosecretory Products

Histological study of *Coccinella septempunctata* retrocerebral–endocrine complex revealed that in this coleopteran, as in Heteroptera and Diptera, the aorta walls act as a neurohemal organ (Ray *et al.*, 1980).

An ultrastructural study of the corpora allata of *Manduca sexta* confirmed that the peripheral area of these organs performs a neurohemal function and may participate in brain hormone release (Sedlak, 1981).

A.2. CONTROL OF ENDOCRINE GLAND ACTIVITY

A.2.1. Control of Brain Hormone Secretion

As already shown in the lepidopterans *Manduca sexta* and *Antheraea*, an endogenous circadian clock was observed to control the timing of the brain hormone secretion that precedes larval ecdysis in *Samia cynthia* (Fujishita and Ishizaki, 1981).

As regards the increase of brain cAMP corresponding to the initiation of metamorphosis in the diapausing silkmoth, a substance with an active site resembling 5-HT was suggested as the activator of brain cyclase (Rasenick and Berry, 1981).

A.2.2. Molting Gland Activation

The role of cAMP in molting gland activation was confirmed in *Galleria mellonella*, in which it induced biphasic changes in the membrane potential similar to those induced by brain hormone (Gersch and Birkenbeil, 1980).

A.2.3. Molting Gland-Innervating Neurons

The presence of neurosecretory axon terminals in the molting gland was reported in *Galleria mellonella* (Benedeczki *et al.,* 1980).

The neurons innervating the molting gland were identified by the cobalt chloride technique in the prothoracic ganglion of the cockroach *Periplaneta americana* (Birkenbeil and Agricola, 1980).

A.2.4. Juvenile Hormone and Molting Gland Stimulation

The activation of molting gland by juvenile hormone analogues, applied to allatectomized prepupae of *Mamestra brassicae*, confirmed the role of juvenile hormone in molting gland stimulation (Hiruma, 1980).

A.2.5. Brain Hormone Purification

Progress in brain hormone purification was achieved in both the silkworm and the tobacco hornworm. Starting from 2,800,000 adult silkworm heads, a "highly purified PTTH" was obtained, 7 ng of which caused adult development in a brainless *Samia* pupa (Nagasawa *et al.*, 1980). In addition, 4400-fold purification was obtained starting from *Manduca* larva heads. The isolated fraction is heat-stable and sensitive to proteolytic enzymes; its molecular weight is about 25,000 (Kingan, 1981).

A.2.6. Ecdysone Production in the Ovary

Ovariectomy in adult *Acheta domesticus* drastically reduced the amount of neurosecretory material stored in the pars intercerebralis, whereas this amount remained unchanged in ovariectomized insects injected with ecdysterone. Thus a regulatory interaction between the brain and ovary is suggested, although ovary implants did not restore a normal situation (Bradley and Simpson, 1981).

A.2.7. Ecdysone and Corpora Allata Activity

In cockroaches, the cyclic activity of the corpora allata was shown to be regulated by a nervous inhibitory pathway and by a humoral factor, which might be ovarian ecdysone. This was confirmed by applying the *in vitro* radiochemical technique to *Periplaneta americana*, in which ovariectomy resulted in the disappearance of the cyclic pattern of juvenile hormone synthesis (Weaver, 1981). Further, 20-hydroxyecdysone was found to inhibit juvenile hormone synthesis by the corpora allata in *Diploptera punctata*. This effect did not appear to be direct but rather indirect (Friedel *et al.*, 1980). In the mosquito *Aedes aegypti*, corpora allata activity also appeared to be regulated by ovarian feedback inhibition (Rossignol *et al.*, 1981).

A.2.8. Juvenile Hormone Esterase Regulation

It was recently shown that the brain and subesophageal ganglion induce the appearance of juvenile hormone esterase in isolated abdomens of *Galleria mellonella*. In *Trichoplusia ni*, the brain and subesophageal ganglion had the same effect, but their activity depended upon the donor's developmental stage. It was therefore proposed that both stimulatory and inhibitory factors are involved in juvenile hormone esterase regulation (Jones *et al.*, 1981).

A.3. DIAPAUSE

A.3.1. Pupal Diapause Regulation

Pupal diapause termination is a complex process. In *Heliothis punctiger* pupae, molting gland activation required not only brain hormone release but also exposure to high temperature (Browning, 1981). The role of ecdysterone and juvenile hormone in pupal diapause was investigated in *Manduca sexta* and *Sarcophaga crassipalpis*. In *Manduca*, both hormones stimulated the neuroendocrine system, but high doses were required, and this might constitute an experimental artifact (Bradfield and Denlinger, 1980). In the flesh fly *Sarcophaga crassipalpis*, juvenile hormone and molting hormone titer determinations indicated that juvenile hormone plays an important role in pupal diapause (Walker and Denlinger, 1980).

A.3.2. Embryonic Diapause

Extraction and purification experiments revealed that diapause hormone occurs in diapause ovaries and eggs of *Bombyx mori*. In these tissues the hormone is probably present in bound form (Kai and Kawai, 1981).

A.4. REPRODUCTION

A.4.1. Vitellogenesis

Recent publications have dealt principally with the role of ovarian ecdysone in vitellogenin synthesis stimulation. In mosquitoes, the initial results were sometimes confirmed (Masler *et al.*, 1980, 1981; Kelly and Fuchs, 1980; Kelly *et al.*, 1981) and sometimes debated, juvenile hormone being suggested to fulfill the role of ecdysone (Borovski, 1981a,b). The complex relationship between juvenile hormone and brain EDNH was investigated in the housefly *Musca domestica* during several gonotropic cycles. It appears that only juvenile hormone is necessary to activate the previtellogenetic stage 3 follicles, whereas EDNH is required later by the stage 4 follicles to initiate vitellogenesis; EDNH is released independently of juvenile hormone, but may be correlated to oviposition. Cyclical egg development seems to be achieved by EDNH (Adams, 1981). The origin of adult ecdysteroids was studied in both a fly and the Colorado beetle. Radioimmunoassay determinations of ecdysone in both sexes of *Sarcophaga bullata*, fed with liver or not fed, suggest that the ovary cannot be the only site of synthesis of the hormone (Briers and de Loof, 1980). In this species ecdysterone injection induced synthesis and release of vitellogenin in both males and females (Huybrechts and de Loof, 1981). In the Colorado beetle *Leptinotarsa decemlineata*, ecdysteroids were found in both males and females, confirming that the ovary is not the only locus for ecdysteroid production (Briers and de Loof, 1981).

A.4.2. Ecdysone and Brain Neurosecretory Activity

In *Rhodnius prolixus*, ovarian ecdysone was suggested to elicit the release of a brain myotropic that causes ovarian contractions (Ruegg *et al.*, 1981).

A.4.3. Origin of Antigonadotropins

By measurement of follicle cell patency *in vitro* after exposure to juvenile hormone or juvenile hormone plus homogenates of recently discovered abdominal neurosecretory organs (see Section 4.5), it was demonstrated that these organs, located in abdominal segments 2 to 5, are the source of antigonadotropins in *Rhodnius prolixus* (Davey and Kuster, 1981).

Oviduct contractions play an important role in oviposition and are known to be increased by substances from the central nervous system. Serotonin and proctolin assays on the oviduct of *Tabanus proximis* show that both substances act upon the oviduct at doses of about 10^{-10}–10^{-9} M. However, their modes of action are different. Serotonin, which works faster, affects particularly the frequency of small phasic contractions; proctolin, on the contrary, modifies tonus and amplitude, and its effects last longer (Cook, 1981).

A.5. FUNCTIONING OF THE VISCERAL MUSCLES

A.5.1. Proctolin Source Cells

The specific neurons synthesizing proctolin were identified by different methods in *Periplaneta americana* and *Gryllus bimaculatus*. This identification was achieved by selective extraction of one pair of conspicuous lateral abdominal neurons, extracts of which revealed proctolinlike activity in a bioassay using locust leg extensor muscle. The proctolinlike factor behaved like proctolin when treated with proteolytic enzymes and in chromatographic experiments (O'Shea and Adams, 1981). Identification of the source cells of proctolin was also carried out by the immunohistochemical technique and axonal $CoCl_2$ iontophoresis in the terminal abdominal ganglion of *Periplaneta*. Six lateral and five to seven ventrocaudal proctolinlike immunoreactive neurons were found to project through the proctodeal nerves into the hindgut, whereas two lateral and several median proctolinlike neurons, arranged in three groups, did not project into the proctodeal nerves (Eckert *et al.*, 1981).

A.5.2. Myotropic Factors

As regards the number of factors involved in visceral muscle contractility, it was shown that proctolin, which acts upon several target tissues, is distinct from the hindgut-stimulating neurohormone. It was also shown that proctolin and serotonin act upon oviduct contractions. However, the alary muscles of *Locusta migratoria*, the contractions of which influence heart amplitude fluctuations, were insensitive to these two substances; they did, however, respond to brain extracts (Miller and Rózsa, 1981).

A.6. MORPHOLOGICAL AND PHYSIOLOGICAL COLOR CHANGE

A.6.1. Insect and Crustacean Chromactivating Factors

Insect adipokinetic hormone is known to concentrate crustacean red chromatophores. It has now been shown that two factors from the central nervous system of the cockroach *Periplaneta americana* exert a melanophore-dispersing effect in the crustacean *Uca pugilator*. These factors are not located in the corpora cardiaca and are water-soluble, heat-stable and trypsin-sensitive; their molecular weights are 3500 and 1000. A third factor, with a molecular weight of 20,000, was isolated from *Apis mellifera* (Dores and Herman, 1981a). It is conceivable that the factors isolated from the cockroach might be identical to the factor inducing color change in insects, since the location and some properties appear to be the same.

A.7. BEHAVIOR AND RHYTHMIC PHENOMENA

A.7.1. Eclosion hormone

The role of eclosion hormone is well-known in the adult *Manduca*. Eclosion hormone is also involved in pupal, larval, and embryonic ecdyses.

However, the origin of eclosion hormone is not the same throughout postembryonic development; in the larvae and pupae it is produced in the ventral nerve cord, whereas in the adult it is produced in the brain and released in the corpora cardiaca. Eclosion hormone activity is found in several insects of the main orders (Truman *et al.*, 1980; Taghert *et al.*, 1980; Truman *et al.*, 1981).

A.8. OSMOREGULATION

A.8.1. Diuretic Hormone

The occurrence of diuretic hormone, which induces a large water loss immediately after eclosion, was reported in the brain and corpora cardiaca of *Acraea horta, Danaus chrysippus,* and *Papilio demodocus* (Nicolson, 1980) and also in the head of *Vanessa cardui* (Herman and Dallmann, 1981). Even though diuretic hormone is present in adults after eclosion, it had no effect on posteclosion *Danaus plexippus.* The loss of sensitivity of the Malpighian tubules appeared to be both temperature-dependent and diuretic hormone-sensitive (Dores and Herman, 1981b).

A.8.2. Origin of Diuretic Hormone

It has been shown in *Locusta migratoria* that a vasopressinlike substance originates in two medioventral cells of the subesophageal ganglion. The vasopressinlike hemolymph titer was then shown to increase when diuresis increases, which suggests that the substance concerned may be involved in diuresis regulation—a hypothesis confirmed by the recent demonstration that the rectum and Malpighian tubules possess a degradative activity (Proux and Rougon-Rapuzzi, 1980).

A.8.3. Ion Regulation

A factor stimulating electrogenic transport of Cl^- in the rectum of *Locusta migratoria* was found in the corpora cardiaca. This chloride transport-stimulating hormone was investigated and purified more than 100-fold. It is destroyed by trypsin and has a molecular weight of 8000–12,000. It is different from locust diuretic, antidiuretic, and adipokinetic hormones (Phillips *et al.*, 1980).

A.8.4. Mode of Action of Diuretic Hormone

In vitro experiments on Malpighian tubules of *Acraea, Danaus,* and *Papilio* demonstrated the stimulating role of cAMP (Nicolson, 1980). As regards the mode of action of diuretic hormone, fluid secretion in *Locusta* was shown to be Na^+-dependent (Morgan and Mordue, 1981).

A.9. METABOLISM

A.9.1. Protease and Amylase Regulation

Midgut amylase activity was shown to increase after brain implantation in a caterpillar, *Hyblaea puera* (Muraleedharan and Prabhu, 1981).

A.9.2. Adipokinetic Hormone Interspecific Activity

The presence of adipokinetic hormone and of red pigment-concentrating hormone was demonstrated in *Danaus plexippus* and *Vanessa cardui*, where they caused significant, dose-dependent elevation of hemolymph lipids (Dallmann *et al.*, 1981). Furthermore, head extracts from *Vanessa cardui* and *Danaus plexippus* mobilized lipids in *Vanessa*, and *Vanessa* extracts elevated blood lipids in *Danaus* (Herman and Dallmann, 1981). In the Colorado beetle *Leptinotarsa decemlineata,* the amino acid proline is used as a main substrate for flight; its synthesis in the fat body was increased by corpora cardiaca extracts and synthetic adipokinetic hormone, which thus plays the same role in flight energetic metabolism, whatever the substrate of muscle activity (Weeda, 1981).

A.9.3. Hypertrehalosemic and Adipokinetic Hormones

The nature of the hypertrehalosemic factor from locust corpora cardiaca is a point of controversy. Sugar hemolymph determinations after injections of AKH I or AKH II isolated from the corpora cardiaca glandular lobe into neck-ligated *Locusta* demonstrated that these hormones behave differently. Whereas AKH I was inactive, AKH II was hypertrehalosemic (Loughton and Orchard, 1981). Other experiments on the same species support this finding. Thus fat body glycogen phosphorylase content increased significantly after injection of corpora cardiaca homogenates, but not after injection of synthetic adipokinetic hormone. Furthermore, activation of fat body phosphorylase may occur independently of lipid mobilization, for instance in stressed insects (Gäde, 1981).

A.9.4. Insulinlike and Glucagonlike Substances in Insects

Insulinlike and glucagonlike immunological reactions have been observed in insects. An insulinlike substance has now been found in the hemolymph and midgut of *Apis mellifica*. This purified substance has a higher molecular weight (18,000) than mammalian insulin (Moreau *et al.*, 1981). A high-molecular-weight form of glucagon was also found in the midgut of another insect, *Manduca sexta*. The factor involved appears to be a peptide of molecular weight 15,000 (Tager and Kramer, 1980).

A.10. NEUROHORMONES AND CUTICLE

A.10.1. A New Factor Involved in Pupariation

A protein with a molecular weight of between 100,000 and 300,000 was isolated from the hemolymph of young *Sarcophaga bullata* pupae. This protein, which immobilizes mature fly larvae, might be identical to the anterior retraction factor. Its function seems to be to shut off muscular activity (Zdarek *et al.*, 1981).

Reviews and Monographs

Berlind, A., 1977, Cellular dynamics in invertebrate neurosecretory systems, *Int. Rev. Cytol.* **49:**171.

Berridge, M. J., and Prince, W. T., 1972, The role of cyclic AMP and calcium in hormone action, in: *Advances in Insect Physiology,* Volume 9 (J. E. Treherne, M. J. Berridge, and V. B. Wigglesworth, eds.), Academic Press, New York, p. 1.

Bodnaryk, R. P., 1978, Structure and function of insect peptides, in: *Advances in Insect Physiology,* Volume 13 (J. E. Treherne, M. J. Berridge, and V. B. Wigglesworth, eds.), Academic Press, New York, p. 69.

Brady, J., 1974, The physiology of insect circadian rhythms, in: *Advances in Insect Physiology,* Volume 10 (J. E. Treherne, M. J. Berridge, and V. B. Wigglesworth, eds.), Academic Press, New York, p. 1.

Chippendale, G. M., 1977, Hormonal regulation of larval diapause, *Annu. Rev. Entomol.* **22:**121.

de Wilde, J., and de Loof, J., 1973, Reproduction—Endocrine control, in: *The Physiology of Insecta,* Volume 1 (M. Rockstein, ed.), Academic Press, New York, p. 97.

Dumser, J. B., 1980, The regulation of spermatogenesis in insects, *Annu. Rev. Entomol.* **25:**341.

Durchon, M., and Joly, P., 1978, *L'endocrinologie des Invertébrés,* Presses Universitaires de France.

Engelmann, F., 1970, *The Physiology of Insect Reproduction* (G. A. Kerkut, ed.), Pergamon Press, New York.

Engelmann, F., 1979, Insect vitellogenin: Identification, biosynthesis, and role in vitellogenesis, in: *Advances in Insect Physiology,* Volume 14 (J. E. Treherne, M. J. Berridge, and V. B. Wigglesworth, eds.), Academic Press, New York, p. 49.

Evans, P. D., 1980, Biogenic amines in the insect nervous system, in: *Advances in Insect Physiology,* Volume 15 (M. J. Berridge, J. E. Treherne, and V. B. Wigglesworth, eds.), Academic Press, New York, p. 317.

Finlayson, L. H., and Osborne, M. P., 1975, Secretory activity of neurons and related electrical activity, *Adv. Comp. Physiol. Biochem.* **6:**165.

Friedman, S., 1978, Trehalose regulation, one aspect of metabolic homeostasis, *Annu. Rev. Entomol.* **23:**389.

Frontali, N., and Gainer, H., 1977, Peptides in invertebrate nervous system, in: *Peptides in Neurobiology* (H. Gainer, ed.), Plenum Press, New York, p. 259.

Gabe, M., 1966, *Neurosécrétion,* Gauthier-Villars, Paris.

Gersch, M., 1964, Vergleichende Endokrinologie der Wirbellosen Tiere, Akad. Verlagsges Leipzig.

Gersch, M., 1975, Prinzipien neurohormonaler und neurohumoraler Steuerung physiologischer Prozesse, Wissenschaftliche Beiträge der Friedrich-Schiller-Universität Jena.

Gilbert, L. I. (ed.), 1976, *The Juvenile Hormones,* Plenum Press, New York.

Goldsworthy, G. J., and Mordue, W., 1974, Neurosecretory hormones in insects, *J. Endocrinol.* **60**:529.

Hagedorn, H. H., and Kunkel, J. G., 1979, Vitellogenin and vitellin in insects, *Annu. Rev. Entomol.* **24**:475.

Highnam, K. C., and Hill, L., 1977, *The Comparative Endocrinology of the Invertebrates* (E. J. W. Barrington and A. J. Willis, eds.), Arnold, London.

Joly, P., 1968, *Endocrinologie des Insectes* (P. P. Grassé, ed.), Masson, Paris.

Joly, P., and Cazal, M., 1969, Données récentes sur les corpora cardiaca, *Bull. Soc. Zool. Fr.* **94**:181.

Keeley, L. L., 1978, Endocrine regulation of fat body development and function, *Annu. Rev. Entomol.* **23**:329.

Klemm, N., 1976, Histochemistry of putative transmitter substances in the insect brain, in: *Progress in Neurobiology,* Volume 7, Pergamon Press, New York, p. 99.

Lafon-Cazal, M., 1978, Les neurotransmetteurs des insectes, *Année Biol.* **17**:489.

Lebrun, D., 1978, Implications hormonales dans la morphogenèse des castes du termite *Kalotermes flavicollis* Fabr., *Bull. Soc. Zool. Fr.* **103**:351.

Lüscher, M., 1976, *Phase and Caste Determination in Insects,* Pergamon Press, Oxford, New York.

Maddrell, S. H. P., 1974, Neurosecretion, in: *Insect Neurobiology* (J. E. Treherne, ed.), Frontiers in Biology Series 35, Elsevier, Amsterdam, p. 307.

Maddrell, S. H. P., and Nordmann, J. J., 1979, *Neurosecretion,* Blackie, Glasgow and London.

Miller, T. A., 1975, Neurosecretion and the control of visceral organs in insects, *Annu. Rev. Entomol.* **20**:133.

Normann, T. C., 1976, Neurosecretion by exocytosis, *Int. Rev. Cytol.* **46**:1.

Novák, V. J. A., 1975, *Insect Hormones,* Chapman & Hall/Constable, London.

Raabe, M., 1975, Les organes périsympathiques, in: *Traité de Zoologie,* Volume VIII (III) (P. P. Grassé, ed.), Masson, Paris, p. 511.

Reynolds, S. E., 1980, Integration of behaviour and physiology in ecdysis, in: *Advances in Insect Physiology,* Volume 15 (J. E. Treherne, M. J. Berridge, and V. B. Wigglesworth, eds.), Academic Press, New York, p. 475.

Rockstein, M., (ed.), 1978, *Biochemistry of Insects,* Academic Press, New York.

Roussel, J. P., 1972b, Physiologie du ganglion frontal des Insectes, *Annee Biol.* **11**:235.

Rowell, H. F., 1976, The cells of the insect neurosecretory system: Constancy, variability, and the concept of the unique identifiable neuron, in: *Advances in Insect Physiology,* Volume 12 (J. E. Treherne, M. J. Berridge, and V. B. Wigglesworth, eds.), Academic Press, New York, p. 63.

Scharrer, B., 1978, Peptidergic neurons: Facts and trends, *Gen. Comp. Endocrinol.* **34**:50.

Steele, J. E., 1976, Hormonal control of metabolism in insects, in: *Advances in Insect Physiology,* Volume 12 (J. E. Treherne, M. J. Berridge, and V. B. Wigglesworth, eds.), Academic Press, New York, p. 239.

Stone, J. V., and Mordue, W., 1980, Isolation of insect neuropeptides, *Insect Biochem.* **10**:229.

Truman, J. W., and Riddiford, L. M., 1974, Hormonal mechanisms underlying insect behaviour, in: *Advances in Insect Physiology,* Volume 10 (J. E. Treherne, M. J. Berridge, and V. B. Wigglesworth, eds.), Academic Press, New York, p. 297.

Wigglesworth, V. B., 1970, *Insect Hormones* (J. E. Treherne, ed.), Oliver & Boyd, Edinburgh.

References

Adams, M. E., Miller, T., and Thomson, W. W., 1973, Fine structure of the alary muscles of the American cockroach, *J. Insect Physiol.* **19:**2199.

Adams, T. S., 1976, The ovaries, ring gland, and neurosecretion during the second gonotrophic cycle in the housefly, *Musca domestica, Gen. Comp. Endocrinol.* **30:**69.

Adams, T. S., 1981, Activation of successive ovarian gonotrophic cycles by the corpus allatum in the house fly, *Musca domestica* (Diptera: Muscidae), *Int. J. Invert. Reprod.* **3:**41.

Adams, T. S., Hintz, A. M., and Pomonis, J. G., 1968, Oöstatic hormone production in houseflies, *Musca domestica*, with developing ovaries, *J. Insect Physiol.* **14:**983.

Adams, T. S., Johnson, J. D., Fatland, C. L., and Olstad, G., 1970, Preparation of a semi-purified extract of the oöstatic hormone and its effect on egg maturation in the house fly, *Ann. Entomol. Soc. Am.* **63:**1565.

Adiyodi, K. G., 1974, Extracerebral cephalic neuroendocrine complex of the blattids, *Periplaneta americana* (L.) and *Neostylopyga rhombifolia* (Stoll): An *in situ* study, *J. Morphol.* **144:**469.

Adiyodi, K. G., and Bern, H. A., 1968, Neuronal appearance of neurosecretory cells in the pars intercerebralis of *Periplaneta americana* (L.), *Gen. Comp. Endocrinol.* **11:**88.

Agui, N., 1975, Activation of prothoracic glands by brains *in vitro*, *J. Insect Physiol.* **21:**903.

Agui, N., and Hiruma, K., 1977a, Ecdysone as a feedback regulator for the neurosecretory brain cells in *Mamestra brassicae*, *J. Insect Physiol.* **23:**1393.

Agui, N., and Hiruma, K., 1977b, *In vitro* activation of neurosecretory brain cells in *Mamestra brassicae* by β-ecdysone, *Gen. Comp. Endocrinol.* **33:**467.

Agui, N., Granger, N. A., Gilbert, L. I., and Bollenbacher, W. E., 1979, Cellular localization of the insect prothoracicotropic hormone: *In vitro* assay of a single neurosecretory cell, *Proc. Natl. Acad, Sci. USA* **76:**5694.

Agui, N., Bollenbacher, W. E., Granger, N. A., and Gilbert, L. I., 1980, Corpus allatum is release site for insect prothoracicotropic hormone, *Nature (London)* **285:**669.

Alexander, N. J., 1970, A regulatory mechanism of ecdysone release in *Galleria mellonella, J. Insect Physiol.* **16:**271.

Alexandrowicz, J. S., 1926, The innervation of the heart of the cockroach (*Periplaneta orientalis*), *J. Comp. Neurol.* **41:**291.

Ali, Z. I., and Pipa, R., 1978, The abdominal perisympathetic neurohemal organs of the cockroach *Periplaneta americana:* Innervation revealed by cobalt chloride diffusion, *Gen. Comp. Endocrinol.* **36:**396.

Altmann, G., 1956, Hormonale Steuerung der Wasserhaushaltes der Honigbiene, *Zool. Anz.* **19:**107.

Altmann, G., 1958, Untersuchungen über den Wasserhaushalt der Honigbiene, *Ann. Univ. Sarav. Sci.* **1.**

Altmann, G., 1960, Die Beeinflussung des Insektenstoffwechsels durch Hormone der Corpora allata, *Verh. Dtsch. Zool. Ges.* **27**:253.

Andrewartha, H. G., Miethke, P. M., and Wells, A., 1974, Induction of diapause in the pupa of *Phalaenoides glycinae* by a hormone from the suboesophageal ganglion, *J. Insect Physiol.* **20**:679.

Anstee, J. H., and Bell, D. M., 1975, Relationship of Na^+-K^+-activated ATPase to fluid production by Malpighian tubules of *Locusta migratoria*, *J. Insect Physiol.* **21**:1779.

Anstee, J. H., and Bowler, K., 1979, Ouabain-sensitivity of insect epithelial tissues, *Comp. Biochem. Physiol. A* **62**:763.

Anstee, J. H., Bell, D. M., and Hyde, D., 1980, Some factors affecting Malpighian tubule fluid secretion and transepithelial potential in *Locusta migratoria* L., *Experientia* **36**:198.

Anwar, I., and Ismail, S., 1979, Neurosecretory centers in the brain of adult *Gryllus bimaculatus* (De Geer) (Orthoptera: Gryllidac), *Int. J. Insect Morphol. Embryol.* **8**:265.

Anwyl, R., and Finlayson, L. H., 1973, The ultrastructure of neurons with both a motor and a neurosecretory function in the insect, *Rhodnius prolixus, Z. Zellforsch. Mikrosk. Anat.* **146**:367.

Anwyl, R., and Finlayson, L. H., 1974, Peripherally and centrally generated action potentials in neurons with both a motor and a neurosecretory function in the insect *Rhodnius prolixus, J. Comp. Physiol.* **91**:135.

Arvy, L., and Gabe, M., 1953, Particularités histophysiologiques des glandes endocrines céphaliques chez *Tenebrio molitor* L., *C. R. Acad. Sci.* **237**:844.

Arvy, L., and Gabe, M., 1962, Histochemistry of the neurosecretory product of the pars intercerebralis of pterygote insects, in: *Proceedings of the 3rd International Symposium on Neurosecretion, Bristol, Mem. Soc. Endocrinol.* **12**:331.

Aston, R. J., 1975, The role of adenosine $3':5'$-cyclic monophosphate in relation to the diuretic hormone of *Rhodnius prolixus, J. Insect Physiol.* **21**:1873.

Aston, R. J., 1979, Studies on the diuretic hormone of *Rhodnius prolixus.* Some observations on the purification and nature of the hormone and the dynamics of its release *in vitro, Insect Biochem.* **9**:163.

Aston, R. J., and White, A. F., 1974, Isolation and purification of the diuretic hormone from *Rhodnius prolixus, J. Insect Physiol.* **20**:1673.

Atzler, M., 1931, Untersuchungen über den morphologischen und physiologischen Farbenwechsel von *Dixippus (Carausius) morosus, Z. Vgl. Physiol.* **13**:509.

Awasthi, V. B., 1968, The functional significance of the nervi corporis allati 1 and nervi corporis allati 2 in *Gryllodes sigillatus, J. Insect Physiol.* **14**:301.

Awasthi, V. B., 1972, Studies on the neurosecretory system and retrocerebral endocrine glands of *Nezara viridula* Linn. (Heteroptera: Pentatomidae), *J. Morphol.* **136**:337.

Awasthi, V. B., 1975, Neurosecretory system of the earwig *Labidura riparia*, and the role of the aorta as a neurohaemal organ, *J. Insect Physiol.* **21**:1713.

Awasthi, V. B., 1976, Ultrastructure of retrocerebral complex of the earwig, *Euborellia annulipes,* and role of aorta as a neurohaemal organ, *J. Insect Physiol.* **22**:1181.

Awasthi, V. B., 1979, The ultrastructure of corpus allatum of the earwig *Euborellia annulipes* Lucas (Dermaptera: Labiduridae), *Z. Mikrosk. Anat. Forsch. (Leipzig)* **93**:982.

Awasthi, V. B., 1980, Neurosecretion in the frontal ganglion of *Achoea janata* Linn. and *Philosamia ricini* (Hutt.) (Lepidoptera), *Z. Mikrosk. Anat. Forsch. (Lepzig)* **94**:825.

Baehr, J. C., 1968, Etude histologique de la neurosécrétion du cerveau et du ganglion sous-oesophagien de *Rhodnius prolixus* Stål (Hémiptère), *C. R. Acad. Sci. Ser. D* **267**:2364.

Baehr, J. C., 1973, Contrôle neuroendocrine du fonctionnement du corpus allatum chez *Rhodnius prolixus, J. Insect Physiol.* **19**:1041.

Baehr, J. C., 1974, Contribution à l'étude des variations naturelles et expérimentales de la protéinémie chez les femelles de *Rhodnius prolixus* (Stål) Insecte Hémiptère Reduviidae, *Gen. Comp. Endocrinol.* **22**: 146.

Baehr, J. C., 1976, Etude du contrôle neuro-endocrine du fonctionnement du corpus allatum chez les larves du quatrième stade de *Rhodnius prolixus*, *J. Insect Physiol.* **22**:73.

Banks, W., 1976, Cardiotropins in the brain and ventral cord ganglia of *Blaberus craniifer* (Orthoptera: Blaberidae), *Trans. Am. Entomol. Soc. (Philadelphia)* **95**:504.

Barker, J. F., 1978, Neuroendocrine regulation of oöcyte maturation in the imported fire ant *Solenopsis invicta*, *Gen. Comp. Endocrinol.* **35**:234.

Barker, J. F., and Herman, W. S., 1973, On the neuroendocrine control of ovarian development in the monarch butterfly, *J. Exp. Zool.* **183**:1.

Barth, R. H., and Sroka, P., 1975, Initiation and regulation of oöcyte growth by the brain and corpora allata of the cockroach, *Nauphoeta cinerea*, *J. Insect Physiol.* **21**:321.

Bar-Zev, A., and Kaulenas, M. S., 1975, The effect of β-ecdysone on *Gromphadorhina* adult female fat body at the transcriptional and translational levels, *Comp. Biochem. Physiol. B* **51**:355.

Bassurmanova, O. K., and Panov, A. A., 1967, Structure of the neurosecretory system in Lepidoptera. Light and electron microscopy of type A'-neurosecretory cells in the brain of normal and starved larvae of the silkworm *Bombyx mori*, *Gen. Comp. Endocrinol.* **9**:245.

Baudry, N., 1968, Etude histologique de la neurosécrétion dans la chaîne nerveuse ventrale de *Rhodnius prolixus* Stål (Hémiptère), *C. R. Acad. Sci. Ser. D* **267**:2356.

Baudry, N., 1969, Recherches histophysiologiques sur la neurosécrétion dans la chaîne nerveuse ventrale de *Rhodnius prolixus* Stål (Hémiptère) au cours de la vie imaginale, *C. R. Acad. Sci. Ser. D* **268**:147.

Baudry, N., 1972a, Comparaison des différents types d'organes périsympathiques de quelques familles d'Hétéroptères (Nepidae, Pyrrhocoridae, Coreidae et Pentatomidae), *C. R. Acad. Sci. Ser. D* **275**:1535.

Baudry, N., 1972b, Les organes périsympathiques des Homoptères. Localisation et structure chez les Cicadidae, *C. R. Acad. Sci. Ser. D* **275**:1673.

Baudry, N., and Baehr, J. C., 1970, Etude histochimique des cellules neurosécrétrices de l'ensemble du système nerveux central de *Rhodnius prolixus* Stål (Hétéroptère, Reduviidae), *C. R. Acad. Sci. Ser. D* **270**:174.

Baudry-Partiaoglou, N., 1978, Anatomie et histologie des organes neurohémaux de quelques Hémiptères, *Int. J. Insect Morphol. Embryol.* **7**:1.

Baudry-Partiaoglou, N., and Grillot, J.P., 1975, Ultrastructure des organes périsympathiques de la glossine, *Glossina fuscipes quanzenis* (Muscidae), *C. R. Acad. Sci. Ser. D* **281**:921.

Beament, J. W. L., 1958, A paralysing agent in the blood of cockroaches, *J. Insect Physiol.* **2**:199.

Bean, D. W., and Beck, S. D., 1980, The role of juvenile hormone in the larval diapause of the European corn borer, *Ostrinia nubilalis*, *J. Insect Physiol.* **26**:579.

Beattie, T. M., 1971, Histology, histochemistry, and ultrastructure of neurosecretory cells in the optic lobe of the cockroach *Periplaneta americana*, *J. Insect Physiol.* **17**:1843.

Beattie, T. M., 1976, Autolysis in axon terminals of a new neurohaemal organ in the cockroach *Periplaneta americana*, *Tissue Cell* **8**:305.

Beaulaton, J., 1968, Etude ultrastructurale et cytochimique des glandes prothoraciques de vers à soie aux quatrième et cinquième âges larvaires. II. Les cellules interstitielles et les fibres nerveuses, *J. Ultrastruct. Res.* **22**:499.

Beenakkers, A. M. Th., 1969, Carbohydrate and fat as a fuel for insect flight. A comparative study, *J. Insect Physiol.* **15**:353.

Bell, R. A., Borg, T. K., and Ittycheriah, P. I., 1974, Neurosecretory cells in the frontal ganglion of the tobacco hornworm, *Manduca sexta*, *J. Insect Physiol.* **20**:669.

Bell, W. J., 1969, Dual role of juvenile hormone in the control of yolk formation in *Periplaneta americana*, *J. Insect Physiol.* **15**:1279.

Benedeczki, I., Malá, J., and Sehnal, F., 1980, Ultrastructural study on the innervation of prothoracic glands in *Galleria mellonella*, *Gen. Comp. Endocrinol.* **41**:400.

Bentz, F., Girardie, A., and Cazal, M., 1970, Etude électrophorétique des variations de la protéinémie chez *Locusta migratoria* pendant la maturation sexuelle, *J. Insect Physiol.* **16**:2257.

Berlind, A., and Maddrell, S. H. P., 1979, Changes in hormone activity of single neurosecretory cell bodies during a physiological secretion cycle, *Brain Res.* **161**:459.

Bernardini-Mosconi, P., and Vecchi, M. L., 1961, Neurosecrezione nel cervello e nei gangli di *Reticulitermes lucifugus rossii* (Rhinotermitidae), *Symp. Genet. Biol. Ital.* **7**:322.

Bernays, E. A., 1980, The post-prandial rest in *Locusta migratoria* nymphs and its hormonal regulation, *J. Insect Physiol.* **26**:119.

Bernstein, A. R., and Mordue, W., 1978, An investigation of diuretic responses to hormones and other physiologically active agents in *Locusta*, in: *9th European Conference on Comparative Endocrinology, Giessen, Gen. Comp. Endocrinol.* **36**:110.

Berridge, M. J., 1966, The physiology of excretion in the cotton stainer, *Dysdercus fasciatus* Signoret. IV. Hormonal control of excretion, *J. Exp. Biol.* **44**:553.

Berridge, M. J., 1970, The role of 5-hydroxytryptamine and cyclic AMP in the control of fluid secretion by isolated salivary glands, *J. Exp. Biol.* **53**:171.

Berridge, M. J., and Patel, N. G., 1968, Insect salivary glands: Stimulation of fluid secretion by 5-hydroxytryptamine and adenosine-3′,5′-monophosphate, *Science* **162**:462.

Berridge, M. J., Lindley, B. D., and Prince, W. T., 1976, Studies on the mechanism of fluid secretion by isolated salivary glands of *Calliphora*, *J. Exp. Biol.* **64**:311.

Berthold, G., 1980, Microtubules in the epidermal cells of *Carausius morosus* (Br.). Their pattern and relation to pigment migration, *J. Insect Physiol.* **26**:421.

Beydon, P., Mauchamp, B., Lhonoré, J., Bouthier, A., and Lafont, R., 1980, The epidermal cell cycle during the last larval instar of *Pieris brassicae* (Lepidoptera), *J. Comp. Physiol.* **136**:21.

Bhakthan, N. M. G., and Gilbert, L. I., 1968, Effects of some vertebrate hormones on lipid mobilization in the insect fat body, *Gen. Comp. Endocrinol.* **11**:186.

Bhargava, S., 1967, Description and distribution of the neurosecretory cells in brain and ventral ganglia of *Lethocerus indicum* Lep. et Serv. (Heteroptera-Belostomatidae), *Zool. Anz.* **179**:272.

Bhargava, S., 1970, Transport, storage and release of neurosecretory products in *Lethocerus indicum* (Heteroptera: Belostomatidae), *J. Nat. Hist.* **4**:377.

Bhaskaran, G., and Jones, G., 1980, Neuroendocrine regulation of corpus allatum activity in *Manduca sexta*: The endocrine basis for starvation-induced supernumerary larval moult, *J. Insect Physiol.* **26**:431.

Bhaskaran, G., DeLeon, G., Looman, B., Shirk, P. D., and Röller, H., 1980, Activity of juvenile hormone acid in brainless, allatectomized diapausing *cecropia* pupae, *Gen. Comp. Endocrinol.* **42**:129.

Bignell, D. E., 1974, The effect of removal of the frontal ganglion on growth and protein synthesis in young adults of *Locusta migratoria*, *Can. J. Zool.* **52**:203.

Birkenbeil, H., 1971, Untersuchungen zur Wirkung von Neurohormon D und von anderen aktiven Substanzen auf das Zentralnervensystem der Schabe *Periplaneta americana* (L.), *Zool. Jahrb. Abt. Allg. Zool. Physiol. Tiere* **75**:493.

Birkenbeil, H., and Agricola, H., 1980, Die Innervierung der Prothorakaldrüse von *Periplaneta americana* L., *Zool. Anz.* **204**:331.

Blazsek, I., and Malá, J., 1978, Steroid transport through the surface of the prothoracic gland cells in *Galleria mellonella* L., *Cell Tissue Res.* **187**:507.

Blazsek, I., Balazs, A., Novák, V. J. A., and Malá, J., 1975, Ultrastructural study of the prothoracic glands of *Galleria mellonella* L. in the penultimate, last larval and pupal stages, *Cell Tissue Res.* **158**:269.

Bloch, B., Thomsen, E., and Thomsen, M., 1966, The neurosecretory system of the adult *Calliphora erythrocephala*. III. Electron microscopy of the medial neurosecretory cells of the brain and some adjacent cells, *Z. Zellforsch. Mikrosk. Anat.* **70**:185.

Blumenfeld, M., and Schneiderman, H. A., 1968, Effect of juvenile hormone on the synthesis and accumulation of a sex-limited blood protein in the polyphemus silkmoth, *Biol. Bull. (Woods Hole, Mass.)* **135**:466.

Bodenstein, D., 1947, Investigations on the reproductive system of *Drosophila*, *J. Exp. Zool.* **104**:101.

Bodenstein, D., 1953a, Studies on the humoral mechanisms in growth and metamorphosis of the cockroach, *Periplaneta americana*. II. The function of the prothoracic gland and the corpus cardiacum, *J. Exp. Zool.* **123**:413.

Bodenstein, D., 1953b, Studies of the humoral mechanism in growth and metamorphosis of the cockroach *Periplaneta americana*. III. Humoral effects on metabolism, *J. Exp. Zool.* **124**:105.

Boer, H. H., Schot, L. P. C., Veenstra, J. A., and Reichelt, D., 1980, Immunocytochemical identification of neural elements in the central nervous system of a snail, some insects, a fish, and a mammal with an antiserum to the molluscan cardio-excitatory tetrapeptide FMRF-amide, *Cell Tissue Res.* **213**:21.

Borg, T. K., and Bell, R. A., 1977, Ultrastructure of the neurosecretory cells in the brain of diapausing pupae of the tobacco hornworm, *Manduca sexta* (L.), *Tissue Cell* **9**:567.

Borg, T. K., Bell, R. A., and Picard, D. J., 1973, Ultrastructure of the neurosecretory cells in the frontal ganglion of the tobacco hornworm, *Manduca sexta* (L.), *Tissue Cell* **5**:259.

Borovski, D., 1981a, Feedback regulation of vitellogenin synthesis in *Aedes aegypti* and *Aedes atropalpus*, *Insect Biochem.* **2**:207.

Borovski, D., 1981b, *In vivo* stimulation of vitellogenesis in *Aedes aegypti* with juvenile hormone, juvenile hormone analogue (ZR 515) and 20-hydroxyecdysone, *J. Insect Physiol.* **6**:371.

Boulétreau-Merle, J., 1976, Destruction de la pars intercerebralis chez *Drosophila melanogaster:* Effet sur la fécondité et sur sa stimulation par l'accouplement, *J. Insect Physiol.* **22**:933.

Bounhiol, J. J., Gabe, M., and Arvy, L., 1953, Données histologiques sur la neurosécrétion chez *Bombyx mori* L. et sur ses rapports avec les glandes endocrines, *Bull. Biol. Fr. Belg.* **87**:323.

Bounias, M., 1980, Cinétiques d'action de la noradrénaline de la dopa et de la dopamine sur la glycémie de l'abeille *in vivo, Arch. Int. Physiol. Biochim.* **88**:93.

Bounias, M., and Pachéco, H., 1979a, Effets hyperglycémiants de l'adrénaline injectée chez l'abeille, *C. R. Acad. Sci. Ser. D* **289**:33.

Bounias, M., and Pachéco, H., 1979b, Sensibilité de la glycémie de l'abeille à l'action de l'insuline et du glucagon injectés *in vivo, C. R. Acad. Sci. Ser. D* **289**:201.

Bouthier, A., and Lafont, R., 1976, Biosynthèse et excrétion de substances à rôle pigmentaire lors de la mue nymphale des Lépidoptères, *Ann. Endocrinol.* **37**:537.

Bowers, B., and Johnson, B., 1966, An electron microscope study of the corpora cardiaca and secretory neurons in the aphid, *Myzus persicae* (Sulz.), *Gen. Comp. Endocrinol.* **6**:213.

Bowers, W. S., and Blickenstaff, C. C., 1966, Hormonal termination of diapause in the alfalfa weevil, *Science* **154**:1673.

Bowers, W. S., and Friedman, S., 1963, Mobilization of fat body glycogen by an extract of corpus cardiacum, *Nature (London)* **198**:685.

Bradfield IV, J. Y., and Denlinger, D. L., 1980, Diapause development in the tobacco hornworm: A role for ecdysone or juvenile hormone?, *Gen. Comp. Endocrinol.* **41**:101.

Bradley, J. T., and Simpson, T. A., 1981, Brain neurosecretion during ovarian development and after ovariectomy in adult *Acheta domesticus* L., *Gen. Comp. Endocrinol.* **44**:117.

Brady, J., 1967a, Control of the circadian rhythm of activity in the cockroach. I. The role of the corpora cardiaca, brain and stress, *J. Exp. Biol.* **47**:153.

Brady, J., 1967b, Control of the circadian rhythm of activity in the cockroach. II. Role of the sub-oesophageal ganglion and ventral nerve cord, *J. Exp. Biol.* **47**:165.

Brady, J., and Maddrell, S. H. P., 1967, Neurohaemal organs in the medial nervous system of insects, *Z. Zellforsch. Mikrosk. Anat.* **76**:389.

Brandt, N. R., and Huber, R. E., 1979, Carbohydrate utilization in the thoraces of honey bees (*Apis mellifera*) during early times of flight, *J. Insect Physiol.* **25**:483.

Bräuer, R., Gersch, M., Böhm, G. A., and Baumann, E., 1977, *In vitro* Stimulierung der Synthese von α-ecdyson in Prothorakaldrüsen von *Periplaneta americana* durch den Aktivationsfaktor I, *Zool. Jahrb. Abt. Allg. Zool. Physiol. Tiere* **81**:1.

Briegel, H., and Lea, A. O., 1975, Relationship between protein and proteolytic activity in the midgut of mosquitoes, *J. Insect Physiol.* **21**:1597.

Briegel, H., and Lea, A. O., 1979, Influence of the endocrine system on tryptic activity in female *Aedes aegypti, J. Insect Physiol.* **25**:227.

Briers, T., and de Loof, A., 1980, The molting hormone activity in *Sarcophaga bullata* in relation to metamorphosis and reproduction, *Int. J. Invert. Reprod.* **2**:363.

Briers, T., and de Loof, A., 1981, Moulting hormone activity in the adult Colorado potato beetle, *Leptinotarsa decemlineata* Say in relation to reproduction and diapause, *Int. J. Invert. Reprod.* **3**:145.

Brookes, V. J., 1969, The maturation of the oöcytes in the isolated abdomen of *Leucophaea maderae, J. Insect Physiol.* **15**:621.

Broomfield, C. E., and Hardy, P. M., 1977, The synthesis of locust adipokinetic hormone, *Tetrahedron Lett.* **25**:2201.

Brousse-Gaury, P., 1967, Généralisation, à divers insectes, de l'innervation deutocérébrale des corpora cardiaca, et rôle neurosécrétoire des nervi corporis cardiaci IV, *C. R. Acad. Sci. Ser. D* **265**:2043.

Brousse-Gaury, P., 1968, Modification de la neurosécrétion au niveau de la pars intercerebralis de *Periplaneta americana* L. en absence de stimuli ocellaires, *Bull. Biol. Fr. Belg.* **102**:481.

Brousse-Gaury, P., 1971a, Influence de stimuli externes sur le comportement neuro-endocrinien de blattes. I. Les organes sensoriels céphaliques, point de départ de réflexes neuro-endocriniens, *Ann. Sci. Nat. Zool. Biol. Anim.* **13**:181.

Brousse-Gaury, P., 1971b, Influence de stimuli externes sur le comportement neuro-endocrinien de blattes. II. Histophysiologie des voies réflexes neuro-endocriniennes, *Ann. Sci. Nat. Zool. Biol. Anim.* **13**:333.

Brown, B. E., 1965, Pharmacologically active constituents of the cockroach corpus cardiacum: Resolution and some characteristics, *Gen. Comp. Endocrinol.* **5**:387.

Brown, B. E., 1967, Neuromuscular transmitter substance in insect visceral muscle, *Science* **155**:595.

Brown, B. E., 1975, Proctolin: A peptide transmitter candidate in insects, *Life Sci.* **17**:1241.

Brown, B. E., 1977, Occurrence of proctolin in six orders of insects, *J. Insect Physiol.* **23**:861.

Brown, B. E., and Starratt, A. N., 1975, Isolation of proctolin, a myotropic peptide, from *Periplaneta americana, J. Insect Physiol.* **21**:1879.

Brown, F. A., Jr., and Meglitsch, A., 1940, Comparison of the chromatophorotropic activity of insect corpora cardiaca with that of crustacean sinus glands, *Biol. Bull (Woods Hole, Mass.)* **79**:409.

Browning, T. O., 1981, Ecdysteroids and diapause in pupae of *Heliothis punctiger, J. Insect Physiol.* **10**:715.

Broza, M., and Pener, M. P., 1969, Hormonal control of the reproductive diapause in the grasshopper, *Oedipoda miniata, Experientia* **25**:414.

Bückmann, D., 1956a, Die Umfärbung der Raupen von *Cerura vinula* unter verschiedenen experimentellen Bedingungen, *Naturwissenschaften* **43**:43.

Bückmann, D., 1956b, Experimentelle Auslösung der Umfärbung bei *Cerura*-Raupen durch Prothorakaldrüsenhormon, *Naturwissenschaften* **43**:44.

Bückmann, D., 1959, Die Auslösung der Umfärbung durch das Häutungshormon bei *Cerura vinula* L. (Lepidoptera, Notodontiadae), *J. Insect Physiol.* **3**:159.

Bückmann, D., 1960, Die Determination der Puppenfärbung bei *Vanessa urticae* L., *Naturwissenschaften* **47**:610.

Bückmann, D., 1971, Melanisierungsverlauf und Melanisierungshemmung bei der Kohlweisslingspuppe *Pieris brassicae* L., *Wilhelm Roux Arch. Entwicklungsmech. Org.* **166**:236.

Bückmann, D., 1974, Die Wirkung von Ecdyson und von Tryptophanüberschuss bei der Ommochrombildung von *Cerura vinula* L. vor der Verpuppung, *Zool. Jahrb. Abt. Allg. Zool, Physiol. Tiere* **78**:257.

Bückmann, D., and Dutsmann, J. H., 1962, Biochemische Untersuchungen über den morphologischen Farbwechsel von *Carausius morosus, Naturwissenschaften* **49**:1.

Burgess, L., 1971, Neurosecretory cells and their axon pathways in *Culiseta inormata* (Williston) (Diptera: Culicidae), *Can. J. Zool.* **49**:889.

Burgess, L., 1973, Axon pathways of the intermediate neurosecretory cells in *Culex tarsalis* Coquillett (Diptera: Culicidae), *Can. J. Zool.* **51**:379.

Burgess, L., and Rempel, J. G., 1966, The stomodeal nervous system, the neurosecretory system, and the gland complex in *Aedes aegypti* (L.) (Diptera: Culicidae), *Can. J. Zool.* **44**:731.

Burov, V. N., Reutskaya, O. E., and Sazonov, A. P., 1972, Activation of diapausing bugs of the species *Eurygaster integriceps* Put. by means of analogues of juvenile hormone, *Dokl. Akad. Nauk SSSR Biol. Sci. Sect.* **204**:295 [English translation].

Busselet, M., 1968, Données histochimiques sur les corps cardiaques d'*Antheraea pernyi* Guer. (Lepidoptera, Attacidae) et de *Rhodnius prolixus* Stal. (Hemiptera, Reduvidae), *C. R. Acad. Sci. Ser. D* **266**:2280.

Busselet, M., 1969, Données histochimiques et ultrastructurales sur les corps allates de *Rhodnius prolixus* Stal et *Antheraea pernyi* Guer., *Bull. Soc. Zool. Fr.* **94**:373.

Butenandt, A., and Karlson, P., 1954, Über die Isolierung eines Metamorphose-Hormons der Insekten in kristallisierter Form, *Z. Zellforsch. Mikrosk. Anat.* **9b**:389.

Butterworth, F. M., and Bodenstein, D., 1969, Adipose tissue of *Drosophila melanogaster*. IV. The effect of the corpus allatum and synthetic juvenile hormone on the tissue of the adult male, *Gen. Comp. Endocrinol.* **13**:68.

Cameron, M. L., 1953, Secretion of an orthodiphenol in the corpus cardiacum of the insect, *Nature (London)* **172**:349.

Candy, D. J., 1974, The control of muscle trehalase activity during locust flight, *Biochem. Soc. Trans.* **2**:1107.

Carlisle, D. B., and Ellis, P. E., 1968, Hormonal inhibition of the prothoracic gland by the brain in locusts, *Nature (London)* **220**:706.

Carlisle, D., Dupont-Raabe, M., and Knowles, F., 1955, Recherches préliminaires relatives à la séparation et à la comparaison des substances chromactives des Crustacés et des Insectes, *C. R. Acad. Sci.* **240**:665.

Carlisle, J. A., and Loughton, B. G., 1979, Adipokinetic hormone inhibits protein synthesis in *Locusta, Nature (London)* **282**:420.

Carlsen, J., Herman, W. S., Christensen, M., and Josefsson, L., 1979, Characterization of a second peptide with adipokinetic and red pigment-concentrating activity from the locust corpora cardiaca, *Insect Biochem.* **9**:497.

Cassier, P., 1966, L'activité des corps allates et la reproduction du criquet migrateur africain *Locusta migratoria migratorioides* R. et F., *Bull. Soc. Zool. Fr.* **91**:133.

Cassier, P., 1970, Influence des conditions d'élevage (isolement et groupement) sur la fécondité des femelles de *Locusta migratoria migratorioides* (R. et F.) et sur les caractéristiques de leur descendance: Données endocrines, *Coll. Int. CNRS* **189**:87.

Cassier, P., and Fain-Maurel, M. A., 1970a, Contribution à l'étude infrastructurale du système neurosécréteur rétrocérébral chez *Locusta migratoria migratorioides* (R. et F.). I. Les corpora cardiaca, *Z. Zellforsch. Mikrosk. Anat.* **111**:471.

Cassier, P., and Fain-Maurel, M. A., 1970b, Contribution à l'étude infrastructurale du système neurosécréteur rétrocérébral chez *Locusta migratoria migratorioides* (R. et F.). II. Le transit des neurosécrétions, *Z. Zellforsch, Mikrosk. Anat.* **111**:483.

Cassier, P., and Fain-Maurel, M. A., 1970c, Contrôle plurifactoriel de l'évolution post-imaginale des glandes ventrales chez *Locusta migratoria* L. Données expérimentales et infrastructurales, *J. Insect Physiol.* **16**:301.

Cassier, P., and Papillon, M., 1968, Effets des implantations de corps allates sur la reproduction des femelles groupées de *Schistocerca gregaria* (Forsk.) et sur le polymorphisme de leur descendance, *C. R. Acad. Sci. Ser. D.* **266**:1048.

Caussanel, C., Breuzet, M., and Karlinsky, A., 1978, Intervention du système neuroendocrine cérébral dans le comportement parental de la femelle de *Labidura riparia* (Insecte, Dermaptère), *C. R. Acad. Sci. Ser. D* **286**:1699.

Cazal, M., 1967, Activité cardiaque de *Locusta migratoria in vitro* et influence des corpora cardiaca. Mise au point expérimentale et premiers résultats, *C. R. Acad. Sci. Ser. D* **264**:842.

Cazal, M., 1969, Action d'extraits de corpora cardiaca sur le péristaltisme intestinal de *Locusta migratoria, Arch. Zool. Exp. Gen.* **110**:83.

Cazal, M., 1971, Action des corpora cardiaca sur la tréhalosémie et la glycémie de *Locusta migratoria* L., *C. R. Acad. Sci. Ser. D* **272**:2596.

Cazal, M., and Girardie, A., 1968, Contrôle humoral de l'équilibre hydrique chez *Locusta, J. Insect Physiol.* **14**:655.

Cazal, M., Joly, L., and Porte, A., 1971, Etude ultrastructurale des corpora cardiaca et de quelques formations annexes chez *Locusta migratoria* L., *Z. Zellforsch. Mikrosk. Anat.* **114**:61.

Cazal, P., 1948, Les glandes endocrines rétro-cérébrales des Insectes. Etude morphologique, *Bull. Biol. Fr. Belg. Suppl.* **32**:1.

Chadha, G. K., and Denlinger, D. L., 1976, Autogeny and endocrine involvement in the reproduction of tropical flesh flies (Diptera: Sarcophagidae), *Entomol. Exp. Appl.* **20**:31.

Chalaye, D., 1965, Recherches histochimiques et histophysiologiques sur la neurosécrétion dans la chaîne nerveuse ventrale du Criquet migrateur, *Locusta migratoria, C. R. Acad. Sci.* **260**:7010.

Chalaye, D., 1966, Recherches sur la destination des produits de neurosécrétion de la chaîne nerveuse ventrale du Criquet migrateur, *Locusta migratoria, C. R. Acad. Sci. Ser. D* **262**:161.

Chalaye, D., 1969, La tréhalosémie et son contrôle neuroendocrine chez le Criquet migrateur, *Locusta migratoria migratorioides.* 2. Rôle des corpora cardiaca et des organes périsympathiques, *C. R. Acad. Sci. Ser. D.* **268**:3111.

Chalaye, D., 1974a, Ultrastructure de la masse ganglionnaire métathoracique de *Locusta migratoria migratorioides* (R. et F.) (Orthoptère). I. Les cellules neurosécrétrices et leurs prolongements dans le neuropile, *Acrida* **3**:19.

Chalaye, D., 1974b, Ultrastructure de la masse ganglionnaire métathoracique de *Locusta migratoria migratorioides* (R. et F.) (Orthoptère). II. Les organes périsympathiques abdominaux et thoraciques, *Acrida* **3**:35.

Chalaye, D., 1975, Etude ultrastructurale des cellules neurosécrétrices du ganglion sous-oesophagien de *Locusta migratoria migratorioides* (R. et F.) (Orthoptère), *Acrida* **4**:19.

Chang, F., 1974a, Effects of vertebrate adipokinetic hormones on the rate of *in vitro* lipid release in insects, *Comp. Biochem. Physiol. B* **49**:567.

Chang, F., 1974b, The effect of oxygen on tanning in ligated tachinid larvae of *Lespesia archippivora* (Diptera: Tachinidae), *Ann. Entomol. Soc. Am.* **67**:427.

Chanussot, B., 1972, Etude histologique et ultrastructurale du ganglion ingluvial de *Blabera craniifer* Burm. (Insecte, Dictyoptère), *Tissue Cell* **4**:85.

Chanussot, B., Dando, J., Moulins, M., and Laverack, M. S., 1969, Mise en évidence d'une amine biogène dans le système nerveux stomatogastrique des Insectes: Etude histochimique et ultrastructurale, *C. R. Acad. Sci. Ser. D* **268**:2101.

Charlet, M., 1969, Etude des cellules neurosécrétrices dans la chaîne nerveuse ventrale d'*Aeschna grandis* L. (Insecta, Odonata), *C. R. Acad. Sci. Ser. D* **269**:1554.

Charlet, M., 1974, Mise en évidence d'un centre neurosécréteur protocérébral intervenant dans l'équilibre hydrique de la larve d'*Aeshna cyanea* Müll. (Insecte, Odonate), *C. R. Acad. Sci. Ser. D* **279**:385.

Charlet, M., and Schaller, F., 1974, Etude des cellules neurosécrétrices cérébrales non associées à la pars intercerebralis chez la larve d'*Aeshna cyanea* (Müller) (Anisoptera: *Aeshnidae*), *Odonatologica (Utrecht)* **3**:93.

Charlet, M., and Schaller, F., 1975a, Effets de la destruction de la pars intercerebralis sur la mue chez la larve d'*Aeshna cyanea* Müll. (Insecte Odonate), *C. R. Acad. Sci. Ser. D* **281**:831.

Charlet, M., and Schaller, F., 1975b, Restauration de la mue chez des larves d'*Aeshna cyanea* Müll. (Insecte Odonate) rendues permanentes par électrocoagulation de la pars intercerebralis, *C. R. Acad. Sci. Ser. D* **281**:1617.

Charlet, M., and Schaller, F., 1976, Blocage de l'exuviation chez la larve d'*Aeshna cyanea* (Insecte Odonate) après électrocoagulation d'un centre neurosécréteur du protocérébron antérieur, *C. R. Acad. Sci. Ser. D* **283**:1539.

Charlet, M., Schaller, F., and Joly, P., 1974, Données sur les phénomènes de neurosérétion chez les Odonates, *Zool. Jahrb. Abt. Allg. Zool. Physiol. Tiere* **78**:279.

Charlet, M., Goltzené, F., and Hoffmann, J. A., 1979, Experimental evidence for a neuroendocrine control of ecdysone biosynthesis in adult females of *Locusta migratoria*, *J. Insect Physiol.* **25**:463.

Chaudhury, M. F. B., and Dhadialla, T. S., 1976, Evidence of hormonal control of ovulation in tsetse flies, *Nature (London)* **260**:243.

Cheeseman, P., and Goldsworthy, G. J., 1979, The release of adipokinetic hormone during flight and starvation in *Locusta*, *Gen. Comp. Endocrinol.* **37**:35.

Cheeseman, P., Jutsum, A. R., and Goldsworthy, G. J., 1976, Quantitative studies on the release of locust adipokinetic hormone, *Physiol. Entomol.* **1**:115.

Cheeseman, P., Goldsworthy, G. J., and Mordue, W., 1977, Studies on the purification of locust adipokinetic hormone, *Life Sci.* **21**:231.

Chen, A. C., and Friedman, S., 1977a, Hormonal regulation of trehalose metabolism in the blowfly, *Phormia regina*: Interaction between hypertrehalosemic and hypotrehalosemic hormones, *J. Insect Physiol.* **23**:1223.

Chen, A. C., and Friedman, S., 1977b, Hormonal regulation of trehalose metabolism in the blowfly *Phormia regina* Meig.: Effects of cardiacectomy and allatectomy at the subcellular level, *Comp. Biochem. Physiol. B* **58**:339.

Chino, H., and Gilbert, L. I., 1964, Diglyceride release from insect fat body: A possible means of lipid transport, *Science* **143**:359.

Chino, H., and Gilbert, L. I., 1965, Lipid release and transport in insects, *Biochim. Biophys. Acta* **98**:94.

Chippendale, G. M., and Kilby, B. A., 1969, Relationship between the proteins of the haemolymph and fat body during development of *Pieris brassicae*, *J. Insect Physiol.* **15**:905.

Chippendale, G. M., and Yin, C. M., 1973, Endocrine activity retained in diapause insect larvae, *Nature (London)* **246**:511.

Chippendale, G. M., and Yin, C. M., 1979, Larval diapause of the European corn borer, *Ostrinia nubilalis:* Further experiments examining its hormonal control, *J. Insect Physiol.* **25**:53.

Claret, J., 1966, Recherche du centre photorécepteur lors de l'induction de la diapause chez *Pieris brassicae* L., *C. R. Acad. Sci. Ser. D* **262**:1464.

Claret, J., Porcheron, P., and Dray, F., 1978, La teneur en ecdysones circulantes au cours du dernier stade larvaire de l'Hyménoptère endoparasite *Pimpla instigator,* et l'entrée en diapause, *C. R. Acad. Sci. Ser. D* **286**:639.

Clarke, K. U., and Anstee, J. H., 1971a, Effects of the removal of the frontal ganglion on the mechanisms of energy production in *Locusta, J. Insect Physiol.* **17**:717.

Clarke, K. U., and Anstee, J. H., 1971b, Effect of the removal of the frontal ganglion on cellular structure in *Locusta, J. Insect Physiol.* **17**:929.

Clarke, K. U., and Baldwin, R. W., 1960, The effect of insect hormones and of 2:4-dinitrophenol on the mitochondrion of *Locusta migratoria* L., *J. Insect Physiol.* **5**:37.

Clarke, K. U., and Gillott, C., 1967a, 1960, Studies on the effects of the removal of the frontal ganglion in *Locusta migratoria* L. I. The effect on protein metabolism, *J. Exp. Biol.* **46**:13.

Clarke, K. U., and Gillott, C., 1967b, Studies on the effects of the removal of the frontal ganglion in *Locusta migratoria* L. II. Ribonucleic acid synthesis, *J. Exp. Biol.* **46**:27.

Clarke, K. U., and Langley, P. A., 1963, Studies on the initiation of growth and moulting in *Locusta migratoria migratorioides* R. & F. IV. The relationship between the stomatogastric nervous system and neurosecretion, *J. Insect Physiol.* **9**:423.

Coles, G. C., 1964, Some effects of decapitation on metabolism in *Rhodnius prolixus* Stål, *Nature (London)* **203**:323.

Coles, G. C., 1965, Studies on the hormonal control of metabolism in *Rhodnius prolixus* Stål. I. The adult female *J. Insect Physiol.* **11**:1325.

Colhoun, E. H., 1963, Synthesis of 5-hydroxytryptamine in the American cockroach, *Experientia* **19**:9.

Collins, C., and Miller, T., 1977, Studies on the action of biogenic amines on cockroach heart, *J. Exp. Biol.* **67**:1.

Conradi-Larsen, E. M., 1970, Influence of neurosecretion and nutrition on O_2-consumption in the bug, *Oncopeltus fasciatus, J. Insect Physiol.* **16**:471.

Cook, B. J., and Eddington, L. C., 1967, The release of triglycerides and free fatty acids from the fat body of the cockroach, *Periplaneta americana, J. Insect Physiol.* **13**:1361.

Cook, B. J., 1981, The action of proctolin and 5-hydroxytryptamine on the oviduct of the horsefly, *Tabanus proximis, Int. J. Invert. Reprod.* **3**:209.

Cook, B. J., and Meola, S., 1978, The oviduct musculature of the horsefly, *Tabanus sulcifrons,* and its response to 5-hydroxytryptamine and proctolin, *Physiol Entomol.* **3**:273.

Cook, B. J., Holman, G. M., and Marks, E. P., 1975, Calcium and cyclic AMP as possible mediators of neurohormone action in the hindgut of the cockroach, *Leucophaea maderae, J. Insect Physiol.* **21**:1807.

Cottrell, C. B., 1962a, The imaginal ecdysis of blowflies. The control of cuticular hardening and darkening, *J. Exp. Biol.* **39**:395.

Cottrell, C. B., 1962b, The imaginal ecdysis of blowflies. Detection of the blood borne darkening factor and determination of some of its properties, *J. Exp. Biol.* **39**:413.

Cottrell, C. B., 1962c, The imaginal ecdysis of blowflies. Observations on the hydrostatic mechanisms involved in digging and expansion, *J. Exp. Biol.* **39**:431.

Cottrell, C. B., 1962d, The imaginal ecdysis of blowflies. Evidence for a change in the mechanical properties of the cuticle at expansion, *J. Exp. Biol.* **39**:449.

Credland, P. F., and Scales, M. D. C., 1976, The neurosecretory cells of the brain and suboesophageal ganglion of *Chironomus riparius, J. Insect Physiol.* **22**:633.

Crossley, A. C., and Waterhouse, D. F., 1969, The ultrastructure of a pheromone-secreting gland in the male scorpion-fly *Harpobittacus australis* (Bittacidae: Mecoptera), *Tissue Cell* **1**:273.

Cymborowski, B., 1970a, Investigations on the neurohormonal factors controlling circadian rhythm of locomotor activity in the house cricket (*Acheta domesticus* L.). I. The role of the brain and suboesophageal ganglion, *Zool. Pol.* **20**:103.

Cymborowski, B., 1970b, Investigations on the neurohormonal factors controlling circadian rhythm of locomotor activity in the house cricket *Acheta domesticus* L. II. Daily histochemical changes in the neurosecretory cells of the pars intercerebralis and suboesophageal ganglion, *Zool. Pol.* **20**:127.

Cymborowski, B., 1970c, The assumed participation of 5-hydroxytryptamine in regulation of the circadian rhythm of locomotor activity in *Acheta domesticus* L., *Comp. Gen. Pharmacol.* **1**:316.

Cymborowski, B., 1973a, Control of the circadian rhythm of locomotor activity in the house cricket, *J. Insect Physiol.* **19**:1423.

Cymborowski, B., 1973b, Les variations diurnes de l'activité des cellules neurosécrétrices du cerveau du grillon (*Acheta domesticus*), *Acrida* **2**:299.

Cymborowski, B., and Dutkowski, A., 1969, Circadian changes in RNA synthesis in the neurosecretory cells of the brain and suboesophageal ganglion of the house cricket, *J. Insect Physiol.* **15**:1187.

Cymborowski, B., and Dutkowski, A., 1970a, Circadian changes in protein synthesis in the neurosecretory cells of the central nervous system of *Acheta domesticus*, *J. Insect Physiol.* **16**:341.

Cymborowski, B., and Dutkowski, A., 1970b, Sex differences in locomotor activity and RNA synthesis in the central nervous system of *Acheta domesticus* L. during the circadian rhythm, *Z. Vgl. Physiol.* **70**:29.

Cymborowski, B., and Flisińska-Bojanowska, A., 1970, The effect of light on the locomotor activity and structure of neurosecretory cells of the brain and suboesophageal ganglion of *Periplaneta americana* L., *Zool. Pol.* **20**:387.

Cymborowski, B., and Stolarz, G., 1979, The role of juvenile hormone during larval–pupal transformation of *Spodoptera littoralis:* Switchover in the sensitivity of the prothoracic gland to juvenile hormone, *J. Insect Physiol.* **25**:939.

Cymborowski, B., Skangiel-Kramska, J., and Dutkowski, A., 1970, Circadian changes of acetylcholinesterase activity in the brain of house-crickets (*Acheta domesticus* L.), *Comp. Biochem. Physiol.* **32**:367.

Dadd, R. M., 1961, Evidence for humoral regulation of digestive secretion in the beetle *Tenebrio molitor*, *J. Exp. Biol.* **38**:259.

Dallmann, S. H., and Herman, W. S., 1978, Hormonal regulation of hemolymph lipid concentration in the monarch butterfly, *Danaus plexippus*, *Gen. Comp. Endocrinol.* **36**:142.

Dallmann, S. H., Herman, W. S., Carlsen, J., and Josefsson, L., 1981, Adipokinetic activity of shrimp and locust peptide hormones in butterflies, *Gen. Comp. Endocrinol.* **43**:256.

Davey, K. G., 1961a, The mode of action of the heart accelerating factor from the corpus cardiacum of insects, *Gen. Comp. Endocrinol.* **1**:24.

Davey, K. G., 1961b, Substances controlling the rate of beating of the heart of *Periplaneta*, *Nature (London)* **192**:284.

Davey, K. G., 1962a, The release by feeding of a pharmacologically active factor from the corpus cardiacum of *Periplaneta americana*, *J. Insect Physiol.* **8**:205.

Davey, K. G., 1962b, The nervous pathway involved in the release by feeding of a pharmacologically active factor from the corpus cardiacum of *Periplaneta*, *J. Insect Physiol.* **8**:579.

Davey, K. G., 1962c, The mode of action of the corpus cardiacum on the hind gut in *Periplaneta americana*, *J. Exp. Biol.* **39**:319.

Davey, K. G., 1963, The release by enforced activity of the cardiac accelerator from the corpus cardiacum of *Periplaneta americana, J. Insect Physiol.* **9**:375.

Davey, K. G., 1965, Copulation and egg-production in *Rhodnius prolixus:* The role of the spermathecae, *J. Exp. Biol.* **42**:373.

Davey, K. G., 1967, Some consequences of copulation in *Rhodnius prolixus, J. Insect Physiol.* **13**:1629.

Davey, K. G., 1978, Hormonal stimulation and inhibition in the ovary of an insect, *Rhodnius prolixus,* in: *Comparative Endocrinology* (P. J. Gaillard and H. H. Boer, eds.), Elsevier/ North-Holland, Amsterdam, p. 13.

Davey, K. G., and Kuster, J. E., 1981, The source of an antigonadotropin in the female of *Rhodnius prolixus* Stål., *Can. J. Zool.* **59**:761.

David, J. C., and Lafon-Cazal, M., 1979, Octopamine distribution in the *Locusta migratoria* nervous system and non-nervous systems, *Comp. Biochem. Physiol. C* **64**:161.

Day, M. F., 1943, The function of the corpus allatum in muscoid Diptera, *Biol. Bull. (Woods Hole, Mass.)* **84**:127.

de Bessé, N., 1965, Recherches histophysiologiques sur la neurosécrétion dans la chaîne nerveuse ventrale d'une blatte, *Leucophaea maderae* (F.), *C. R. Acad. Sci.* **260**:7014.

de Bessé, N., 1966, Recherche des organes neurohémaux associés à la chaîne nerveuse ventrale de deux blattes, *Leucophaea maderae* et *Periplaneta americana, C. R. Acad. Sci. Ser. D* **263**:404.

de Bessé, N., 1975, Effets de la cautérisation de la pars intercerebralis sur la reproduction de *Leucophaea maderae* (F.), *C. R. Acad. Sci. Ser. D* **280**:729.

de Bessé, N., 1978, Mise en évidence des cellules neurosécrétrices protocérébrales impliquées dans la régulation de l'équilibre hydrique et de la sécrétion salivaire chez les adultes de *Leucophaea maderae, C. R. Acad. Sci. Ser. D* **286**:1695.

de Bessé, N., and Cazal, M., 1968, Action des extraits d'organes périsympathiques et de corpora cardiaca sur la diurèse de quelques insectes, *C. R. Acad. Sci. Ser. D* **266**:615.

Delachambre, J., 1971, Le tannage de la cuticule adulte de *Tenebrio molitor* mise en évidence d'une action hormonale induite par la région céphalique, *J. Insect Physiol.* **17**:2481.

Delachambre, J., 1976, Contrôle du métabolisme de la tyrosine par l'hormone de tannage (bursicon) chez *Tenebrio molitor* L. (Insecta, Col.) étude *in vitro* de l'activité tyrosine hydroxylasique, *Coll. Int. CNRS* **251**:493.

Delachambre, J., Provansal, A., and Grillot, J. P., 1972, Mise en évidence de la libération d'un facteur de tannage assimilable à la bursicon par les organes pèrisympathiques chez *Tenebrio molitor* L. (Ins. Col.), *C. R. Acad. Sci. Ser. D* **275**:2703.

Delachambre, J., Delbecque, J. P., Provansal, A., Grillot, J. P., de Reggi, M. L., and Cailla, H. L., 1979a, Total and epidermal cyclic AMP levels related to the variations of ecdysteroids and bursicon during the metamorphosis of the mealworm *Tenebrio molitor* L., *Insect Biochem.* **9**:95.

Delachambre, J., Delbecque, J. P., Provansal, A., de Reggi, M. L., and Cailla, H., 1979b, Induction of epidermal cyclic AMP by bursicon in mealworm, *Tenebrio molitor, Experientia* **35**:701.

Delépine, Y., 1965, Recherches sur la neurosécrétion dans l'ensemble du système nerveux central d'un Lépidoptère, *Galleria mellonella, Bull. Soc. Zool. Fr.* **90**:525.

de Lerma, B., 1956, Corpora cardiaca et neurosécrétion protocérébrale chez le Coléoptère *Hydrous piceus* L., *Ann. Sci. Nat. Zool. Biol. Anim.* **18**:235.

de Loof, A., Van Loon, J., and Vanderroost, C., 1979, Influence of ecdysterone, precocene and compounds with juvenile hormone activity on induction, termination and maintenance of diapause in the parasitoid wasp, *Nasonia vitripennis, Physiol. Entomol.* **4**:319.

de Loof, A., and de Wilde, J., 1970, Hormonal control of synthesis of vitellogenic female protein in the Colorado beetle, *Leptinotarsa decemlineata, J. Insect Physiol.* **16**:1455.

de Loof, A., and Lagasse, A., 1970, Juvenile hormone and ultrastructural properties of the fat body of the adult Colorado beetle. *Leptinotarsa decemlineata* Say, *Z. Zellforsch. Mikrosk. Anat.* **106**:439.

Delphin, F., 1963, Histology and possible functions of neurosecretory cells in the ventral ganglia of *Schistocerca gregaria* Forsk., *Nature (London)* **200**:913.

Delphin, F., 1965, The histology and possible functions of neurosecretory cells in the ventral ganglia of *Schistocerca gregaria* Forskål (Orthoptera: Acrididae), *Trans. R. Entomol. Soc. London* **117**:167.

Delye, G., 1972, Le système endocrine céphalique de *Camponotus vagus* Scopoli (Hyménoptère Formicidae). Une disposition particulière des fibres neurosécrétrices, *C. R. Acad. Sci. Ser. D* **274**:2065.

Denlinger, D. L., Chaudhury, M. F. B., and Dhadialla, T. S., 1978, Cyclic AMP is a likely mediator of ovulation in the tsetse fly, *Experientia* **34**:1296.

Descamps, M., 1978, Rôle des centres endocrines céphaliques dans la régulation de la spermatogenèse chez *Lithobius forficatus* L. (Myriapode Chilopode), *Bull. Soc. Zool. Fr.* **103**:367.

de Roberts, S. K. F., 1966, Circadian activity rhythms in cockroaches. III. The role of endocrine and neural factors, *J. Cell. Physiol.* **67**:473.

de Wilde, J., 1954, Aspects of diapause in adults insects, with special regard to the Colorado beetle, *Leptinotarsa decemlineata* Say, *Arch. Neerl. Zool.* **10**:375.

de Wilde, J., 1966, Endocrine aspects of nutrition and excretion in the Colorado beetle, *J. Endocrinol.* **37**:xi.

de Wilde, J., and de Boer, J. A., 1969, Humoral and nervous pathways in photoperiodic induction of diapause in *Leptinotarsa decemlineata*, *J. Insect Physiol.* **15**:661.

de Wilde, J., and Stegwee, D., 1958, Two major effects of the corpus allatum in the adult Colorado beetle, *Arch. Neerl. Zool.* **13**:277.

de Wilde, J., Staal, G. B., de Kort, C. A. D., de Loof, A., and Baard, A., 1968, Juvenile hormone titer in the haemolymph as a function of photoperiodic treatment in the adult Colorado beetle (*Leptinotarsa decemlineata* Say), *Proc. R. Neth. Acad. Sci. Ser. C* **71**:321.

Divakar, J., and Němec, V., 1973, Hormonally induced changes in haemolymph saccharid concentration in diapausing *Pyrrhocoris apterus* (L.) (Heteroptera), *Acta Entomol. Bohemoslov,* **70**:371.

Doerr-Schott, J., Joly, L., and Dubois, M. P., 1978, Sur l'existence dans la pars intercerebralis d'un Insecte (*Locusta migratoria* R. et F.) de cellules neurosécrétrices fixant un antisérum antisomatostatine, *C. R. Acad. Sci. Ser. D* **286**:93.

Dogra, G. S., 1967a, Neurosecretory system of Heteroptera (Hemiptera) and role of the aorta as a neurohaemal organ, *Nature (London)* **215**:199.

Dogra, G. S., 1967b, Studies on the neurosecretory system of *Ranatra elongata* Fabricius (Hemiptera: Nepidae) with reference to the distal fate of NCCI and II, *J. Morphol.* **121**:223.

Dogra, G. S., 1969, Studies *in situ* on the neuroendocrine system of the giant water bug, *Belostoma indica* (Lep. and Serv.) (Heteroptera: Belostomatidae), *Acta Anat.* **72**:429.

Dogra, G. S., 1975, An autoradiographic study of neurosecretory cell activity of allatectomized females of the grasshopper, *Melanoplus sanguinipes* (Fab.), *Z. Naturforsch. Teil C* **30**:511.

Dogra, G. S., and Ewen, A. B., 1969, The effects of salt stress on the cerebral neurosecretory system of the grasshopper, *Melanoplus sanguinipes* (F.) (Orthoptera), *Experientia* **25**:940.

Dogra, G. S., and Ewen, A. B., 1970, Histology of the neurosecretory system and the retrocerebral endocrine glands of the adult migratory grasshopper, *Melanoplus sanguinipes* (Fab.) (Orthoptera: Acrididae), *J. Morphol.* **130**:451.

Dogra, G. S., and Gillott, C., 1971, Neurosecretory activity and protease synthesis in relation to feeding in *Melanoplus sanguinipes* (Fab.), *J. Exp. Zool.* **177**:41.

Dogra, G. S., and Tandan, B. K., 1965, Ontogenic fate of the neurosecretory cells in the larval brain of *Sarcophaga ruficornis* (Fabricius, 1774) (Diptera: Cyclorrhapha), *Experientia* **21**:216.

Dores, R. M., and Herman, W. S., 1981a, Effect of age, ambient temperature, and exposure to hormone on the posteclosion diuretic response of the monarch butterfly, *Experientia* **37**:529.

Dores, R. M., and Herman, W. S., 1981b, Insect chromatophorotropic factors: The isolation of polypeptides from *Periplaneta americana* and *Apis mellifera* with melanophore-dispersing activity in the crustacean, *Uca pugilator*, *Gen. Comp. Endocrinol.* **43**:76.

Dores, R. M., Dallmann, S. H., and Herman, W. S., 1979, The regulation of post-eclosion and post-feeding diuresis in the monarch butterfly, *Danaus plexippus*, *J. Insect Physiol.* **25**:895.

Dorn, A., 1978, Neurosecretion in the frontal ganglion of the stick insect: Electron microscopic study and extirpation experiments, in: *7th International Symposium on Neurosecretion, Leningrad*, Springer-Verlag, Berlin, Heidelberg, New York, p. 361.

Dortland, J. F., 1978, Synthesis of vitellogenins and diapause proteins by the fat body of *Leptinotarsa*, as a function of photoperiod, *Physiol. Entomol.* **3**:281.

Dortland, J. F., 1979, The hormonal control of vitellogenin synthesis in the fat body of the female Colorado potato beetle, *Gen. Comp. Endocrinol.* **38**:332.

Downer, R. G. H., 1972, Interspecificity of lipid-regulating factors from insect corpus cardiacum, *Can. J. Zool.* **50**:63.

Downer, R. G. H., 1979, Induction of hypertrehalosemia by excitation in *Periplaneta americana*, *J. Insect Physiol.* **25**:59.

Downer, R. G. H., and Parker, G. H., 1979, Glycogen utilization during flight in the American cockroach, *Periplaneta americana* L., *Comp. Biochem. Physiol. A* **64**:29.

Downer, R. G. H., and Steele, J. E., 1969, Hormonal control of lipid concentration in fat body and hemolymph of the American cockroach, *Periplaneta americana*, *Proc. Entomol. Soc. Ont.* **100**:113.

Downer, R. G. H., and Steele, J. E., 1972, Hormonal stimulation of lipid transport in the American cockroach, *Periplaneta americana*, *Gen. Comp. Endocrinol.* **19**:259.

Downer, R. G. H., and Steele, J. E., 1973, Haemolymph lipase activity in the American cockroach, *Periplaneta americana*, *J. Insect Physiol.* **19**:523.

Dumser, J. B., 1980, *In vitro* effects of ecdysterone on the spermatogonial cell cycle in *Locusta*, *Int. J. Invert. Reprod.* **2**:165.

Dumser, J. B., and Davey, K. G., 1975, The *Rhodnius* testis: Hormonal effects on germ cell division, *Can. J. Zool.* **53**:1682.

Dupont-Raabe, M., 1949, Les chromatophores de la larve de Corèthre, *Arch. Zool. Exp. Gen.* **86**:32.

Dupont-Raabe, M., 1951, Etude expérimentale de l'adaptation chromatique chez le phasme, *Carausius morosus* Br., *C. R. Acad. Sci.* **232**:886.

Dupont-Raabe, M., 1952a, Substances chromactives de crustacés et d'insectes activité réciproque, répartition, différences qualitatives, *Arch. Zool. Exp. Gen.* **89**:102.

Dupont-Raabe, M., 1952b, Contribution à l'étude du rôle endocrine du cerveau et notamment de la pars intercerebralis chez les Phasmides, *Arch. Zool. Exp. Gen.* **89**:128.

Dupont-Raabe, M., 1954, Répartition des activités chromatiques dans le ganglion susoesophagien des Phasmides: Mise en évidence d'une région sécrétoire dans la partie deuto et tritocérébrale, *C. R. Acad. Sci.* **238**:950.

Dupont-Raabe, M., 1956a, Quelques données relatives aux phénomènes de neurosécrétion chez les Phasmides, *Ann. Sci. Nat. Zool. Biol. Anim.* **18**:293.

Dupont-Raabe, M., 1956b, Rôle des différents éléments du système nerveux central dans la variation chromatique des Phasmides, *C. R. Acad. Sci.* **243**:1358.

Dupont-Raabe, M., 1956c, Les mécanismes de l'adaptation chromatique chez les insectes, *Année Biol.* **32**:247.

Dupont-Raabe, M., 1957, Les mécanismes de l'adaptation chromatique chez les insectes, *Arch. Zool. Exp. Gen.* **94**:61.

Dupont-Raabe, M., 1958, Quelques aspects des phénomènes de neuro-sécrétion chez les Phasmides, in: *Zweites International Symposium über Neurosecretion,* Springer-Verlag, Berlin, Göttingen, Heidelberg, p. 85.

Dürnberger, H., Pohlhammer, K., and Weinbörmair, G., 1978, The paramedial neurosecretory cells of the subesophageal ganglion of the cricket, *Teleogryllus commodus* (Walk.). I. Effect of ovariectomy on stainability and incorportion of cysteine, *Cell Tissue Res.* **187**:489.

Dustmann, J. H., 1964, Die Redoxpigmente von *Carausius morosus* und ihre Bedeutung für den morphologischen Farbwechsel, *Z. Vgl. Physiol.* **49**:28.

Dutkowski, A. B., and Cymborowski, B., 1973, The rôle of the suboesophageal ganglion in regulation of RNA synthesis in some insect tissues, *J. Insect Physiol.* **19**:1533.

Dutkowski, A. B., Cymborowski, B., and Przełęcka, A., 1971, Circadian changes in the ultrastructure of neurosecretory cells of the pars intercerebralis of the house cricket, *J. Insect Physiol.* **17**:1763.

Dutrieu, J., 1962, Rôle du froid et de l'anaérobiose sur le taux de tréhalose de *Bombyx mori, C. R. Soc. Biol.* **156**:2020.

Dutrieu, J., 1963, Quelques données biochimiques sur la métamorphose de *Calliphora erythrocephala* en fonction de divers facteurs, *C. R. Acad. Sci.* **256**:304.

Dutrieu, J., and Gourdoux, L., 1967, Le contrôle neuroendocrinien de la tréhalosémie de *Carausius morosus, C. R. Acad. Sci. Ser. D* **265**:1067.

Dutrieu, J., and Gourdoux, L., 1974, Rôle des corps cardiaques et des corps allates dans le métabolisme du glycogène et du tréhalose au cours du développement du Coléoptère *Tenebrio molitor, Ann. Endocrinol.* **35**:375.

Duve, H., 1978, The presence of a hypoglycemic and hypotrehalosemic hormone in the neurosecretory system of the blowfly *Calliphora erythrocephala, Gen. Comp. Endocrinol.* **36**:102.

Duve, H., and Thorpe, A., 1979, Immunofluorescent localization of insulin-like material in the median neurosecretory cells of the blowfly, *Calliphora vomitoria* (Diptera), *Cell Tissue Res.* **200**:187.

Duve, H., and Thorpe, A., 1980, Localisation of pancreatic polypeptide (PP)-like immunoreactive material in neurones of the brain of the blowfly, *Calliphora erythrocephala* (Diptera), *Cell Tissue Res.* **210**:101.

Duve, H., and Thorpe, A., 1981, Gastrin/cholecystokinin (CCK)-like immunoreactive neurones in the brain of the blowfly, *Calliphora erythrocephala* (Diptera), *Gen. Comp. Endocrinol.* **43**:381.

Duve, H., Thorpe, A., and Lazarus, N. R., 1979, Isolation of material displaying insulin-like immunological and biological activity from the brain of the blowfly *Calliphora vomitoria, Biochem. J.* **184**:221.

Eckert, M., Gersch, M., and Wagner, M., 1971, Immunologische Untersuchungen des Neuroendokrinen Systems von Insekten. II. Nachweis von Gewebeantigenen des Gehirns und der Corpora cardiaca von *Periplaneta americana* mit fluorescein und peroxydasemarkierten Antikörpern, *Zool. Jahrb. Abt. Allg. Zool. Physiol. Tiere* **76**:29.

Eckert, M., Agricola, H., and Penzlin, H., 1981, Immunocytochemical identification of proctolinlike immunoreactivity in the terminal ganglion and hindgut of the cockroach *Periplaneta americana* (L.), *Cell Tissue Res.* **217**:633.

Eidmann, H., 1956, Uber rhythmische Erscheinungen bei der Stabheuschrecke *Carausius morosus* Br., *Z. Vgl. Physiol.* **38**:370.

Ejezie, G. C., and Davey, K. G., 1974, Changes in the neurosecretory cells, corpus cardiacum and corpus allatum during pregnancy in *Glossina austeni* Newst. (Diptera, Glossinidae), *Bull. Entomol. Res.* **64**:247.

Ejezie, G. C., and Davey, K. G., 1976, Some effects of allactectomy in the female tsetse, *Glossina austeni, J. Insect Physiol.* **22**:1743.

Ejezie, G. C., and Davey, K. G., 1977, Some effects of mating in female tsetse *Glossina austeni* Newst., *J. Exp. Zool.* **200**:303.

El-Ibrashy, M. T., and Boctor, I. Z., 1970, Effect of allatectomy upon lipid metabolism of the female moth of *Spodoptera littoralis* Boid., *Z. Vgl. Physiol.* **68**:111.

Elliott, H. J., 1976, Structural analysis of the corpus allatum of an aphid, *Aphis craccivora, J. Insect Physiol.* **22**:1275.

Elliott, R. H., and Gillott, C., 1977, Changes in the protein concentration and volume of the haemolymph in relation to yolk deposition, ovariectomy, allatectomy, and cautery of the median neurosecretory cells in *Melanoplus sanguinipes, Can. J. Zool.* **55**:97.

Elliott, R. H., and Gillott, C., 1978, The neuro-endocrine control of protein metabolism in the migratory grasshopper *Melanoplus sanguinipes, J. Insect Physiol.* **24**:119.

Elliott, R. H., and Gillott, C., 1979, An electrophoretic study of proteins of the ovary, fat body, and haemolymph in the migratory grasshopper *Melanoplus sanguinipes, J. Insect Physiol.* **25**:405.

Ellis, P. E., and Carlisle, D. B., 1961, The prothoracic gland and colour change in locusts, *Nature (London)* **190**:368.

El-Salhy, M., 1981, Immunohistochemical localization of pancreatic polypeptide (PP) in the brain of the larval instar of the hoverfly, *Eristalis aenus* (Diptera), *Experientia* **37**:1009.

El-Salhy, M., Abou-El-Ela, R., Falkmer, S., Grimelius, L., and Wilander, E., 1980, Immunohistochemical evidence of gastro-entero-pancreatic neurohormonal peptides of vertebrate type in the nervous system of the larva of a dipteran insect, the hoverfly, *Eristalis aeneus, Regulatory Peptides* **1**:187.

Enders, E., 1955, Die hormonal Steuerung rhythmischer Bewegungen von Insekten-Ovidukten, *Verh. Dtsch. Zool. Ges.* 19:113.

Engelmann, F., 1957, Die Steuerung der Ovarfunktion bei der ovoviviparen Schabe *Leucophaea maderae, J. Insect Physiol.* **1**:257.

Engelmann, F., 1959, The control of reproduction in *Diploptera punctata* (Blattaria), *Biol. Bull. (Woods Hole, Mass.)* **116**:406.

Engelmann, F., 1965, The mode of regulation of the corpus allatum in adult insects, *Arch. Anat. Microsc. Morphol. Exp.* **54**:387.

Engelmann, F., 1968, Endocrine control of reproduction in insects, *Annu. Rev. Entomol.* **13**:1.

Engelmann, F., 1969, Food-stimulated synthesis of intestinal proteolytic enzymes in the cockroach *Leucophaea maderae, J. Insect Physiol.* **15**:217.

Engelmann, F., and Barth, R. H., Jr., 1968, Endocrine control of female receptivity in *Leucophaea maderae* (Blattaria), *Ann. Entomol. Soc. Am.* **61**:503.

Engelmann, F., and Lüscher, M., 1957, Die hemmende Wirkung des Gehirns auf die Corpora allata bei *Leucophaea maderae* (Orthoptera), *Verh. Dtsch. Zool. Ges.* **1956**:215.

Engelmann, F., and Müller, H. P., 1966, Fat body respiration as influenced by previously isolated corpora cardiaca, *Naturwissenschaften* **53**:388.

Engelmann, F., and Penney, D., 1966, Studies on the endocrine control of metabolism in *Leucophaea maderae* (Blattaria). 1. The hemolymph proteins during egg maturation, *Gen. Comp. Endocrinol.* **7**:314.

Engelmann, F., and Wilkens, J. L., 1969, Synthesis of digestive enzyme in the fleshfly *Sarcophaga bullata* stimulated by food, *Nature (London)* **222**:798.

Engelmann, F., Hill, L., and Wilkens, J. L., 1971, Juvenile hormone control of female specific protein synthesis in *Leucophaea maderae, Schistocerca vaga,* and *Sarcophaga bullata, J. Insect Physiol.* **17**:2179.

Engels, W., and Bier, K., 1967, Zur Glykogenspeicherung während der Oogenese und ihren vorzeitigen Auslösung durch Blochierung der RNS-Versorgung (Untersuchungen an *Musca domestica* L.), *Wilhelm Roux Arch. Entwicklungsmech. Org.* **158**:64.

Evans, P. D., 1978, Octopamine distribution in the insect nervous system, *J. Neurochem.* **30**:1009.

Ewen, A. B., 1962, Histophysiology of the neurosecretory system and retrocerebral endocrine glands of the alfalfa plant bug, *Adelphocoris lineolatus* (Goeze) (Hemiptera: Miridae), *J. Morphol.* **3**:255.

Ewen, A. B., 1966, A possible endocrine mechanism for inducing diapause in the eggs of *Adelphocoris lineolatus* (Goeze) (Hemiptera: Miridae), *Experientia* **22**:470.

Fain, M. J., and Riddiford, L. M., 1976, Reassessment of the critical periods for prothoracicotropic hormone and juvenile hormone secretion in the larval molt of the tobacco hornworm *Manduca sexta*, *Gen. Comp. Endocrinol.* **30**:131.

Fain-Maurel, M. A., and Cassier, P., 1968, Etude infrastructurale des glandes de mue de *Locusta migratoria migratorioides* (R. et F.), *Arch. Zool. Exp. Gen.* **109**:445.

Fallon, A. M., Hagedorn, H. H., Wyatt, G. R., and Laufer, H., 1974, Activation of vitellogenin synthesis in the mosquito *Aedes aegypti* by ecdysone, *J. Insect Physiol.* **20**:1815.

Farley, R. D., and Evans, S. J., 1972, Neurosecretion in the terminal ganglion of the cockroach, *Periplaneta americana*, *J. Insect Physiol.* **18**:289.

Faruqui, S. A., 1974, Neurosecretory system and retrocerebral endocrine glands of *Sphaerodema rusticum* Fabr. (Heteroptera-Belostomatidae), *Zool. Jahrb. Abt. Allg. Zool. Physiol. Tiere* **92**:416.

Faruqui, S. A., 1977a, Histochemical observations on the neuroendocrine complex of Hemiptera. I. The neurosecretory cells, *Zool. Jahrb. Abt. Allg. Zool. Physiol. Tiere* **98**:491.

Faruqui, S. A., 1977b, Histochemical observations on the neuroendocrine complex of Heteroptera. II. The endocrine glands, *Folia Histochem. Cytochem.* **15**:259.

Fifield, S. M., and Finlayson, L. H., 1978, Peripheral neurons and peripheral neurosecretion in the stick insect, *Carausius morosus*, *Proc. R. Soc. London Ser. B.* **200**:63.

Fingerman, A. M., Lago, A. D., and Lowe, M. E., 1958, Rhythms of locomotor activity and O$_2$-consumption of the grasshopper *Romalea microptera*, *Am. Midl. Nat.* **59**:58.

Finlayson, L. H., 1956, Normal and induced degeneration of abdominal muscles during metamorphosis in the Lepidoptera, *Q. J. Microsc. Sci.* **97**:215.

Finlayson, L. H., and Osborne, M. P., 1968, Peripheral neurosecretory cells in the stick insect (*Carausius morosus*) and the blowfly larva (*Phormia terrae-novae*), *J. Insect Physiol.* **14**:1793.

Finlayson, L. H., and Osborne, M. P., 1970, Electrical activity of neurohaemal tissue in the stick insect, *Carausius morosus*, *J. Insect Physiol.* **16**:791.

Firstenberg, D. E., and Silhacek, D. L., 1973, Juvenile hormone regulation of oxidative metabolism in isolated insect mitochondria, *Experientia* **29**:1420.

Fisk, F. W., 1950, Studies on proteolytic digestion in adult *Aedes aegypti* mosquitoes, *Ann. Entomol. Soc. Am.* **43**:555.

Fletcher, B. S., 1969, The diversity of cell types in the neurosecretory system of the beetle *Blaps mucronata*, *J. Insect Physiol.* **15**:119.

Fogal, W., and Fraenkel, G., 1969, The rôle of bursicon in melanization and endocuticle formation in the adult fleshfly, *Sarcophaga bullata*, *J. Insect Physiol.* **15**:1235.

Formigoni, A., 1956, Neurosécrétion et organes endocrines chez *Apis mellifica* L., *Ann. Sci. Nat. Zool. Biol. Anim.* **11**:283.

Foster, W. A., 1972, Influence of medial neurosecretory cells on reproduction in female *G. austeni*, *Trans. R. Soc. Trop. Med. Hyg.* **66**:322.

Fourche, J., 1969, Le métabolisme respiratoire au cours des métamorphoses. Essai d'interprétation de la courbe en U chez *Drosophila melanogaster* et *Bombyx mori*, *Bull. Biol. Fr. Belg.* **103**:225.

Fraenkel, G., 1935, A hormone causing pupation in the blowfly, *Calliphora erythrocephala*, *Proc. R. Soc. London Ser. B* **118**:1.

Fraenkel, G., and Hsiao, C., 1962, Hormonal and nervous control of tanning in the fly, *Science* **138**:27.

Fraenkel, G., and Hsiao, C., 1963, Tanning in the adult fly: A new function of neurosecretion in the brain, *Science* **141**:1057.

Fraenkel, G., and Hsiao, C., 1965, Bursicon, a hormone which mediates tanning of the cuticle in the adult fly and other insects, *J. Insect Physiol.* **11**:513.

Fraenkel, G., Hsiao, C., and Seligman, M., 1966, Properties of bursicon: An insect hormone that controls cuticular tanning, *Science* **151**:91.

Fraenkel, G., Zdarek, J., and Sivasubramanian, P., 1972, Hormonal factors in the CNS and hemolymph of pupariating fly larvae which accelerate puparium formation and tanning, *Biol. Bull. (Woods Hole, Mass.)* **143**:127.

Fraenkel, G., Blechl, A., Blechl, J., Herman, P., and Seligman, M. I., 1977, 3':5'-cyclic AMP and hormonal control of puparium formation in the fleshfly *Sarcophaga bullata, Proc. Natl. Acad. Sci. USA* **74**:2182.

Fraser, A., 1958, Humoral control of metamorphosis and diapause in the larvae of certain *Calliphoridae* (Diptera: Cyclorrhapha), *Proc. R. Soc. Edinburgh Sect. B* **67**:127.

Fraser, A., 1959, Neurosecretory cells in the abdominal ganglia of larvae of *Lucilia caesar* (Diptera), *Q. J. Microsc. Sci.* **100**:395.

Fraser, A., and Pipa, R., 1977, Corpus allatum regulation during the metamorphosis of *Periplaneta americana:* Axon pathways, *J. Insect Physiol.* **23**:975.

Fréon, G., 1964, Contribution à l'étude de la neurosécrétion dans la chaîne nerveuse ventrale du criquet migrateur, *Locusta migratoria* (L.), *Bull. Soc. Zool. Fr.* **89**:819.

Friedel, T., 1974, Endocrine control of vitellogenesis in the harlequin bug, *Dindymus versicolor, J. Insect Physiol.* **20**:717.

Friedel, T., and Gillott, C., 1976, Male accessory gland substance of *Melanoplus sanguinipes:* An oviposition stimulant under the control of the corpus allatum, *J. Insect Physiol.* **22**:489.

Friedel, T., Feyereisen, R., Mundall, E. C., and Tobe, S. S., 1980, The allatostatic effect of 20-hydroxyecdysone on the adult viviparous cockroach, *Diploptera punctata, J. Insect Physiol.* **26**:665.

Friedel, T., Orchard, I., and Loughton, B. G., 1981, Release of neurosecretory protein from insect neurohaemal tissue following electrical stimulation, *Brain Res.* **208**:451.

Friedman, S., 1967, The control of trehalose synthesis in the blowfly, *Phormia regina* Meig., *J. Insect Physiol.* **13**:397.

Frontali, N., 1968, Histochemical localization of catecholamines in the brain of normal and drug-treated cockroaches, *J. Insect Physiol.* **14**:881.

Fuchs, M. S., Sundland, B. R., and Kang, S. H., 1980, *In vivo* induction of ovarian development in *Aëdes atropalpus* by a head extract from *Aëdes aegypti, Int. J. Invert. Reprod.* **2**:121.

Fujishita, M., and Ishizaki, H., 1981, Circadian clock and prothoracicotropic hormone secretion in relation to the larval–larval ecdysis rhythm of the Saturniid *Samia cynthia ricini, J. Insect Physiol.* **27**:121.

Fukaya, M., and Mitsuhashi, J., 1957, The hormonal control of larval diapause in the rice stem borer, *Chilo suppressalis.* I. Some factors in the head maintaining larval diapause, *Jpn. J. Appl. Entomol. Zool.* **1**:145.

Fukaya, M., and Mitsuhashi, J., 1961, Larval diapause in the rice stem borer with special reference of its hormonal mechanism, *Bull. Natl. Inst. Agric. Sci. Tokyo C* **13**:1.

Fukuda, S., 1951, The production of the diapause eggs by transplanting the suboesophageal ganglion in the silkworm, *Proc. Jpn. Acad.* **27**:672.

Fukuda, S., 1952, Function of the pupal brain and suboesophageal ganglion in the production of non-diapause and diapause eggs in the silkworm, *Annot. Zool. Jpn.* **25**:149.

Fukuda, S., 1953, Alteration of voltinism in the silkworm following transection of pupal oesophageal connectives, *Proc. Jpn. Acad.* **29**:389.

Fukuda, S., 1963, Déterminisme hormonal de la diapause chez le ver à soie, *Bull. Soc. Zool. Fr.* **88**:151.

Fukuda, S., and Takeuchi, S., 1967, Studies on the diapause factor-producing cells in the suboesophageal ganglion of the silkworm, *Bombyx mori* L., *Embryologia* **9**:333.

Füller, H. B., 1960, Morphologische und experimentelle Untersuchungen über die neurosekretorischen Verhältnisse im Zentralnervensystem von Blattiden und Culiciden, *Zool. Jahrb. Abt. Allg. Zool. Physiol. Tiere* **69**:223.

Furtado, A. F., 1971a, Etude des cellules neurosécrétrices, de leur site de décharge et des corpora cardiaca chez une punaise vivipare, *Stilbocoris natalensis* (Hétéroptères, Lygéidés), *C. R. Acad. Sci. Ser. D* **272**:2364.

Furtado, A. F., 1971b, Recherches sur le contrôle endocrine cérébal de la vitellogenèse et de la parturition chez une Punaise vivipare, *Stilbocoris natalensis* (Hétéroptères, Lygéidés), *C. R. Acad. Sci. Ser. D* **272**:2468.

Furtado, A., 1976a, Contrôle endocrine de l'ovogenèse au cours du cinquième stade nymphal de *Panstrongylus megistus* (Hemiptera, Heteroptera: Reduviidae), *C. R. Acad. Sci. Ser. D* **282**:561.

Furtado, A., 1976b, Etude histophysiologique des cellules neurosécrétrices de la pars intercerebralis de larves femelles de *Panstrongylus megistus* (*Heteroptera:* Reduviidae), *C. R. Acad. Sci. Ser. D* **283**:163.

Furtado, A., 1976c, Rôle de la pars intercerebralis au cours de la cinqième intermue de *Panstrongylus megistus* (Heteroptera: Reduviidae), *C. R. Acad. Sci. Ser. D* **283**:527.

Furtado, A., 1977a, Dualité d'action de la pars intercerebralis de *Panstrongylus megistus* (Heteroptera, Reduviidae) dans le contrôle de l'ovogenèse et de la mue, *C. R. Acad. Sci. Ser. D* **284**:659.

Furtado, A., 1977b, Hormones cérébrales, ecdysones et leurs implications dans le contrôle des mitoses goniales et de la méiose chez la femelle de *Panstrongylus megistus* (Hemiptera Reduviidae), *C. R. Acad. Sci. Ser. D* **284**:2377.

Furtado, A. F., 1979, The hormonal control of mitosis and meiosis during oögenesis in a blood-sucking bug *Panstrongylus megistus*, *J. Insect Physiol.* **25**:561.

Fuzeau-Braesch, S., 1968, Action de l'hormone juvénile de synthèse dans la morphogenese et la pigmentogenèse de *Gryllus bimaculatus* (Orthoptères), *C. R. Soc. Biol.* **162**:1086.

Gabe, M., 1955, Signification histochimique de certaines affinités tinctoriales du produit de neurosécrétion hypothalamique, *C. R. Soc. Biol.* **149**:462.

Gäde, G., 1977, Effect of corpus cardiacum extract on cyclic AMP concentration in the fat body of *Periplaneta americana*, *Zool. Jahrb. Abt. Allg. Zool. Physiol. Tiere* **81**:245.

Gäde, G., 1979a, Studies on the influence of synthetic adipokinetic hormone and some analogs on cyclic AMP levels in different arthropod systems, *Gen. Comp. Endocrinol.* **37**:122.

Gäde, G., 1979b, Adipokinetic and hyperglycaemic factor(s) in the corpora cardiaca/corpora allata complex of the stick insect, *Carausius morosus*. I. Initial characteristics, *Physiol. Entomol.* **4**:131.

Gäde, G., 1980, Further characteristics of adipokinetic and hyperglycaemic factor(s) of stick insects, *J. Insect Physiol.* **26**:351.

Gäde, G., 1981, Activation of fat body glycogen phosphorylase in *Locusta migratoria* by corpus cardiacum extract and synthetic adipokinetic hormone, *J. Insect Physiol.* **27**:155.

Gäde, G., and Beenakkers, A. M. Th., 1977, Adipokinetic hormone-induced lipid mobilization and cyclic AMP accumulation in the fat body of *Locusta migratoria* during development, *Gen. Comp. Endocrinol.* **32**:481.

Gäde, G., and Holwerda, D. A., 1976, Involvement of adenosine $3':5'$ cyclic monophosphate in lipid mobilization in *Locusta migratoria*, *Insect Biochem.* **6**:535.

Gaude, H., 1975, Histologische Untersuchungen zur Struktur und Funktion des neurosekretorischen Systems der Hausgrille *Acheta domesticus* L., *Zool. Anz.* **194**:151.

Gaude, H., and Weber, W., 1966, Untersuchungen zur Neurosekretion bei *Acheta domesticus* L., *Experientia* **22**:296.

Gavin, J. A., and Williamson, J. F., 1976, Juvenile hormone-induced vitellogenesis in *Apetrous 4*, a non-vitellogenic mutant in *Drosophila melanogaster*, *J. Insect Physiol.* **22**:1737.

Gee, J. D., 1975, The control of diuresis in the tsetse fly *Glossina austeni:* A preliminary investigation of the diuretic hormone, *J. Exp. Biol.* **63**:391.

Gee, J. D., 1976, Active transport of sodium by the Malpighian tubules of the tsetse fly *Glossina morsitans*, *J. Exp. Biol.* **64**:357.

Gee, J. D., Whitehead, D. L., and Koolman, J., 1977, Steroids stimulate secretion by insect Malpighian tubules, *Nature (London)* **269**:238.

Geldiay, S., 1959, Neurosecretory cells in ganglia of the roach, *Blaberus craniifer*, *Biol. Bull. (Woods Hole, Mass.)*, **117**:267.

Geldiay, S., 1967, Hormonal control of adult reproductive diapause in the Egyptian grasshopper, *Anacridium aegyptium* L., *J. Endocrinol.* **37**:63.

Geldiay, S., 1970, Photoperiodic control of neurosecretory cells in the brain of the Egyptian grasshopper, *Anacridium aegyptium* L., *Gen. Comp. Endocrinol.* **14**:35.

Geldiay, S., and Edwards, J. S., 1973, The protocerebral neurosecretory system and associated cerebral neurohemal area of *Acheta domesticus*, *Z. Zellforsch, Mikrosk, Anat.* **145**:1.

Geldiay, S., and Edwards J. S., 1976, Neurosecretion and water balance in the house cricket *Acheta domesticus* L., *Gen. Comp. Endocrinol.* **28**:163.

Geldiay, S., and Karaçali, S., 1980, The neurosecretory system of the adult *Melanogryllus desertus* Pall. (Orthoptera, Gryllidae). II. Cerebral neurohemal area, *Cell Tissue Res.* **211**:235.

Gersch, M., 1955, Untersuchungen über Auslösung und Steuerung der Darmbewegungen bei der Larve von *Chaoborus (Corethra)*, *Biol. Zentralbl.* **74**:603.

Gersch, M., 1956, Untersuchungen zur Frage der hormonalen Beeinflussung der Melanophoren bei der *Corethra*-Larve, *Z. Vgl. Physiol.* **39**:190.

Gersch, M., 1958, Neurohormonale Beeinflussung der Hertztätigkeit bei der Larve von *Corethra*, *J. Insect Physiol.* **2**:281.

Gersch, M., 1962, The activation hormone of the metamorphosis of insects, *Gen. Comp. Endocrinol. Suppl.* **1**:322.

Gersch, M., 1967, Experimental examinations on the hormonal control of the water balance and excretion of the larva of *Corethra (Chaoborus)*, *Gen. Comp. Endocrinol.* **9**:453.

Gersch, M., 1969, Experimentelle Untersuchungen zur endocrinen Regulation des Wasserhaushaltes der Larve von *Corethra (Chaoborus)*, *Zool. Jahrb. Abt. Allg. Zool. Physiol. Tiere* **75**:1.

Gersch, M., 1972, Experimentelle Untersuchungen zum Freisetzungsmechanismus von Neurohormonen nach elektrischer Reizung der Corpora cardiaca von *Periplaneta americana in vitro*, *J. Insect Physiol.* **18**:2425.

Gersch, M., 1974a, Selektive Freisetzung des hyperglykämischen Faktors aus den Corpora cardiaca von *Periplaneta americana in vivo*, *Experientia* **30**:767.

Gersch, M., 1974b, Experimentelle Untersuchungen zur Ausschüttung von Neurohormonen aus Ganglien des Bauchmarks von *Periplaneta americana* nach elektrischer Reizung *in vitro*, *Zool. Jahrb. Abt. Allg. Zool. Physiol. Tiere* **78**:138.

Gersch, M., 1976, Krontrolle der circadianen Rhythmik des Trehalosegehaltes der Haemolymphe von *Periplaneta americana*, *Zool. Jahrb. Abt. Allg. Zool. Physiol. Tiere* **80**:1.

Gersch, M., and Birkenbeil, H., 1973, Das Membranpotential der Prothoracaldrüsenzellen von *Galleria mellonella* in Beziehung zum Entwicklungsstadium und nach Einwirkung von Neurohormonen, *Zool. Jahrb. Abt. Allg. Zool. Physiol. Tiere* **77**:1.

Gersch, M., and Birkenbeil, H., 1980, C-AMP als möglicher Mediator der neurohormonalen Beeinflussung des Membranpotentials der Prothoracaldrüsenzellen von *Galleria mellonella*, *Zool. Jahrb. Abt. Allg. Zool. Physiol. Tiere* **84**:58.

Gersch, M., and Bräuer, R., 1974, *In-vitro*-Stimulation der Prothoracaldrüsen von Insekten als Testsystem (Prothoracaldrüsentest), *J. Insect Physiol.* **20**:735.

Gersch, M., and Mothes, G., 1956, Neurohormonaler Wirkungsantogonismus beim Farbwechsel von *Dixippus morosus*, *Naturwissenschaften* **23**:542.

Gersch, M., and Richter, K., 1963, Auslösung von Nervenimpulsen durch ein Neurohormon bei *Periplaneta americana*, *Zool. Jahrb. Abt. Allg. Zool. Physiol. Tiere* **70**:301.

Gersch, M., and Stürzebecher, J., 1967, Zur Frage der Identität und des Vorkommens von Neurohormon D in verschiedenen Bereichen des Zentralnervensystems von *Periplaneta americana*, *Z. Naturforsch. Teil C* **22b**:563.

Gersch, M., and Stürzebecher, J., 1968, Weitere Untersuchungen zur Kennzeichnung des Aktivationshormons der Insektenhäutung, *J. Insect Physiol.* **14**:87.

Gersch, M., and Stürzebecher, J., 1970, Experimentelle Stimulierung der Zellulären Aktivität der Prothorakaldrüsen von *Periplaneta americana* durch den Aktivationsfaktor, *J. Insect Physiol.* **16**:1813.

Gersch, M., Unger, H., and Fischer, F., 1957, Die Isolierung eines Neurohormons aus dem Nervensystem von *Periplaneta americana* und einige biologische Test verfahren, *Wiss. Z. Friedrich-Schiller Univ. Jena* **6**:125.

Gersch, M., Fischer, F., Unger, H., and Koch, H., 1960, Die Isolierung neurohormonalen Faktoren aus dem Nervensystem der Küchenschabe *Periplaneta americana*, *Z. Naturforsch. Teil C* **15**:319.

Gersch, M., Fischer, F., Unger, H., and Kabitza, W., 1961, Vorkommen von Serotonin im Nervensystem von *Periplaneta americana* L. (Insecta), *Z. Naturforsch. Teil C* **16**:351.

Gersch, M., Richter, K., Böhm, G. A., and Stürzebecher, J., 1970, Selektive Ausschüttung von Neurohormonen nach elektrischer Reizung der Corpora cardiaca von *Periplaneta americana in vitro*, *J. Insect Physiol.* **16**:1991.

Gersch, M., Bräuer, R., and Birkenbeil, H., 1973, Experimentelle Untersuchungen zur Wirkungsmechanismus der beiden entwicklungsphysiologischaktiven Fraktionen des "Gehirnhormons" der Insekten (Aktivationsfaktor I und II) auf die Prothoracaldrüse, *Experientia* **29**:425.

Gersch, M., Hentschel, E., and Ude, J., 1974, Aminerge Substanzen im lateralen Herznerven und im stomatogastrischen Nervensystem der Schabe *Blaberus craniifer* Burm., *Zool. Jahrb. Abt. Allg. Zool. Physiol. Tiere* **78**:1.

Gersch, M., Eckert, M., Baumann, E., and Birkenbeil, H., 1977, Immunologische Untersuchungen des neuroendokrinen Systems der Insekten. V. Zur Kennzeichung des Aktivationsfaktors II mit immunologischen, biochemischen und physiologischen Methoden, *Zool. Jahrb. Abt. Allg. Zool. Physiol. Tiere* **81**:153.

Ghosh, K. K., and Faruqui, S. A., 1977, The retrocerebral neuroendocrine complex of a reduvid *Oncopeltus sp.*, (Hemiptera), with special reference to the neurohaemal organs, *Zool. Jahrb. Abt. Allg. Zool. Physiol. Tiere* **97**:523.

Gibbs, D., and Riddiford, L. M., 1977, Prothoracicotropic hormone in *Manduca sexta*: Localization by a larval assay, *J. Exp. Biol.* **66**:255.

Giersberg, H., 1928, Über den morphologischen and physiologischen Farbwechsel der Stabheuschrecke *Dixippus* (*Carausius morosus*), *Z. Vgl. Physiol.* **7**:657.

Gi">Gieryng, R., 1976, Untersuchungen über neurosekretorische Zellen im Gehirn von Fliegen (Diptera), *Zool. Anz.* **197**:300.

Gilbert, L. I., 1962, Maintenance of the prothoracic gland by the juvenile hormone in insects, *Nature (London)* **193**:1205.

Gilbert, L. I., 1967, Changes in lipid content during the reproductive cycle of *Leucophaea maderae* and effects of the juvenile hormone on lipid metabolism *in vitro*, *Comp. Biochem. Physiol.* **21**:237.

Gilbert, L. I. (ed.), 1976, *The Juvenile Hormone*, Plenum Press, New York.

Gillett, J. D., 1957, Variation in the time of release of the ovarian development hormone in *Aëdes aegypti, Nature (London)* **180**:656.

Gillott, C., 1964, The role of the frontal ganglion in the control of protein metabolism in *Locusta migratoria, Helgol. Wiss. Meeresunters.* **9**:41.

Gillott, C., and Elliott, R. H., 1976, Reproductive growth in normal, allatectomized, median-neurosecretory-cell-cauterized, and ovariectomized females of *Melanoplus sanguinipes, Can. J. Zool.* **54**:162.

Gillott, C., Dogra, G. S., and Ewen, A. B., 1970, An autoradiographic study of endocrine activity following frontal ganglionectomy in virgin females of *Melanoplus sanguinipes* (Orthoptera: Acrididae), *Can. Entomol.* **102**:1083.

Girardie, A., 1964, Action de la pars intercerebralis sur le développement de *Locusta migratoria* L., *J. Insect Physiol.* **10**:599.

Girardie, A., 1966a, Contribution à l'étude du contrôle du métabolisme de l'eau chez *Gryllus bimaculatus*. Fonction diurétique de la pars intercerebralis, *C. R. Acad. Sci. Ser. D* **262**:1361.

Girardie, A., 1966b, Contrôle de l'activité génitale chez *Locusta migratoria*. Mise en évidence d'un facteur gonadotrope et d'un facteur allatotrope dans la *pars intercerebralis, Bull. Soc. Zool. Fr.* **91**:423.

Girardie, A., 1967, Contrôle neuro-hormonal de la métamorphose et de la pigmentation chez *Locusta migratoria cinerascens* (Orthoptère), *Bull. Biol. Fr. Belg.* **101**:79.

Girardie, A., 1970, Mise en évidence, dans le protocérébron de *Locusta migratoria migratorioïdes* et de *Schistocerca gregaria,* de nouvelles cellules neurosécrétrices contrôlant le métabolisme hydrique, *C. R. Acad. Sci. Ser. D* **271**:504.

Girardie, A., 1974, Recherches sur le rôle physiologique des cellules neurosécrétrices latérales du protocérébron de *Locusta migratoria migratorioides* (Insecte Orthoptère), *Zool. Jahrb. Abt. Allg. Zool. Physiol. Tiere* **178**:310.

Girardie, A., and Cazal, M., 1965, Rôle de la pars intercerebralis et des corpora cardiaca sur la mélanisation chez *Locusta migratoria* (L.), *C. R. Acad. Sci.* **261**:4325.

Girardie, A., and de Reggi, M., 1978, Moulting and ecdysone release in response to electrical stimulation of protocerebral neurosecretory cells in *Locusta migratoria, J. Insect Physiol.* **24**:797.

Girardie, A., and Girardie, J., 1966, Mise en évidence d'une activité neurosécrétrice des cellules C de la pars intercerebralis de *Locusta migratoria* L. par étude comparative histologique et ultrastructurale, *C. R. Acad. Sci. Ser. D* **263**:1119.

Girardie, A., and Girardie, J., 1967, Etude histologique, histochimique et ultrastructurale de la pars intercerebralis chez *Locusta migratoria* L. (Orthoptère), *Z. Zellforsch. Mikrosk. Anat.* **78**:54.

Girardie, A., and Lafon-Cazal, M., 1972, Contrôle endocrine des contractions de l'oviducte isolé de *Locusta migratoria migratorioides* (R. et F.), *C. R. Acad. Sci. Ser. D* **274**:2208.

Girardie, A., Moulins, M., and Girardie, J., 1974, Rupture de la diapause ovarienne d'*Anacridium aegyptium* par stimulation électrique des cellules neurosécrétrices médianes de la pars intercerebralis, *J. Insect Physiol.* **20**:2261.

Girardie, J., 1973, Aspects histologique, histochimique et ultrastructural des péricaryones neurosécréteurs latéraux du protocérébron de *Locusta migratoria migratorioides* (Insecte: Orthoptère), *Z. Zellforsch. Mikrosk. Anat.* **141**:75.

Girardie, J., 1975, Recherche en microscopie photonique et électronique des éléments neurosécréteurs tritocérébraux de *Locusta migratoria* (Insecte Orthoptère), *Arch. Anat. Microsc. Morphol. Exp.* **64**:223.

Girardie, J., and Girardie, A., 1972, Evolution de la radioactivité des cellules neurosécrétrices de la pars intercerebralis chez *Locusta migratoria migratorioides* (Insecte Orthoptère) après injection de cystéine S35, *Z. Zellforsch. Mikrosk. Anat.* **128**:212.

Girardie, J., and Girardie, A., 1977, Libération provoquée *in vitro* du produit de neurosécrétion des cellules protocérébrales médianes chez le criquet migrateur, *J. Physiol. (Paris)* **73**:707.

Girardie, J., and Rémy, C., 1980, Particularités histo-cytologiques des prolongements distaux des 2 cellules A "vasopressine-neurophysine-like" du criquet migrateur, *J. Physiol. (Paris)* **76**:265.

Girardie, J., Girardie, A., and Moulins, M., 1975, Preuves radiochimiques et physiologiques d'une activation par électrostimulation des cellules neurosécrétrices de la pars intercerebralis chez *Locusta migratoria* (Insecte, Orthoptère), *Gen. Comp. Endocrinol.* **25**:416.

Girardie, J., Girardie, A., and Moulins, M., 1976, Etude radiochimique après électrostimulation de la dynamique fonctionnelle des cellules neurosécrétrices protocérébrales médianes de *Locusta migratoria* (Insecte Orthoptère), *Gen. Comp. Endocrinol.* **30**:410.

Goldbard, G. A., Sauer, J. R., and Mills, R. R., 1970, Hormonal control of excretion in the American cockroach. II. Preliminary purification of a diuretic and antidiuretic hormone, *Comp. Gen. Pharmacol.* **1**:82.

Goldsworthy, G. J., 1969, Hyperglycaemic factors from the corpus cardiacum of *Locusta migratoria, J. Insect Physiol.* **15**:2131.

Goldsworthy, G. J., 1970, The action of hyperglycaemic factors from the corpus cardiacum of *Locusta migratoria* on glycogen phosphorylase, *Gen. Comp. Endocrinol.* **14**:78.

Goldsworthy, G. J., 1971, The effects of removal of the cerebral neurosecretory cells on haemolymph and tissue carbohydrate in *Locusta migratoria migratorioides* R. et F., *J. Endocrinol.* **50**:237.

Goldsworthy, G. J., Johnson, R. A., and Mordue, W., 1972a, *In vivo* studies on the release of hormones from the corpora cardiaca of locusts, *J. Comp. Physiol.* **79**:85.

Goldsworthy, G. J., Mordue, W., and Guthkelch, J., 1972b, Studies on insect adipokinetic hormones, *Gen. Comp. Endocrinol.* **18**:545.

Goldsworthy, G. J., Coupland, A. J., and Mordue, W., 1973, The effects of corpora cardiaca on tethered flight in the locust, *J. Comp. Physiol.* **82**:339.

Goldsworthy, G. J., Lee, S. S., and Jutsum, A. R., 1977, Cerebral neurosecretory cells and flight in the locust, *J. Insect Physiol.* **23**:717.

Goltzené, F., and Porte, A., 1978, Endocrine control by neurosecretory cells of the pars intercerebralis and the corpora allata during the earlier phases of vitellogenesis in *Locusta migratoria migratorioides* R and F (Orthoptera), *Gen. Comp. Endocrinol.* **35**:35.

Goltzené, F., Lagueux, M., Charlet, M., and Hoffman, J. A., 1978, The follicle cell epithelium of maturing ovaries of *Locusta migratoria*: A new biosynthetic tissue for ecdysone, *Hoppe Seyler's Z. Physiol. Chem.* **359**:1427.

Goltzené-Bentz, F., Joly, L., Brehelin, M., and Hoffmann, J. A., 1972, Influence des corpora allata sur la protéinémie de larves de stade V et d'imagos femelles de *Locusta migratoria* L. (Insecte Orthoptère), *C. R. Acad. Sci. Ser. D* **274**:1059.

Gooding, R. H., 1974, Digestive processes of haematophagous insects: Control of trypsin secretion in *Glossina morsitans, J. Insect Physiol.* **20**:957.

Goosey, M. W., and Candy, D. J., 1980, The D-octopamine content of the haemolymph of the locust, *Schistocerca americana gregaria* and its elevation during flight, *Insect Biochem.* **10**:393.

Gordon, R., 1968, Observations on the effect of the neuroendocrine system of *Blatta orientalis* L. on the midgut protease activity of the adult female and the level of infestation with the nematode *Hammerschmidtiella diesingi* (Hammerschmidt, 1838), *Gen. Comp. Endocrinol.* **11**:284.

Gosbee, J. L., Milligan, J. V., and Smallman, B. N., 1968, Neural properties of the pro-tocerebral neurosecretory cells of the adult cockroach *Periplaneta americana, J. Insect Physiol.* **14**:1785.

Govardhan, T. L., Shyamasundari, K., and Rao, K. H., 1978, The structure and cytochemistry of the neurosecretory cells in the cerebral and ventral ganglia of *Hydrous triangularis* Say (Insecta: Coleoptera), *Boll. Zool.* **45**:307.

Granger, N. A., 1978, Innervation of the prothoracic glands in *Galleria mellonella* larvae (Lepidoptera: Pyralidae), *Int. J. Insect Morphol. Embryol.* **7**:315.

Granger, N. A., and Sehnal, F., 1974, Regulation of larval corpora allata in *Galleria mellonella, Nature (London)* 251:415.

Grillot, J.-P., 1968, Description d'organes neurohémaux métamériques associés à la chaîne nerveuse ventrale chez deux coléoptères: *Chrysocarabus auronitens* Fabr. (Carabidae) et *Oryctes rhinoceros* L. *(Scarabaeidae), C. R. Acad. Sci. Ser. D* **267**:772.

Grillot, J.-P., 1971a, Les organes périsympathiques latéraux des Coléoptères. Organes existant en l'absence d'organes médians, *C. R. Acad. Sci. Ser. D* **273**:1210.

Grillot, J.-P., 1971b, Les organes périsympathiques latéraux des Coléoptères. Organes coexistant avec des organes médians, *C. R. Acad. Sci. Ser. D* **273**:1318.

Grillot, J.-P., 1976a, Les organes périsympathiques des insectes. Essai sur leur évolution, données structurales et physiologiques chez les Coléoptères et les Diptéres, *Ann. Sci. Nat. Zool. Biol. Anim.* **18**:311.

Grillot, J.-P., 1976b, Types morphologiques et phylogénie des organes périsympathiques des insectes, *Bull. Biol. Fr. Belg.* **110**:143.

Grillot, J.-P., 1977, Les organes périsympathiques des Diptères, *Int. J. Insect Morphol. Embryol.* **6**:303.

Grillot, J.-P., and Raabe, M., 1973, Recherches préliminaires sur les organes périsympathiques des Diptères, *C. R. Acad. Sci. Ser. D* **277**:425.

Grillot, J.-P., Provansal, A., Baudry, N., and Raabe, M., 1971a, Les organes périsympathiques des Insectes Ptérygotes. Les principaux types morphologiques, *C. R. Acad. Sci. Ser. D* **273**:2126.

Grillot, J.-P., Delachambre, J., and Provansal, A., 1971b, Les organes périsympathiques de *Tenebrio molitor* L. (Coléoptère). Description. Rôle possible dans la décharge d'un facteur de tannage, *C. R. Acad. Sci. Ser. D* **273**:2599.

Grillot, J.-P., Delachambre, J., and Provansal, A., 1976, Rôle des organes périsympathiques et dynamique de la sécrétion de la bursicon chez *Tenebrio molitor, J. Insect Physiol.* **22**:763.

Guilvard, E., Raabe, M., and Rioux, J. A., 1976, Autogenèse et neurosécrétion cérébrale chez *Aedes detritus* (Haliday, 1833) (Diptera, Culicidae), *C. R. Acad. Sci. Ser. D* **283**:1217.

Gundel, M., and Penzlin, H., 1978, The neuronal connections of the frontal ganglion of the cockroach *Periplaneta americana, Cell Tissue Res.* **193**:353.

Gundel, M., and Penzlin, H., 1980, Identification of neuronal pathways between the stomatogastric nervous system and the retrocerebral complex of the cockroach *Periplaneta americana* (L.), *Cell Tissue Res.* **208**:283.

Gupta, B. L., and Berridge, M. J., 1966, Fine structural organization of the rectum in the blowfly, *Calliphora erythrocephala* (Meig.), with special reference to connective tissue, tracheae and neurosecretory innervation in the rectal papillae, *J. Morphol.* **120**:23.

Gupta, B. L., and Smith, D. S., 1969, Fine structural organization of the spermatheca in the cockroach, *Periplaneta americana, Tissue Cell* **1**:295.

Gupta, D. P., 1971, Histochemical observations on the neurosecretory cells of *Dysdercus, Acta Histochem.* **41**:79.

Gwadz, R. W., 1969, Regulation of blood meal size in the mosquito, *J. Insect Physiol.* **15**:2039.

Gwadz, R. W., and Spielman, A., 1973, Corpus allatum control of ovarian development in *Aedes aegypti, J. Insect Physiol.* **19**:1441.

Hadorn, E., and Frizzi, G., 1949, Experimentelle Untersuchungen zur Melanophoren-reaktion von *Corethra, Rev. Suisse Zool.* **56**:5.

Hagedorn, H. H., 1974, The control of vitellogenesis in the mosquito, *Aedes aegypti, Am. Zool.* **14**:1207.

Hagedorn, H. H., Shapiro, J. P., and Hanaoka, K., 1979, Ovarian ecdysone secretion is controlled by a brain hormone in an adult mosquito, *Nature (London)* **282:**92.

Hanaoka, K., and Takahashi, S. Y., 1976, Effect of a hyperglycemic factor on haemolymph trehalose and fat body carbohydrates in the American cockroach, *Insect Biochem.* **6:**621.

Hanaoka, K., and Takahashi, S. Y., 1977, Adenylate cyclase system and the hyperglycemic factor in the cockroach, *Periplaneta americana, Insect Biochem.* **7:**95.

Handler, A. M., and Konopka, R. J., 1979, Transplantation of a circadian pacemaker in *Drosophila, Nature (London)* **279:**236.

Handler, A. M., and Postlethwait, J. H., 1977, Endocrine control of vitellogenesis in *Drosophila melanogaster*: Effects of brain and corpus allatum, *J. Exp. Zool.* **202:**389.

Hanström, B., 1937, Inkretorische Organe und Hormonfunktion bei den Wirbellosen, *Ergeb. Biol.* **14:**143.

Hanström, B., 1938, Zwei Probleme betreffs der hormonalen Lokalisation im Insektenkopf, *K. Fysiogr. Sällsk. Handl. Lund Arsb., Lunds Univ. Arsskr. N. F.* **49:**3.

Hanström, B., 1940, Die chromatophoraktivierende Substanz des Insektenkopfes, *K. Fysiogr. Sällsk. Lund* **51:**3.

Hardie, J., 1980, Juvenile hormone mimics the photoperiodic apterization of the alate gynopara of aphid, *Aphis fabae, Nature (London)* **286:**602.

Harker, J. E., 1956, Factors controlling the diurnal rhythm of activity of *Periplaneta americana* L., *J. Exp. Biol.* **33:**224.

Harker, J. E., 1960, Endocrine and nervous factors in insect circadian rhythms, *Cold Spring Harbor Symp. Quant. Biol.* **25:**279.

Hart, D. E., and Steele, J. E., 1973, The glycogenolytic effect of the corpus cardiacum on the cockroach nerve cord, *J. Insect Physiol.* **19:**927.

Hasegawa, K., 1951, Studies on the voltinism in the silkworm, *Bombyx mori* L., with special reference to the organs concerning determination of voltinism, *Proc. Jpn. Acad.* **27:**667.

Hasegawa, K., 1952, Studies on the voltinism in the silkworm, *Bombyx mori* L., with special reference to the organs concerning determination of voltinism, *J. Fac. Agric. Tottori Univ.* **1:**83.

Hasegawa, K., 1957, The diapause hormone of the silkworm, *Bombyx mori, Nature (London)* **179:**1300.

Hasegawa, K., and Yamashita, O., 1965, Studies on the mode of action of the diapause hormone in the silkworm, *Bombyx mori* L. VI. The target organ of the diapause hormone, *J. Exp. Biol.* **43:**271.

Hasegawa, K., Isobe, M., and Goto, T., 1972, Highly purified diapause hormone from the silkworm, *Naturwissenschaften* **59:**364.

Hasegawa, K., Isobe, M., Kubota, I., and Goto, T., 1974, Activity of two species of the diapause hormone in the silkworm, *Bombyx mori, Zool. Jahrb. Abt. Allg. Zool. Physiol. Tiere* **78:**327.

Hashiguchi, T., 1960, On the relations between pupal color and ligature of a black pupa strain in the silkworm, *Bombyx mori, Jpn. J. Genet.* **35:**269.

Hashiguchi, T., 1962, Hormone-like factor controlling the manifestation of black pupa in silkworm, *Bombyx mori, Jpn. J. Genet.* **37:**91.

Hashiguchi, T., 1964, Source of the hormone-like factor controlling the manifestation of the black pupa in the silkworm, *Bombyx mori* L., *Mem. Fac. Agric. Kagoshima Univ.* **5:**33.

Hashiguchi, T., Yoshitake, N., and Takahashi, N., 1965, Hormone determining the black pupal colour in the silkworm, *Bombyx mori, Nature (London)* **206:**215.

Heinzeller, T., 1976, Circadiane Änderungen im endokrinen system der Honigbiene *Apis mellifera*: Effekt von Haft und Ocellenblendung, *J. Insect Physiol.* **22:**315.

Hentschel, E., 1972, Ovulation und aminerges neurosekretorisches System bei *Periplaneta americana* (L.) (Blattoidea, Insecta), *Zool. Jahrb. Abt. Allg. Zool. Physiol. Tiere* **76:**356.

Herlant-Meewis, H., and Paquet, L., 1956, Neurosécrétion et mue chez *Carausius morosus* Brdt, *Ann. Sci. Nat. Zool. Biol. Anim.* **18**:163.

Herman, W. S., and Dallmann, S. H., 1981, Endocrine biology of the painted lady butterfly *Vanessa cardui, J. Insect Physiol.* **27**:163.

Herman, W. S., and Gilbert, L. I., 1966, The neuroendocrine system of *Hyalophora cecropia* (L.) (Lepidoptera: Saturniidae). 1. The anatomy and histology of the ecdysial glands, *Gen. Comp. Endocrinol.* **7**:275.

Herman, W. S., Carlsen, J. B., Christensen, M., and Josefsson, L., 1977, Evidence for an adipokinetic function of the RPCH activity present in the desert locust neuroendocrine system, *Biol. Bull (Woods Hole, Mass.)* **153**:527.

Hertel, W., 1971, Untersuchungen zur neurohormonalen Steuerung des Herzens der Amerikanischen Schabe *Periplaneta americana* (L.), *Zool. Jahrb. Abt. Allg. Zool. Physiol. Tiere* **76**:152.

Hertel, W., 1975, Untersuchungen zur Beeinflussung des Herzschlages von *Periplaneta americana* L. unter den Bedingungen der Kurzzeitkultur von Herzsegment-präparaten *in vitro, Zool. Jahrb. Abt. Allg. Zool. Physiol. Tiere* **79**:170.

Hidaka, T., 1956, Recherches sur le déterminisme hormonal de la coloration pupale chez les Lépidoptères. I. Les effets de la ligature, de l'ablation des ganglions et de l'incision des nerfs chez prépupes et larves âgées de quelques Papilionides, *Annot. Zool. Jpn.* **29**:69.

Hidaka, T., 1961, Recherches sur le mécanisme endocrine de l'adaptation chromatique morphologique chez les nymphes de *Papilio xuthus* L., *J. Fac. Sci. Tokyo Univ.* **9**:223.

Hidaka, T., and Ohtaki, T., 1963, Effet de l'hormone juvénile et du farnésol sur la coloration tégumentaire de la nymphe de *Pieris rapae crucivora* Boisd., *C. R. Soc. Biol.* **157**:928.

Highnam, K. C., 1961a, The histology of the neurosecretory system of the adult female desert locust, *Schistocerca gregaria, Q. J. Microsc. Sci.* **102**:27.

Highnam, K. C., 1961b, Induced changes in the amounts of material in the neurosecretory system of the desert locust, *Nature (London)* **191**:199.

Highnam, K. C., 1962, Neurosecretory control of ovarian development in the desert locust, *Schistocerca gregaria, Mem. Soc. Endocrinol.* **12**:379.

Highnam, K. C., 1976, L'activité neurosécrétrice chez les Acridiens migrateurs, *Acrida* **5**:3.

Highnam, K. C., and Goldsworthy, G. J., 1972, Regenerated corpora cardiaca and hyperglycemic factor in *Locusta migratoria, Gen. Comp. Endocrinol.* **18**:83.

Highnam, K. C., and Lusis, O., 1962, The influence of mature males on the neurosecretory control of ovarian development in the desert locust, *Q. J. Microsc. Sci.* **103**:73.

Highnam, K. C., and Mordue, A. J., 1970, Estimates of neurosecretory activity by an autoradiographic method in adult female *Schistocerca gregaria* (Forsk), *Gen. Comp. Endocrinol.* **15**:38.

Highnam, K. C., and Mordue A. J., 1974, Induced changes in neurosecretory activity of adult female *Schistocerca gregaria* in relation to feeding, *Gen. Comp. Endocrinol.* **22**:519.

Highnam, K. C., and West, M. W., 1971, The neuropilar neurosecretory reservoir of *Locusta migratoria migratorioides* R. and F., *Gen. Comp. Endocrinol.* **16**:574.

Highnam, K. C., Lusis, O., and Hill, L., 1963, The role of the corpora allata during oöcyte growth in the desert locust, *Schistocerca gregaria* Forsk, *J. Insect Physiol.* **9**:587.

Highnam, K. C., Hill, L., and Gingell, D. J., 1965, Neurosecretion and water balance in the male desert locust (*Schistocerca gregaria*), *J. Zool.* **147**:201.

Highnam, K. C., Hill, L., and Mordue, W., 1966, The endocrine system and oöcyte growth in *Schistocerca* in relation to starvation and frontal ganglionectomy, *J. Insect Physiol.* **12**:977.

Hill, L., 1962, Neurosecretory control of haemolymph protein concentration during ovarian development in the desert locust, *J. Insect Physiol.* **8**:609.

Hill, L., 1963, Endocrine control of protein synthesis in female desert locusts (*Schistocerca gregaria*), *J. Endocrinol.* **26**:17.

Hill, L., and Goldsworthy, G. J., 1968, Growth, feeding activity, and the utilization of reserves in larvae of *Locusta*, *J. Insect Physiol.* **14**:1085.

Hill, L., and Izatt, M. E. G., 1974, The relationships between corpora allata and fat body and haemolymph lipids in the adult female desert locust, *J. Insect Physiol.* **20**:2143.

Hinks, C. F., 1967, Relationship between serotonin and the circadian rhythm in some nocturnal moths, *Nature (London)* **214**:386.

Hinks, C. F., 1975, Peripheral neurosecretory cells in some Lepidoptera, *Can. J. Zool.* **53**:1035.

Hintze-Podufal, C., 1970, The innervation of the prothoracic glands of *Cerura vinula* L. (Lepidoptera), *Experientia* **26**:1269.

Hiruma, K., 1980, Possible roles of juvenile hormone in the prepupal stage of *Mamestra brassicae*, *Gen. Comp. Endocrinol.* **41**:392.

Hiruma, K., and Agui, N., 1977, Relationship between histological changes and functions of the neurosecretory cells in the brain of the cabbage armyworm, *Mamestra brassicae* L., *Appl. Entomol. Zool.* **12**:42.

Hiruma, K., Yagi, S., and Agui, N., 1978, Action of juvenile hormone on the cerebral neurosecretory cells of *Mamestra brassicae in vivo* and *in vitro*, *Appl. Entomol. Zool.* **13**:149.

Hiruma, K., Shimada, H., and Yagi, S., 1978, Activation of the prothoracic gland by juvenile hormone and prothoracicotropic hormone in *Mamestra brassicae*, *J. Insect Physiol.* **24**:215.

Hodgson, E. S., and Geldiay, S., 1959, Experimentally induced release of neurosecretory materials from roach corpora cardiaca, *Biol. Bull. (Woods Hole, Mass.)*, **117**:275.

Hodková, M., 1976, Nervous inhibition of corpora allata by photoperiod in *Pyrrhocoris apterus*, *Nature (London)* **263**:521.

Hodková, M., 1977, Function of the neuroendocrine complex in diapausing *Pyrrhocoris apterus* females, *J. Insect Physiol.* **23**:23.

Hodková, M., 1979, Hormonal and nervous inhibition of reproduction by brain in diapausing females of *Pyrrhocoris apterus* L. (Hemiptera), *Zool. Jahrb. Abt. Allg. Zool. Physiol. Tiere* **83**:126.

Holman, G. M., and Cook, B. J., 1970, Pharmacological properties of excitatory neuromuscular transmission in the hindgut of the cockroach, *Leucophaea maderae*, *J. Insect Physiol.* **16**:1891.

Holman, G. M., and Cook, B. J., 1972, Isolation, partial purification and characterization of a peptide which stimulates the hindgut of the cockroach, *Leucophaea maderae* (Fabr.), *Biol. Bull (Woods Hole, Mass.)* **142**:446.

Holman, G. M., and Cook, B. J., 1979, Evidence for proctolin and a second myotropic peptide in the cockroach, *Leucophaea maderae*, determined by bioassay and HPLC analysis, *Insect Biochem.* **9**:149.

Holman, G. M., and Marks, E. P., 1974, Synthesis, transport, and release of a neurohormone by cultured neuroendocrine glands from the cockroach, *Leucophaea maderae*, *J. Insect Physiol.* **20**:479.

Holwerda, D. A., Van Doorn, J., and Beenakkers, A. M. Th., 1977a, Characterization of the adipokinetic and hyperglycaemic substances from the locust corpus cardiacum, *Insect Biochem.* **7**:151.

Holwerda, D. A., Weeda, E., and Van Doorn, J. M., 1977b, Separation of the hyperglycemic and adipokinetic factors from the cockroach corpus cardiacum, *Insect Biochem.* **7**:477.

Houben, N. M. D., and Beenakkers, A. M. Th., 1974, The influence of haemolymph carbohydrate concentration on the release of adipokinetic hormone during locust flight, *J. Endocrinol.* **64**:66.

Hrubešová, H., and Sláma, K., 1967, The effect of hormones on the intestinal proteinase activity of adult *Pyrrhocoris apterus* L. (Hemiptera), *Acta Entomol. Bohemoslov.* **64**:175.

Hsiao, C., and Fraenkel, G., 1966, Neurosecretory cells in the central nervous system of the adult blowfly *Phormia regina* Meigen. (Diptera: Calliphoridae), *J. Morphol.* **119**:21.

Huebner, E., and Davey, K. G., 1973, An antigonadotropin from the ovaries of the insect *Rhodnius prolixus* Stål., *Can. J. Zool.* **51**:113.

Hughes, L., 1977, High molecular weight forms of diuretic hormone from *Rhodnius prolixus, Biochem. Soc. Trans.* **5**:1060.

Hughes, L., 1979, Further investigations of the isolation of diuretic hormone from *Rhodnius prolixus, Insect Biochem.* **9**:247.

Huignard, J., 1964, Recherches histophysiologiques sur le contrôle hormonal de l'ovogenèse chez *Gryllus domesticus* L., *C. R. Acad. Sci.* **259**:1557.

Huybrechts, R., and de Loof, A., 1981, Effect of ecdysterone on vitellogenin concentration in haemolymph of male and female *Sarcophaga bullata, Int. J. Invert. Reprod.* **3**:157.

Ichikawa, M., and Ishizaki, H., 1961, Brain hormone of the silkworm, *Bombyx mori, Nature (London)* **191**:933.

Ichikawa, M., and Ishizaki, H., 1963, Protein nature of the brain hormone of insects, *Nature (London)* **198**:308.

Ichikawa, M., and Nishiitsutsuji-Uwo, J., 1959, Studies on the role of the corpus allatum in the eri-silkworm, *Philosamia cynthia ricini, Biol. Bull. (Woods Hole, Mass.)* **116**:88.

Ichimasa, Y., 1976, Sterol accumulation in developing ovaries of the silkworm *Bombyx mori* in relation to the diapause hormone, *J. Insect Physiol.* **22**:1071.

Ichimasa, Y., and Hasegawa, K., 1973, Studies on the mode of action of the diapause hormone with special reference to lipid metabolism in the silkworm, *Bombyx mori* L. I. Effect of the hormone on lipid in pupal ovaries and matured eggs, *J. Seric. Sci. Jpn.* **42**:379.

Irving, S. N., and Miller, T. A., 1980, Octopamine and proctolin mimic spontaneous membrane depolarisations in *Lucilia* larvae, *Experientia* **36**:566.

Ishay, J., Gitter, S., Galun, R., Doron, M., and Laron, Z., 1976, The presence of insulin in and some effects of exogenous insulin on Hymenoptera tissues and body fluids, *Comp. Biochem. Physiol. A* **54**:203.

Ishizaki, H., 1969, Changes in titer of the brain during development of the silkworm, *Bombyx mori, Dev. Growth Differ.* **11**:1.

Ishizaki, H., and Ichikawa, M., 1967, Purification of the brain hormone of the silkworm *Bombyx mori, Biol. Bull. (Woods Hole, Mass.)* **133**:355.

Ishizaki, H., Suzuki, A., Isogai, A., Nagasawa, H., and Tamura, S., 1977, Enzymatic and chemical inactivation of partially purified prothoracicotropic (brain) hormone of the silkworm, *Bombyx mori, J. Insect Physiol.* **23**:1219.

Isobe, M., Hasegawa, K., and Goto, T., 1973, Isolation of the diapause hormone from the silkworm A, *Bombyx mori, J. Insect Physiol.* **19**:1221.

Isobe, M., Hasegawa, K., and Goto, T., 1975, Further characterization of the silkworm diapause hormone A, *J. Insect Physiol.* **21**:1917.

Ito, T., and Horie, Y., 1957, Glycogen in the ligated silkworm pupa (*Bombyx mori*), *Nature (London)* **179**:1136.

Ivanović, J. P., Janković-Hladni, M. I., Stanić, V., Milanović, M. P., and Nenadović, V., 1978, The midgut of coleopteran larvae, the possible target organ for the action of neurohormones, in: *7th International Symposium on Neurosecretion, Leningrad,* Springer-Verlag, Berlin, Heidelberg, New York, p. 373.

Janda, V., 1936, Sur les changements de coloration de l'épiderme transplanté et des fragments du corps artificiellement réunis de *Dixippus morosus* (Br. et Redt.), *Bull. Int. Acad. Sci. Bohême* **1**:1.

Janda, V., and Sláma, K., 1965, Über den Einfluss von Hormonen auf den Glykogen-Fett und

Stickstoffmetabolismus bei den Imagines von *Pyrrhocoris apterus* L. (Hemiptera), *Zool. Jahrb. Abt. Allg. Zool. Physiol. Tiere* **71**:345.

Janković-Hladni, M. I., Stanić, V., and Ivanović, J. P., 1976, The effect of temperature and crowding on midgut amylolytic activity in *Tenebrio molitor* adults, *J. Insect Physiol.* **22**:851.

Janković-Hladni, M., Ivanović, J., Stanić, V., and Milanović, M., 1978, Possible role of hormones in the control of midgut amylolytic activity during adult development of *Tenebrio molitor*, *J. Insect Physiol.* **24**:61.

Jarial, M. S., and Scudder, G. G. E., 1971, Neurosecretion and water balance in *Cenocorixa bifida* (Hung.) (Hemiptera, Corixidae), *Can. J. Zool.* **49**:1369.

Johansson, A. S., 1958, Relation of nutrition to endocrine–reproductive functions in the milkweed bug *Oncopeltus fasciatus* (Dallas) (Heteroptera: Lygaeidae), *Nytt Mag. Zool. (Oslo)* **7**:1.

Johnson, B., 1963, A histological study of neurosecretion in aphids, *J. Insect Physiol.* **9**:727.

Johnson, B., 1966a, Ultrastructure of probable sites of release of neurosecretory materials in an insect, *Calliphora stygia* Fabr. (Diptera), *Gen. Comp. Endocrinol.* **6**:99.

Johnson, B., 1966b, Fine structure of the lateral cardiac nerves of the cockroach *Periplaneta americana* L., *J. Insect Physiol.* **12**:645.

Johnson, B., and Bowers, B., 1963, Transport of neurohormones from the corpora cardiaca in insects, *Science* **141**:264.

Johnson, R. A., and Hill, L., 1973, Quantitative studies on the activity of the corpora allata in adult male *Locusta* and *Schistocerca*, *J. Insect Physiol.* **19**:2459.

Joly, L., 1954, Résultats d'implantations systématiques de corpora allata à de jeunes larves de *Locusta migratoria* L., *C. R. Soc. Biol.* **148**:579.

Joly, L., and Joly, P., 1975, Interaction entre l'hormone juvénile et l'ecdysone chez *Locusta migratoria* L., *3rd Colloque Physiologie des Insectes*, Strasbourg (May 2–3, 1975).

Joly, L., Joly, P., Porte, A., and Girardie, A., 1969, Analyse ultrastructurale de l'activité des corpora allata de *Locusta migratoria* L. et ses conséquences sur la structure que l'on doit attribuer au mécanisme humoral contrôlant la métamorphose, *Arch. Zool. Exp. Gen.* **110**:617.

Joly, P., 1951, Déterminisme endocrine de la pigmentation chez *Locusta migratoria* L., *C. R. Soc. Biol.* **145**:1362.

Joly, P., 1952, Déterminisme de la pigmentation chez *Acrida turrita* L. (Insecte orthoptéröide), *C. R. Acad. Sci.* **235**:1054.

Joly, P., 1970, Voies physiologiques d'action des facteurs externes sur l'ovaire, *Coll. Int. CNRS* **189**:401.

Joly, P., and Joly, L., 1953, Résultats de greffes de corpora allata chez *Locusta migratoria* L., *Ann. Sci. Nat. Zool. Biol. Anim.* **15**:21.

Joly, P., Joly, L., and Halbwachs, M., 1956, Contrôle humoral du développement chez *Locusta migratoria* L., *Ann. Sci. Nat. Zool. Biol. Anim.* **18**:257.

Jones, G., Wing, K. D., Jones, D., and Hammock, B. D., 1981, The source and action of head factors regulating juvenile hormone esterase in larvae of the cabbage looper, *Trichoplusia ni*, *J. Insect Physiol.* **27**:85.

Jones, P., Stone, J. V., and Mordue, W., 1977, The hyperglycaemic activity of locust adipokinetic hormone, *Physiol. Entomol.* **2**:185.

Juberthie, C., and Caussanel, C., 1980, Release of brain neurosecretory products from the neurohaemal part of the aorta during egg-laying and egg care in *Labidura riparia* (Insecta, Dermaptera), *J. Insect Physiol.* **26**:427.

Judy, K. J., 1972, Diapause termination and metamorphosis in brainless tobacco hornworms (Lepidoptera), *Life Sci.* **11**:605.

Kai, H., and Kawai, T., 1981, Diapause hormone in *Bombyx* eggs and adult ovaries, *J. Insect Physiol.* **27**:623.

Kaiser, H., 1979, Nachweis von Neurosekretspeichern bei der Eintagsfliege *Ephemera danica* Müll. (Ephemeroptera: Ephemeridae), *Eur. J. Cell Biol.* **19**:145.

Kallapur, V. L., and Basalingappa, S., 1977, Influence of corpora allata and brain extract on the lipid release from the fat body of termite queen *Odontotermes assmuthi, Experientia* **33**:99.

Kambysellis, M. P., and Heed, W. B., 1974, Juvenile hormone induces ovarian development on diapausing cave-dwelling *Drosophila* species, *J. Insect Physiol.* **20**:1779.

Kambysellis, M. P., and Williams, C. M., 1971, *In vitro* development of insect tissue. II. The role of ecdysone in the spermatogenesis of silkworms, *Biol. Bull. (Woods Hole, Mass.)* **141**:541.

Karlson, P., and Sekeris, C., 1962a, *N*-acetyldopamin as sclerotizing agent of the insect cuticle, *Nature (London)* **195**:183.

Karlson, P., and Sekeris, C., 1962b, Zum Tyrosinstoffwechsel der Insekten. IX. Kontrolle des Tyrosinstoffwechsels durch Ecdyson, *Biochim. Biophys. Acta* **63**:489.

Karlson, P., and Shaaya, E., 1964, Der Ecdysontiter während der Insektenentwicklung. I. Eine Methode zur Bestimmung des Ecdysongehalts, *J. Insect Physiol.* **10**:797.

Kater, S. B., 1967, Release of a cardioaccelerator substance by stimulation of nerves to the corpora cardiaca in *Periplaneta americana, Am. Zool.* **7**:53.

Kater, S. B., 1968, Cardioaccelerator release in *Periplaneta americana* (L.), *Science* **160**:765.

Keeley, L. L., 1971, Endocrine effects on the biochemical properties of fat body mitochondria from the cockroach, *Blaberus discoidalis, J. Insect Physiol.* **17**:1501.

Keeley, L. L., and Friedman, S., 1967, Corpus cardiacum as a metabolic regulator in *Blaberus discoidalis* Serville (Blattidae). I. Long-term effects of cardiatectomy on whole body and tissue respiration and on trophic metabolism, *Gen. Comp. Endocrinol.* **8**:129.

Keeley, L. L., and Friedman, S., 1969, Effects of long-term cardiactectomy-allatectomy on mitochondrial respiration in the cockroach, *Blaberus discoidalis, J. Insect Physiol.* **15**:509.

Keeley, L. L., and Van Waddill, H., 1971, Insect hormones: Evidence for a neuroendocrine factor affecting respiratory metabolism, *Life Sci.* **10**:737.

Kelly, T. J., and Fuchs, M. S., 1980, *In vivo* induction of ovarian development in decapitated *Aedes atropalpus* by physiological levels of 20-hydroxyecdysone, *J. Exp. Zool.* **213**:25.

Kelly, T. J., and Telfer, W. H., 1977, Antigenic and electrophoretic variants of vitellogenin in *Oncopeltus* blood and their control by juvenile hormone, *Dev. Biol.* **61**:58.

Kelly, T. J., Fuchs, M. S., and Kang, S. H., 1981, Induction of ovarian development in autogenous *Aedes atropalpus* by juvenile hormone and 20-hydroxyecdysone, *Int. J. Invert. Reprod.* **3**:101.

Key, K. H. L., and Day, M. F., 1954a, A temperature-controlled physiological colour response in the grasshopper *Kosciuscola tristis* Sjost. (Orthoptera: Acrididae), *Aust. J. Zool.* **2**:309.

Key, K. H. L., and Day, M. F., 1954b, The physiological mechanism of colour change in the grasshopper *Kosciuscola tristis* Sjost. (Orthoptera: Acrididae), *Aust. J. Zool.* **2**:340.

Khan, T. R., 1976a, Stomodeal nervous system and retrocerebral complex of the cockroaches, *Periplaneta americana* (L.) and *Blatta orientalis* (L.), *Zool. Anz.* **197**:105.

Khan, T. R., 1976b, Neurosecretory cells in the brain and frontal ganglion of the cockroaches, *Periplaneta americana* (L.) and *Blatta orientalis, Zool. Anz.* **197**:117.

Khan, T. R., Singh, S. B., Singh, A. K., and Singh, T. K., 1978a, Neurosecretory control of corpora allata activity in cockroach, *Periplaneta americana* L., *Experientia* **34**:49.

Khan, T. R., Singh, S. B., Singh, R., Singh, R. K., and Singh, T. K., 1978b, Ontogenic fate of cerebral neurosecretory cells in *Calliphora erythrocephala* (Meig.) (Cyclorrhapha: Diptera), *Folia Morphol. (Prague)* **26**:117.

Khattar, N., 1972, Neurosecretory cells in the ventral ganglia of *Schizodactylus monstrosus* (Drury) (Orthoptera), *Bull. Soc. Zool. Fr.* **97**:67.

Kiguchi, K., 1972, Hormonal control of the coloration of larval body and the pigmentation of larval markings in *Bombyx mori* (1) Endocrine organs affecting the coloration of larval body and the pigmentation of markings, *J. Seric. Sci. Jpn.* **41**:407.

Kimura, S., and Kobayashi, M., 1975, Prothoracotropic action of ecdysone analogues in *Bombyx mori, J. Insect Physiol.* **21**:417.

Kind, T. V., 1965, Neurosecretion and voltinism in the moth *Orgyia antiqua* L. (Lepidoptera, Lymantriidae), *Entomol. Rev. Entomol. Obozr.* **44**:326 [English translation].

Kind, T. V., 1969, The role of neurosecretory cells of the suboesophageal ganglion in determination of embryonic diapause in *Orgyia antiqua* L., *Dokl. Akad. Nauk SSSR Biol. Sci. Sect.* **187**:517 [English translation].

Kind, T. V., 1978, Neurosecretory cells of the brain of *Mamestra brassicae* (Lepidoptera, Noctuidae) during metamorphosis and diapause development, in: *7th International Symposium on Neurosecretion, Leningrad,* Springer-Verlag, Berlin, Heidelberg, New York, p. 378.

Kind, T. V., 1977, Dynamics of the brain endocrine activity under the reactivation of the *Barathra brassicae* diapausing pupae and subsequent imaginal development, *Zool. Zh.* **56**:881.

Kind, T. V., 1978, Dynamics of activity of the brain neurosecretory cells in *Pieris brassicae* under the photoperiodic control of development and diapause of pupae, *Zool. Zh.* **57**:1668.

Kind, T. V., and Vaghina, N. P., 1976, The neurosecretory system of the brain in diapausing and developing pupae of *Acronycta rumicis* L. (Lepidoptera, Noctuidae), *Entomol. Rev. Entomol. Obozr.* **55**:15 [English translation].

King, R. C., Aggarwal, S. K., and Bodenstein, D., 1966, The comparative submicroscopic cytology of the corpus allatum–corpus cardiacum complex of wild type and *fes* adult female *Drosophila melanogaster, J. Exp. Zool.* **161**:151.

Kingan, T. G., 1981, Purification of the prothoracicotropic hormone from the tobacco hornworm *Manduca sexta, Life Sci* **28**:2585.

Kirchner, E., 1962, Untersuchungen über neurohormonale Faktoren bei *Melolontha vulgaris, Zool. Jahrb. Abt. Allg. Zool. Physiol. Tiere* **70**:43.

Klemm, N., 1968a, Monoaminhaltige Strukturen im Zentralnervensystem der Trichoptera (Insecta), *Z. Zellforsch. Mikrosk. Anat.* **92**:487.

Klemm, N., 1968b, Monoaminerge Zellelemente im stomatogastrischen Nervensystem der Trichopteren (Insecta), *Z. Naturforsch. Teil C* **23b**:1279.

Klemm, N., 1971, Monoaminhaltige Zellelemente im stomatogastrischen Nervensystem und in den Corpora cardiaca von *Schistocerca gregaria* Forskål (Insecta, Orthoptera), *Z. Naturforsch. Teil C* **26b**:1085.

Klemm, N., 1972, Monoamine-containing fibres in foregut and salivary gland of the desert locust, *Schistocerca gregaria* Forskål (Orthoptera, Acrididae), *Comp. Biochem. Physiol. A* **43**:207.

Klemm, N., 1974, Vergleichend-histochemische Untersuchungen über die Verteilung monoaminhaltiger Strukturen im Oberschlundganglion von Angehörigen verschiedener Insekten-Ordnungen, *Entomol. Germ.* **1**:21.

Klemm, N., and Axelsson, S., 1973, Detection of dopamine, noradrenaline, and 5-hydroxytryptamine in the cerebral ganglion of desert locust, *Schistocerca gregaria* Forsk. (Insecta, Orthoptera), *Brain Res.* **57**:289.

Klemm, N., and Falck, B., 1978, Monoamines in the pars intercerebralis–corpus cardiacum complex of locusts, *Gen. Comp. Endocrinol.* **34**:180.

Klemm, N., and Schneider, L., 1975, Selective uptake of indolamine into nervous fibres in the brain of the desert locust, *Schistocerca gregaria* Forskål (Insecta). A fluorescence and electron microscopic investigation, *Comp. Biochem. Physiol. C* **50**:177.

Klug, H., 1958, Neurosekretion und Aktivitätsperiodik bei Carabiden, *Naturwissenschaften* **45**:141.

Knowles, F., and Bern, H. A., 1966, The function of neurosecretion in endocrine regulation, *Nature (London)* **210**:271.

Knowles, F. G. W., Carlisle, D. B., and Dupont-Raabe, M., 1955, Studies on pigment activating substances in animals. I. The separation by paper electrophoresis of chromactivating substances in arthropods, *J. Mar. Biol. Assoc. U. K.* **34**:611.

Knowles, F. G. W., Carlisle, D. B., and Dupont-Raabe, M., 1956, Inactivation enzymatique d'une substance chromative des Insectes et des Crustacés, *C. R. Acad. Sci.* **242:**825.

Kobayashi, M., 1957, Studies on the neurosecretion in the silkworm, *Bombyx mori* L., *Bull. Seric. Exp. Stn. (Tokyo)* **15:**181.

Kobayashi, M., and Kimura, S., 1967, Action of ecdysone on the conversion of ^{14}C-glucose in dauer pupa of the silkworm, *Bombyx mori, J. Insect Physiol.* **13:**545.

Kobayashi, M., and Kirimura, J., 1958, The "brain" hormone in the silkworm, *Bombyx mori, Nature (London)* **181:**1217.

Kobayashi, M., Fukaya, M., and Mitsuhashi, J., 1960, Imaginal differentiation of "dauer-pupae" in the silkworm, *Bombyx mori, J. Seric. Sci. Jpn.* **24:**337.

Kobayashi, M., Kirimura, J., and Saito, M., 1962, The "brain" hormone in an insect, *Bombyx mori* L., *Mushi* **36:**85.

Kobayashi, M., Kimura, S., and Yamazaki, M., 1967, Action of insect hormones on the fate of ^{14}C-glucose in the diapausing, brainless pupa of *Samia cynthia pryeri* (Lepidoptera: Saturniidae), *Appl. Entomol. Zool.* **2:**79.

Kobayashi, M., Ishitoya, Y., and Yamazaki, M., 1968, Action of proteinic brain hormone to the prothoracic gland in an insect, *Bombyx mori* L. (Lepidoptera: Bombycidae), *Appl. Entomol. Zool.* **3:**150.

Koeppe, J., and Ofengand, J., 1976, Juvenile hormone-induced biosynthesis of vitellogenin in *Leucophaea maderae, Arch. Biochem. Biophys.* **173:**100.

Koller, G., 1948, Rhythmische Bewegung und hormonale Steuerung bei den Malpighischen Gefässen der Insekten, *Biol. Zentralbl.* **67:**201.

Koller, G., 1954, Zur Frage der hormonalen Steuerung bei rhythmischen Eingeweidebewegungen von Insekten, *Verh. Dtsch. Zool. Ges.* **27:**417.

Kono, Y., 1973, Light and electron microscopic studies on the neurosecretory control of diapause incidence in *Pieris rapae crucivora, J. Insect Physiol.* **19:**255.

Kopec, S., 1922, Studies on the necessity of the brain for the inception of insect metamorphosis, *Biol. Bull. (Woods Hole, Mass.)* **42:**322.

Kopenec, A., 1949, Farbwechsel der Larve von *Corethra plumicornis, Z. Vgl. Physiol.* **31:**490.

Köpf, H., 1957, Uber Neurosekretion bei *Drosophila*. I. Zur Topographie und Morphologie neurosekretorischer Zentren bei der Imago von *Drosophila, Biol. Zentralbl.* **76:**28.

Kramer, K. J., 1978, Diabetes . . . A means of controlling insects?, *Agric. Res* **26:**3.

Kramer, K. J., Tager, H. S., Childs, C. N., and Speirs, R. D., 1977, Insulin-like hypoglycemic and immunological activities in honeybee royal jelly, *J. Insect Physiol.* **23:**293.

Kramer, K. J., Tager, H. S., and Childs, C. N., 1980, Insulin-like and glucagon-like peptides in insect hemolymph, *Insect Biochem.* **10:**179.

Kramer, S. J., 1978, Regulation of the activity of JH-specific esterases in the Colorado potato beetle, *Leptinotarsa decemlineata, J. Insect Physiol.* **24:**743.

Krishnakumaran, A., and Schneiderman, H. A., 1965, Prothoracotrophic activity of compounds that mimic juvenile hormone, *J. Insect Physiol.* **11:**1517.

Krishnakumaran, A., Oberlander, H., and Schneiderman, H. A., 1965, Rates of DNA and RNA synthesis in various tissues during a larval moult cycle of *Samia cynthia ricini* (Lepidoptera), *Nature (London)* **205:**1131.

Krogh, I. M., 1973, Light microscopy of living neurosecretory cells of the corpus cardiacum of *Schistocerca gregaria, Acta. Zool. (Stockholm)* **54:**73.

Krolak, J. M., Zimmermann, M. L., and Mills, R. R., 1977, Cardioaccelerating factors from the terminal abdominal ganglion of the American cockroach, *J. Insect Physiol.* **23:**1343.

Kubota, I., Isobe, M., Goto, T., and Hasegawa, K., 1976, Molecular size of the diapause hormone of the silkworm *Bombyx mori, Z. Naturforsch. Teil C* **31c:**132.

Kubota, I., Isobe, M., Imai, K., Goto, T., Yamashita, O., and Hasegawa, K., 1979, Characterization of the silkworm diapause hormone B, *Agric. Biol. Chem.* **43:**1075.

Kuster, J. E., and Davey, K. G., 1981, Fine structure of the abdominal neurosecretory organs of *Rhodnius prolixus* Stål., *Can. J. Zool.* **59**:765.

Lafon-Cazal, M., 1976, Radioautographic detection of monoamines in a neuroendocrine gland (corpora cardiaca) of locusts (Insects), *J. Microsc. Biol. Cell.* **27**:257.

Lafon-Cazal, M., and Arluison, M., 1976, Localization of monoamines in the corpora cardiaca and the hypocerebral ganglion of locusts, *Cell Tissue Res.* **172**:517.

Lafon-Cazal, M., and Roussel, J. P., 1971, Modifications métaboliques après ablation simultanée des corpora allata et des corpora cardiaca chez *Schistocerca gregaria* et *Locusta migratoria*, *C. R. Acad. Sci. Ser. D* **273**:2613.

Lafon-Cazal, M., Calas, A., and Bosc, S., 1973, Capture et rétention de monoamines tritiées dans les corpora cardiaca de *Locusta migratoria* L. Etude *in vitro* par radioautographie à haute résolution, *J. Microsc. (Paris)* **17**:223.

Langley, P. A., 1967a, Effect of ligaturing on puparium formation in the larva of the tsetse fly, *Glossina morsitans* Westwood, *Nature (London)* **214**:389.

Langley, P. A., 1967b, Experimental evidence for a hormonal control of digestion in the tsetse fly, *Glossina morsitans* Westwood: Study of the larva, pupa, and teneral adult fly, *J. Insect Physiol.* **13**:1921.

Lanzrein, B., 1974, Influence of a juvenile hormone analogue on vitellogenin synthesis and oögenesis in larvae of *Nauphoeta cinerea*, *J. Insect Physiol.* **20**:1871.

Lanzrein, B., 1975, Programming, induction, or prevention of the breakdown of the prothoracic gland in the cockroach, *Nauphoeta cinerea*, *J. Insect Physiol.* **21**:367.

Lauverjat, S., 1977, L'évolution post-imaginale du tissu adipeux femelle de *Locusta migratoria* et son contrôle endocrine, *Gen. Comp. Endocrinol.* **33**:13.

Laverdure, A. M., 1970, Etude électrophorétique des variations du protéinogramme au cours du développement postembryonnaire de *Tenebrio molitor* (Coleoptère). Etude comparée du cas de *Galleria mellonella* (Lepidoptère), *Ann. Endocrinol.* **31**:504.

Laverdure, A. M., 1972, L'évolution de l'ovaire chez la femelle adulte de *Tenebrio molitor*. La prévitellogenèse, *J. Insect Physiol.* **18**:1477.

Lazarovici, P., and Pener, M. P., 1978, The relations of the pars intercerebralis, corpora allata, and juvenile hormone to oöcyte development and oviposition in the african migratory locust, *Gen. Comp. Endocrinol.* **35**:375.

Lea, A. O., 1967, The medial neurosecretory cells and egg maturation in mosquitoes, *J. Insect Physiol.* **13**:419.

Lea, A. O., 1969, Egg maturation in mosquitoes not regulated by the corpora allata, *J. Insect Physiol.* **15**:537.

Lea, A. O., 1972, Regulation of egg maturation in the mosquito by the neurosecretory system: The role of the corpus cardiacum, *Gen. Comp. Endocrinol. Supp.* **3**:602.

Lea, A. O., 1975, Evidence that the ovaries of *Musca domestica* do not maintain cyclicity by regulating the corpus allatum, *J. Insect Physiol.* **21**:1747.

Lea, A. O., and Van Handel, E., 1970, Suppression of glycogen synthesis in the mosquito by a hormone from the medial neurosecretory cells, *J. Insect Physiol.* **16**:319.

Lees, A. D., 1978, Endocrine aspects of photoperiodism in aphids, in: *Comparative Endocrinology* (P. J. Gaillard and H. H. Boer, eds.), Elsevier/North-Holland, Amsterdam, p. 165.

L'Hélias, C., 1953, Rôle des corpora allata dans le métabolisme des glucides, de l'azote et des lipides chez le phasme *Dixippus morosus*, *C. R. Acad. Sci.* **236**:2164.

L'Hélias, C., 1955, Recherches préliminaires sur les hormones du cerveau et du complexe post-cérébral des Phasmes *Carausius morosus* et *Cuniculina cuniculina*, *C. R. Acad. Sci.* **240**:1141.

Linzen, B., and Bückmann, D., 1961, Biochemische und histologische Untersuchungen zur Umfärbung der Raupe von *Cerura vinula* L., *Z. Naturforsch. Teil C* **16b**:6.

Liu, T. P., 1973, The effect of allatectomy on the blood trehalose in the female housefly, *Musca domestica* L., *Comp. Biochem. Physiol. A* **46**:109.

Liu, T. P., 1974, The effect of allactectomy on glycogen metabolism in the fat body of the female housefly, *Musca domestica* L., *Comp. Biochem. Physiol. B* **47**:79.

Liu, T. P., and Davey, K. G., 1974, Partial characterization of a proposed antigonadotropin from the ovaries of the insect *Rhodnius prolixus* Stål, *Gen. Comp. Endocrinol.* **24**:405.

Locke, M., and Collins, J. V., 1968, Protein uptake into multivesicular bodies and storage granules in the fat body of an insect, *J. Cell Biol.* **36**:453.

Loher, W., 1961, The chemical acceleration of the maturation process and its hormonal control in the male of the desert locust, *Proc. R. Soc. London Ser. B* **153**:380.

Loher, W., 1966, Die Steuerung sexueller Verhaltensweisen und der Oocytenentwicklung bei *Gomphocerus rufus* L., *Z. Vgl. Physiol.* **53**:277.

Loher, W., 1974, Circadian control of spermatophore formation in the cricket *Teleogryllus commodus* Walker, *J. Insect Physiol.* **20**:1155.

Loher, W., 1979, The influence of Prostaglandin E_2 on oviposition in *Teleogryllus commodus, Entomol. Exp. Appl.* **25**:107.

Loughton, B. G., and Orchard, I., 1981, The nature of the hyperglycaemic factor from the glandular lobe of the corpus cardiacum of *Locusta migratoria, J. Insect Physiol.* **27**:383.

Lüscher, M., 1965, The influence of hormones on tissue respiration in the insect, *Leucophaea maderae, Gen. Comp. Endocrinol.* **5**:72.

Lüscher, M., 1968a, Hormonal control of respiration and protein synthesis in the fat body of the cockroach *Nauphoeta cinerea* during oöcyte growth, *J. Insect Physiol.* **14**:499.

Lüscher, M., 1968b, Oöcyte protection—A function of a corpus cardiacum hormone in the cockroach, *Nauphoeta cinerea, J. Insect Physiol.* **14**:685.

Lüscher, M., and Leuthold, R., 1965, Über die hormonale Beeinflussung des repiratorischen Stoffwechsels bei der Schabe *Leucophaea maderae* (F.), *Rev. Suisse Zool.* **72**:618.

Lüscher, M., Moesh, K., Scheurer, R., and Wyss-Huber, M., 1971, Hormonal control of protein synthesis in the fat body of *Leucophaea maderae, Endocrinol. Exp.* **5**:69.

McCaffery, A. R., 1976, Effects of electrocoagulation of cerebral neurosecretory cells and implantation of corpora allata on oöcyte development in *Locusta migratoria, J. Insect Physiol.* **22**:1081.

McCaffery, A. R., and Highnam, K. C., 1975a, Effects of corpora allata on the activity of the cerebral neurosecretory system of *Locusta migratoria* R. and F., *Gen. Comp. Endocrinol.* **25**:358.

McCaffery, A. R., and Highnam, K. C., 1975b, Effects of corpus allatum hormone and its mimics on the activity on the cerebral neurosecretory system of *Locusta migratoria migratorioides* R. & F., *Gen. Comp. Endocrinol.* **25**:373.

McCaleb, D. C., and Kumaran, A. K., 1980, Control of juvenile hormone esterase activity in *Galleria mellonella* larvae, *J. Insect Physiol.* **26**:171.

McCarthy, R., and Ralph, C. L., 1962, The effects of corpora allata and cardiaca extracts on hemolymph sugars of the cockroach, *Am. Zool.* **2**:429.

McDaniel, C. N., and Berry, S. J., 1967, Activation of the prothoracic glands of *Antheraea polyphemus, Nature (London)* **214**:1032.

Maddrell, S. H. P., 1963, Excretion in the blood-sucking bug, *Rhodnius prolixus* Stål. I. The control of diuresis, *J. Exp. Biol.* **40**:247.

Maddrell, S. H. P., 1964a, Excretion in the blood-sucking bug, *Rhodnius prolixus* Stål. II. The normal course of diuresis and the effect of temperature, *J. Exp. Biol.* **41**:163.

Maddrell, S. H. P., 1964b, Excretion in the blood-sucking bug, *Rhodnius prolixus* Stål. III. The control of the release of diuretic hormone, *J. Exp. Biol.* **41**:459.

Maddrell, S. H. P., 1965, Neurosecretory supply to the epidermis of an insect, *Science* **150**:1033.

Maddrell, S. H. P., 1966, The site of release of the diuretic hormone in *Rhodnius*—A new neurohaemal system in insects, *J. Exp. Biol.* **45**:499.

Maddrell, S. H. P., 1969, Secretion by the Malpighian tubules of *Rhodnius*. The movements of ions and water, *J. Exp. Biol.* **51**:71.

Maddrell, S. H. P., 1972, The functioning of insect Malpighian tubules, in: *Role of Membranes in Secretory Processes* (L. Bolis, R. D. Keynes, and W. W. Wilbrant, eds.), North-Holland, Amsterdam, p. 338.

Maddrell, S. H. P., and Gee, J. D., 1974, Potassium-induced release of the diuretic hormones of *Rhodnius prolixus* and *Glossina austeni:* Ca dependence, time course and localization of neurohaemal areas, *J. Exp. Biol.* **61**:155.

Maddrell, S. H. P., and Klunsuwan, S., 1973, Fluid secretion by *in vitro* preparations of the Malpighian tubules of the desert locust *Schistocerca gregaria, J. Insect Physiol.* **19**:1369.

Maddrell, S. H. P., and Phillips, J. E., 1975, Secretion of hypo-osmotic fluid by the lower Malpighian tubules of *Rhodnius prolixus, J. Exp. Biol.* **62**:671.

Maddrell, S. H. P., and Phillips, J. E., 1976, Regulation of absorption in insect excretory systems, in: *Perspectives in Experimental Biology,* Volume 1 (P. Spencer Davies, ed.), Pergamon Press, New York, p. 179.

Maddrell, S. H. P., and Phillips, J. E., 1978, Induction of sulphate transport and hormonal control of fluid secretion by Malpighian tubules of larvae of the mosquito *Aedes teaniorhynchus, J. Exp. Biol.* **72**:181.

Maddrell, S. H. P., Pilcher, D. E. M., and Gardiner, B. O. C., 1969, Stimulatory effect of 5-hydroxytryptamine (serotonin) on secretion by Malpighian tubules of insects, *Nature (London)* **222**:784.

Maddrell, S. H. P., Pilcher, D. E. M., and Gardiner, B. O. C., 1971, Pharmacology of the Malpighian tubules of *Rhodnius* and *Carausius:* The structure–activity relationship of tryptamine analogues and the role of cyclic AMP, *J. Exp. Biol.* **54**:779.

Mahon, D. C., and Nair, K. K., 1975, A comparison of aldehyde fuchsin and Alcian blue staining of neurosecretory material in *Oncopeltus fasciatus, Cell Tissue Res.* **161**:477.

Malá, J., Granger, N. A., and Sehnal, F., 1977, Control of prothoracic gland activity in larvae of *Galleria mellonella, J. Insect Physiol.* **23**:309.

Mancini, G., and Frontali, N., 1970, On the ultrastructural localization of catecholamines in the beta lobes (corpora pedunculata) of *Periplaneta americana, Z. Zellforsch. Mikrosk. Anat.* **103**:341.

Marks, E. P., and Reinecke, J. P., 1965, Regenerating tissues from the cockroach *Leucophaea maderae*: Effects of endocrine glands *in vitro, Gen. Comp. Endocrinol.* **5**:241.

Marks, E. P., Reinecke, J. P., and Leopold, R. A., 1968, Regenerating tissues from the cockroach, *Leucophaea maderae,* nerve regeneration *in vitro, Biol. Bull (Woods Hole, Mass.)* **135**:520.

Marks, E. P., Ittycheriah, P. I., and Leloup, A. M., 1972, The effect of β-ecdysone on insect neurosecretion *in vitro, J. Insect Physiol.* **18**:847.

Maslennikova, V. A., 1970, Hormonal regulation of diapause in *Pieris brassicae* L., *Dokl. Akad. Nauk SSSR Biol. Sci. Sect.* **192**:412 [English translation].

Maslennikova, V. A., 1973, Hormonal control of insect diapause, in: *Hormonal Control of Growth and Differentiation,* Academic Press, New York.

Masler, E. P., Fuchs, M. S., Sage, B., and O'Connor, J. D., 1980, Endocrine regulation of ovarian development in the autogenous mosquito, *Aedes atropalpus, Gen. Comp. Endocrinol.* **41**:250.

Masler, E. P., Fuchs, M. S., Sage, B., and O'Connor, J. D., 1981, A positive correlation between oocyte production and ecdysteroid levels in adult *Aedes, Physiol. Entomol.* **6**:45.

Mason, C. A., 1973, New features of the brain–retrocerebral neuroendocrine complex of the locust *Schistocerca vaga* (Scudder), *Z. Zellforsch. Mikrosk. Anat.* **114**:19.

Matsumoto, S., Suzuki, A., Yamamoto, H., Ogura, N., and Ikemoto, H., 1979, Cyclic nucleotides and hormonal control of cuticular melanization in the armyworm larva, *Leucania separata* (Lepidoptera: Noctuidae), *Appl. Entomol. Zool.* **14**:159.

Matthews, J. R., and Downer, R. G. H., 1974, Origin of trehalose in stress-induced hyperglycaemia in the American cockroach, *Periplaneta americana, Can. J. Zool.* **52**:1005.

Mayer, R. J., and Candy, D. J., 1967, Changes in haemolymph lipoproteins during locust flight, *Nature (London)* **215**:987.

Mayer, R. J., and Candy, D. J., 1969, Control of haemolymph lipid concentration during locust flight: An adipokinetic hormone from the corpora cardiaca, *J. Insect Physiol.* **15**:611.

Meneses, P., and Ortiz, M., 1975, A protein extract from *Drosophila melanogaster* with insulin-like activity, *Comp. Biochem. Physiol. A* **51**:483.

Menon, M., 1965, Endocrine influences on yolk deposition in insects, *J. Anim. Morphol. Physiol.* **12**:76.

Meola, R. W., and Adkisson, P. L., 1977, Release of prothoracicotropic hormone and potentiation of developmental ability during diapause in the bollworm, *Heliothis zea, J. Insect Physiol.* **23**:683.

Meola, R., and Lea, A. O., 1972, Humoral inhibition of egg development in mosquitoes, *J. Med Entomol.* **9**:99.

Mesnier, M., 1972a, Etude du contrôle neuroendocrine de l'oviposition chez *Clitumnus extradentatus* (Phasmides-Chéleutoptères), *C. R. Acad. Sci. Ser. D* **274**:564.

Mesnier, M., 1972b, Recherches sur le déterminisme de la ponte chez *Galleria mellonella* (Lépidoptères), *C. R. Acad. Sci. Ser. D* **274**:708.

Mesnier, M., and Provansal, A., 1975, Rôle des organes périsympathiques dans l'induction de la ponte chez *Galleria mellonella* (Lépidoptère), *C. R. Acad. Sci. Ser. D* **281**:905.

Michel, R., 1972, Influence des corpora cardiaca sur les possibilités de vol soutenu du criquet pélerin *Schistocerca gregaria, J. Insect Physiol.* **18**:1811.

Michel, R., 1973, Variations de la tendance au vol soutenu de criquet pélerin *Schistocerca gregaria* après implantations de corpora cardiaca, *J. Insect Physiol.* **19**:1317.

Michel, R., and Bernard, A., 1973, Influence de la pars intercerebralis sur l'induction au vol soutenu chez le criquet pélerin *Schistocerca gregaria, Acrida* **2**:139.

Michel, R., and Lafon-Cazal, M., 1978, Ergastoplasmic granules, cytophysiological adaptation of the locusts corpora cardiaca to migratory flights?, *Experientia* **34**:812.

Milburn, N. S., and Roeder, K. D., 1962, Control of efferent activity in the cockroach terminal abdominal ganglion by extracts of corpora cardiaca, *Gen. Comp. Endocrinol.* **2**:70.

Milburn, N., Weiant, E. A., and Roeder, K. D., 1960, The release of efferent nerve activity in the roach *Periplaneta americana,* by extracts of the corpus cardiacum, *Biol. Bull. (Woods Hole, Mass.)* **118**:111.

Miller, T., 1968, Role of cardiac neurons in the cockroach heartbeat, *J. Insect Physiol.* **14**:1265.

Miller, T., 1973, Regulation of the heartbeat of the American cockroach, in: *Neurobiology of Invertebrates* (J. Salanki, ed.), Akadémiai Kiadó, Budapest, p. 195.

Miller, T., and Adams, M. E., 1974, Ultrastructure and electrical properties of the hyperneural muscle of *Periplaneta americana, J. Insect Physiol.* **20**:1925.

Miller, T., and Thomson, W. W., 1968, Ultrastructure of cockroach cardiac innervation, *J. Insect Physiol.* **14**:1099.

Miller, T., and Usherwood, P. N. R., 1971, Studies of cardioregulation in the cockroach, *Periplaneta americana, J. Exp. Biol.* **54**:329.

Miller, T. A., and Rózsa, K. S., 1981, Control of the alary muscles of locust dorsal diaphragm, *Physiol Entomol.* **6**:51.

Mills, R. R., 1965, Hormonal control of tanning in the American cockroach. II. Assay for the hormone and the effect of wound healing, *J. Insect Physiol.* **11**:1269.

Mills, R. R., 1966, Hormonal control of tanning in the American cockroach. III. Hormone stability and post-ecdysial changes in hormone titre, *J. Insect Physiol.* **12**:275.

Mills, R. R., 1967, Hormonal control of excretion in the American cockroach. I. Release of a diuretic hormone from the terminal abdominal ganglion, *J. Exp. Biol.* **46**:35.

Mills, R. R., and Lake, C. R., 1966, Hormonal control of tanning in the American cockroach. IV. Preliminary purification of the hormone, *J. Insect Physiol.* **12**:1395.

Mills, R. R., and Nielsen, D. J., 1967, Hormonal control of tanning in the American cockroach. V. Some properties of the purified hormone, *J. Insect Physiol.* **13**:273.

Mills, R. R., and Whitehead, D. L., 1970, Hormonal control of tanning in the American cockroach: Changes in blood cell permeability during ecdysis, *J. Insect Physiol.* **16**:331.

Mills, R. R., Lake, C. R., and Alworth, W. L., 1967, Biosynthesis of *N*-acetyl dopamine by the American cockroach, *J. Insect Physiol.* **13**:1539.

Mills, R. R., Androuny, S., and Fox, F. R., 1968, Correlation of phenoloxydase activity with ecdysis and tanning hormone release in the American cockroach, *J. Insect Physiol.* **14**:603.

Minks, A. K., 1965, Hemolymph protein and amino-acid composition as influenced by the corpus allatum in *Locusta migratoria migratorioides* (R. and F.) *Proc. K. Ned. Akad. Wet. Ser. C* **68**:320.

Minks, A. K., 1967, Biochemical aspects of juvenile hormone action in the adult *Locusta migratoria, Arch. Neerl. Zool.* **17**:175.

Mitsuhashi, J., 1963, Histological studies on the neurosecretory cells of the brain and on the corpus allatum during diapause in some lepidopterous insects, *Bull. Natl. Inst. Agric. Sci. Ser. C* **16**:67.

Mitsuhashi, J., and Fukaya, M., 1960, The hormonal control of larval diapause in the rice stem borer, *Chilo suppressalis:* Histological studies on the neurosecretory cells of the brain and the secretory cells of the corpora allata during diapause and host diapause, *Jpn. J. Appl. Entomol. Zool.* **4**:127.

Mitsui, T., Riddiford, L. M., and Bellamy, G., 1979, Metabolism of juvenile hormone by the epidermis of the tobacco hornworm, *Manduca sexta, Insect Biochem.* **9**:637.

Mjeni, A. M., and Morrison, P. E., 1976, Juvenile hormone analogue and egg development in the blowfly, *Phormia regina* (Meig.), *Gen. Comp. Endocrinol.* **28**:17.

Mokia, G. G., 1941, Contribution to the study of hormones in insects, *Dokl. Akad. Nauk SSSR* **30**:368.

Mordue, A. J., and Highnam, K. C., 1973, Incorporation of cysteine into the cerebral neurosecretory system of adult desert locusts, *Gen. Comp. Endocrinol.* **20**:351.

Mordue, W., 1965a, Studies on oöcyte production and associated histological changes in the neuro-endocrine system in *Tenebrio molitor* L., *J. Insect Physiol.* **11**:493.

Mordue, W., 1965b, The neuro-endocrine control of oöcyte development in *Tenebrio molitor* L., *J. Insect Physiol.* **11**:505.

Mordue, W., 1965c, Neuro-endocrine factors in the control of oöcyte production in *Tenebrio molitor* L., *J. Insect Physiol.* **11**:617.

Mordue, W., 1967, The influence of feeding upon the activity of the neuroendocrine system during oöcyte growth in *Tenebrio molitor, Gen. Comp. Endocrinol.* **9**:406.

Mordue, W., 1969, Hormonal control of Malpighian tube and rectal function in the desert locust, *Schistocerca gregaria, J. Insect Physiol.* **15**:273.

Mordue, W., 1970, Evidence for the existence of diuretic and anti-diuretic hormones in locusts, *J. Endocrinol.* **46**:119.

Mordue, W., 1972, Hormones and excretion in locusts, *Gen. Comp. Endocrinol. Suppl.* **3**:289.

Mordue, W., and Goldsworthy, G. J., 1969, The physiological effects of corpus cardiacum extracts in locusts, *Gen. Comp. Endocrinol.* **12**:360.

Mordue, W., and Stone, J. V., 1976, Comparison of the biological activities of an insect and a crustacean neurohormone that are structurally similar, *Nature (London)* **264**:287.

Mordue, W., and Stone, J. V., 1977, Relative potencies of locust adipokinetic hormone and prawn red pigment-concentrating hormone in insect and crustacean systems, *Gen. Comp. Endocrinol.* **33**:103.

Mordue, W., Highnam, K. C., Hill, L., and Luntz, A. J., 1970, Environmental effects upon endocrine-mediated processes in locusts, *Mem. Soc. Endocrinol.* **18**:111.

Moreau, R., Raoelison, C., and Sutter, B. Ch. J., 1981, An intestinal insulin-like molecule in *Apis mellifica* L. (Hymenoptera), *Comp. Biochem. Physiol. A* **69**:79.

Moreteau-Levita, B., 1972a, Rôle des lobes optiques dans la réalisation de l'homochromie d'*Oedipoda coerulescens* L. (Acridien, Orthoptère), *C. R. Acad. Sci. Ser. D* **274**:2690.

Moreteau-Levita, B., 1972b, Rôle de la pars intercerebralis dans la réalisation de l'homochromie d'*Oedipoda coerulescens* L. (Acridien, Orthoptère), *C. R. Acad. Sci. Ser. D* **274**:3277.

Moreteau-Levita, B., 1972c, Rôle des corpora cardiaca dans la pigmentation d'*Oedipoda coerulescens* L. (Acridien, Orthoptère), *C. R. Acad. Sci. Ser. D* **275**:2699.

Morgan, P. J., and Mordue, W., 1981, Stimulated fluid secretion is sodium dependent in the Malpighian tubules of *Locusta migratoria, J. Insect Physiol.* **27**:271.

Morohoshi, S., and Fugo, H., 1977, The control of growth and development in *Bombyx mori*. XXXVII. Some aspects on a hormone controlling adult eclosion, *Proc. Jpn. Acad.* **53**:75.

Morohoshi, S., and Ohkuma, T., 1968, Change in the heartbeat in the larva of *Bombyx mori* by injection of extracts of the corpus allatum and suboesophageal ganglion and adrenalin and insulin, *J. Seric. Sci. Jpn.* **37**:375.

Morohoshi, S., and Oshiki, T., 1969, Effect of the brain on the suboesophageal ganglion and determination of voltinism in *Bombyx mori, J. Insect Physiol.* **15**:167.

Morris, G. P., and Steel, C. G. H., 1975, Ultrastructure of neurosecretory cells in the pars intercerebralis of *Rhodnius prolixus* (Hemiptera), *Tissue Cell* **7**:73.

Mothes, G., 1960, Weitere Untersuchungen über den physiologischen Farbwechsel von *Carausius morosus* (Br.), *Zool. Jahrb. Abt. Allg. Zool. Physiol. Tiere* **69**:133.

Mouton, J., 1968, Effet de la castration sur les cellules neurosécrétrices ventrales du ganglion infra-oesophagien chez le Phasme *Carausius morosus* (Phasmides-Orthoptères), *C. R. Acad. Sci. Ser. D* **266**:2120.

Mouton, J., 1969, Données histochimiques sur la neurosecrétion de *Carausius morosus* (Phasmides-Orthoptères), *Ann. Endocrinol.* **30**:839.

Mouton, J., 1971, Influence de la neuro-sécrétion sur la reproduction du phasme *Carausius morosus* (Chéleutoptère), *Ann. Endocrinol.* **32**:709.

Müller, H. P., and Engelmann, F., 1968, Studies on the endocrine control of metabolism in *Leucophaea maderae* (Blattaria). II. Effect of the corpora cardiaca on fat-body respiration, *Gen. Comp. Endocrinol.* **11**:43.

Mundall, E., 1978, Oviposition in *Triatoma protracta:* Role of mating and relationship to egg growth. *J. Insect Physiol.* **24**:321.

Mundall, E., and Engelmann, F., 1977, Endocrine control of vitellogenin synthesis and vitellogenesis in *Triatoma protracta, J. Insect Physiol.* **23**:825.

Muraleedharan, D., and Prabhu, V. K. K., 1979, Role of the median neurosecretory cells in secretion of protease and invertase in the red cotton bug, *Dysdercus cingulatus, J. Insect Physiol.* **25**:237.

Muraleedharan, D., and Prabhu, V. K. K., 1981, Hormonal influence on feeding and digestion in a plant bug, *Dysdercus cingulatus,* and a caterpillar, *Hyblaea puera, Physiol. Entomol.* **6**:183.

Murphy, T. A., and Wyatt, G. R., 1965, The enzymes of glycogen and trehalose synthesis in silkmoth fat body, *J. Biol. Chem.* **240**:1500.

Muszyńska-Pytel, M., and Cymborowski, B., 1978a, The role of serotonin in regulation of the circadian rhythms of locomotor activity in the cricket (*Acheta domesticus* L.). I. Circadian variations in serotonin concentration in the brain and hemolymph, *Comp. Biochem. Physiol. C* **59**:13.

Muszyńska-Pytel, M., and Cymborowski, B., 1978a, The role of serotonin in regulation of the circadian rhythms of locomotor activity in the cricket (*Acheta domesticus* L.). I. Circadian variations in serotonin concentration in the brain and hemolymph, *Comp. Biochem. Physiol. C* **59**:13.

Mwangi, R. W., and Goldsworthy, G. J., 1977a, Age-related changes in the response to adipokinetic hormone in *Locusta, Physiol. Entomol.* **2**:37.

Mwangi, R. W., and Goldsworthy, G. J., 1977b, Diglyceride-transporting lipoproteins in *Locusta, J. Comp. Physiol.* **114**:177.

Nagasawa, H., Isogai, A., Suzuki, A., Tamura, S., and Ishizaki, H., 1979, Purification and properties of the prothoracicotropic hormone of the silkworm, *Bombyx mori, Dev. Growth Differ.* **21**:29.

Nagasawa, H., Fu, G., Xiangchen, Z., Bangying, X., Zongshun, W., Xujia, Q., Dingyi, W., Eying, C., Jingze W., Suzuki, A., Isogai, A., Hori, Y., Tamura, S., and Ishizaki, H., 1980, Large-scale purification of prothoracicotropic hormone of the silkworm *(Bombyx mori)*, *Scientia Sinica* **23**:1053.

Nagy, F., 1978, Ultrastructure of a peripheral neurosecretory cell in the proctodaeal nerve of the larva of *Oryctes nasicornis* L. (Coleoptera: Scarabaeidae), *Int. J. Insect Morphol. Embryol.* **7**:325.

Nair, K. S. S., 1974, Studies on the diapause of *Trogoderma granarium:* Effects of juvenile hormone analogues on growth and metamorphosis, *J. Insect Physiol.* **20**:231.

Naisse, J., 1965, Contrôle endocrinien de la différentiation sexuelle chez les Insectes, *Arch. Anat. Microsc. Morphol. Exp.* **54**:417.

Naisse, J., 1969, Rôle des neurohormones dans la différentiation sexuelle de *Lampyris noctiluca, J. Insect Physiol.* **15**:877.

Naisse, J., and Mouton, J., 1965, Phénomènes neuro-endocrines au niveau de la chaîne nerveuse ventrale de *Carausius morosus* (Phasmides-Orthoptères), *C. R. Acad. Sci.* **261**:3887.

Natalizi, G. M., and Frontali, N., 1966, Purification of insect hyperglycaemic and heart accelerating hormones, *J. Insect Physiol.* **12**:1279.

Natalizi, G. M., Pansa, M. C., D'Ajello, V., Casaglia, O., Bettini, S., and Frontali, N., 1970, Physiologically active factors from corpora cardiaca of *Periplaneta americana, J. Insect Physiol.* **16**:1827.

Nayar, K. K., 1954, The neurosecretory system of the fruitfly *Chaetodacus cucurbitae* Coq. I. Distribution and description of the neurosecretory cells in the adult fly, *Proc. Indian Acad. Sci. B.* **40**:138.

Nayar, K. K., 1955, Studies on the neurosecretory system of *Iphita limbata* Stal. I. Distribution and structure of the neurosecretory cells of the nerve ring, *Biol. Bull. (Woods Hole, Mass.)* **108**:296.

Nayar, K. K., 1956a, Studies on the neurosecretory system of *Iphita limbata* Stal. (Hemiptera). Part III. The endocrine glands and the neurosecretory pathways in the adult, *Z. Zellforsch. Mikrosk. Anat.* **44**:697.

Nayar, K. K., 1956b, Studies on the neurosecretory system of *Iphita limbata* Stål. (Pyrrhocoridae: Hemiptera). Part IV. Observations on the structure and functions of the corpora cardiaca of the adult insect, *Proc. Natl. Inst. Sci. India Part B* **22**:171.

Nayar, K. K., 1957, Water content and release of neurosecretory products in *Iphita limbata* Stal, *Curr. Sci.* **26**:25.

Nayar, K. K., 1958, Studies on the neurosecretory system of *Iphita limbata* Stal. Part V. Probable endocrine basis of oviposition in the female insect, *Proc. Indian Acad. Sci. B.* **47**:233.

Nayar, K. K., 1960, Studies on the neurosecretory system of *Iphita limbata* Stal. VI. Structural changes in the neurosecretory cells induced by changes in water content, *Z. Zellforsch. Mikrosk., Anat.* **51**:320.

Nayar, K. K., 1962, Effects of injecting juvenile hormone extracts on the neurosecretory system of adult male cockroaches *(Periplaneta americana)* in: *Proceedings of the 3rd International Symposium on Neurosecretion, Bristol, Mem. Soc. Endocrinol.* **12**:371.

Nicolson, S. W., 1976, The hormonal control of diuresis in the cabbage white butterfly *Pieris brassicae, J. Exp. Biol.* **65**:565.

Nicolson, S. W., 1980, Diuresis and its hormonal control in butterflies, *J. Insect Physiol.* **26**:841.

Nijhout, H. F., 1975, Axonal pathways in the brain–retrocerebral neuroendocrine complex of *Manduca sexta* (L.) (Lepidoptera: Sphingidae), *Int. J. Insect. Morphol. Embryol.* **4**:529.

Nijhout, H. F., 1976, The role of ecdysone in pupation of *Manduca sexta*, *J. Insect Physiol.* **22**:453.

Nijhout, H. F., and Carrow, G. M., 1978, Diuresis after a bloodmeal in female *Anopheles freeborni*, *J. Insect Physiol.* **24**:293.

Nijhout, H. F., and Williams, C. M., 1974a, Control of moulting and metamorphosis in the tobacco hornworm, *Manduca sexta* (L.): Growth of the last instar larva and the decision to pupate, *J. Exp. Biol.* **61**:481.

Nijhout, H. F., and Williams, C. M., 1974b, Control of moulting and metamorphosis in the tobacco hornworm, *Manduca sexta* (L.): Cessation of juvenile hormone secretion as a trigger for pupation, *J. Exp. Biol.* **61**:493.

Nishiitsutsuji-Uwo, J., 1961, Electron microscopic studies on the neurosecretory system in Lepidoptera, *Z. Zellforsch. Mikrosk. Anat.* **54**:613.

Nishiitsutsuji-Uwo, J., 1972, Purification and some properties of insect brain hormone extracted from silkworm: heads, *Botyu-Kagaku* **37**:93.

Nishiitsutsuji-Uwo, J., and Pittendrigh, C. S., 1968, Central nervous system control of circadian rhythmicity in the cockroach. III. The optic lobes, locus of the driving oscillation?, *Z. Vgl. Physiol.* **58**:14.

Nishiitsutsuji-Uwo, J., Petropulos, S. F., and Pittendrigh, C. S., 1967, Central nervous system control of circadian rhythmicity in the cockroach. I. Role of the pars intercerebralis, *Biol. Bull. (Woods Hole, Mass.)* **133**:679.

N'Kouka, E., 1976, Variations de la neurosecrétion cérébrale au cours du cycle génital d'un insecte vivipare, *Glossina fuscipes fuscipes* (Diptére, Muscidae), *C. R. Acad. Sci. Ser. D* **282**:557.

N'Kouka, E., 1977a, Recherches sur le contrôle endocrine cérébral de la reproduction chez un insecte vivipare, *Glossina fuscipes fuscipes* (Diptère, Muscidae), *C. R. Acad. Sci. Ser. D* **284**:2381.

N'Kouka, E., 1977b, Contrôle neuro-endocrine de l'ovulation et de la parturition chez un insecte vivipare *Glossina fuscipes fuscipes* (Diptera, Muscidae), *C. R. Acad. Sci. Ser. D* **285**:77.

Noirot, C., 1957, Neurosécrétion et sexualité chez le Termite à cou jaune *"Calotermes flavicollis* F.,*" C. R. Acad. Sci.* **245**:743.

Noirot, C., 1977, Various aspects of hormone action in social insects, in: *Proceedings of the 8th International Congress of IUSSI, Wageningen, The Netherlands*, p. 12.

Noirot, C., and Noirot-Timothée, C., 1976, Fine structure of the rectum in cockroaches (Dictyoptera): General organization and intercellular junctions, *Tissue Cell* **8**:345.

Nopp-Pammer, E., and Nopp, H., 1967, Zur hormonalen Steuerung des Spinnverhaltens von *Philosamia cynthia*, *Naturwissenschaften* **54**:592.

Normann, T. C., 1965, The neurosecretory system of the adult *Calliphora erythrocephala*. I. The fine structure of the corpus cardiacum with some observations on adjacent organs, *Z. Zellforsch. Mikrosk. Anat.* **67**:461.

Normann, T. C., 1969, Experimentally induced exocytosis of neurosecretory granules, *Exp. Cell Res.* **55**:285.

Normann, T. C., 1970, The mechanism of hormone release from neurosecretory axon endings in the insect *Calliphora erythrocephala*, in: *Aspects of Neuroendocrinology* (W. Bargmann and B. Scharrer, eds.), Springer-Verlag, New York, p. 30.

Normann, T. C., 1972, Heart activity and its control in the adult blowfly, *Calliphora erythrocephala*, *J. Insect Physiol.* **18**:1793.

Normann, T. C., 1973, Membrane potential of the corpus cardiacum neurosecretory cells of the blowfly, *Calliphora erythrocephala*, *J. Insect Physiol.* **19**:303.

Normann, T. C., 1974, Calcium dependence of neurosecretion by exocytosis, *J. Exp. Biol.* **61**:401.

Normann, T. C., 1975, Neurosecretory cells in insect brain and production of hypoglycaemic hormone, *Nature (London)* **254**:259.

Normann, T. C., and Duve, H., 1969, Experimentally induced release of a neurohormone influencing hemolymph trehalose level in *Calliphora erythrocephala* (Diptera), *Gen. Comp. Endocrinol.* **12**:449.

Novák, V. J. A., and Rohdendorf, E., 1959, The influence of the implanted corpus allatum on the corpus allatum of host, in: *Acta Symposii de Evolutione Insectorum, Prague*, p. 157.

Nowosielski, J. W., and Patton, R. L., 1963, Studies on circadian rhythm of the house cricket *Gryllus domesticus*, *J. Insect Physiol.* **9**:401.

Nunez, J. A., 1956, Untersuchungen über die Regelung das Wasserhaushaltes bei *Anisotarsus cupripennis Germ.*, *Z. Vgl. Physiol.* **36**:341.

Oberlander, H., Berry, S. J., Krishnakumaran, A., and Schneiderman, H. A., 1965, RNA and DNA synthesis during activation and secretion of the prothoracic glands of saturniid moths, *J. Exp. Zool.* **159**:15.

Odhiambo, T. R., 1966, The metabolic effects of the corpus allatum hormone in the male desert locust. I. Lipid metabolism, *J. Exp. Biol.* **45**:45.

O'Farrell, A. F., 1964, On physiological colour change in some Australian Odonata, *J. Entomol. Soc. Aust.* **1**:1.

Ogura, N., 1975, Hormonal control of larval coloration in the armyworm, *Leucania separata, J. Insect Physiol.* **21**:559.

Ogura, N., and Mitsuhashi, J., 1978, *In vitro* cultures of the endocrine organs secreting melanization and reddish coloration hormone in the common armyworm, *Leucania separata* (Lepidoptera: Noctuidae), *Appl. Entomol. Zool.* **13**:274.

Ogura, N., and Mitsuhashi, J., 1979, Melanization of integuments cultured *in vivo* and *in vitro*, *Appl. Entomol. Zool.* **14**:118.

Ogura, N., and Saito, T., 1973, Induction of embryonic diapause in the silkworm, *Bombyx mori* L. (Lepidoptera: Bombycidae), by implantation of ganglia of the common armyworm larvae, *Leucania separata* Walker (Lepidoptera: Noctuidae), *Appl. Entomol. Zool.* **8**:46.

Ogura, N., Yagi, S., and Fukaya, M., 1971, Hormonal control of larval coloration in the common armyworm, *Leucania separata* Walker, *Appl. Entomol. Zool.* **6**:93.

Ohnishi, E., and Hidaka, T., 1957, Effet des facteurs du milieu ambiant sur la détermination des types de chrysalides de certains papillons diurnes nommés "queue d'hirondelle," *Bull. Soc. Hist. Nat. Toulouse* **92**:177.

Ohtaki, T., 1960, Humoral control of pupal coloration in the cabbage white butterfly, *Pieris rapae crucivora, Annot. Zool. Jpn.* **33**:97.

Ohtaki, T., 1963, Further studies on the development of pupal coloration in the cabbage white butterfly, *Pieris rapae crucivora, Annot. Zool. Jpn.* **36**:78.

Ohtaki, T., and Ohnishi, E., 1967, Pigments in the pupal integuments of two colour types of cabbage white butterfly, *Pieris rapae crucivora, J. Insect Physiol.* **13**:1569.

Okelo, O., 1971, Physiological control of oviposition in the female desert locust *Schistocerca gregaria* Forsk, *Can. J. Zool.* **49**:969.

Oltmer, A., 1968, Die Steuerung des Melanineibaus in das Farbmuster der Kohlweisslingspuppe *Pieris brassicae* L., *Wilhelm Roux Arch. Entwicklungsmech. Org.* **160**:401.

Opoczyńska-Sembratowa, Z., 1936, Recherches sur l'anatomie et l'innervation du coeur de *Carausius morosus, Bull. Int. Acad. Pol. Sci. Lett. Cl. Sci. Math. Nat. Ser. B* **1936**:411.

Orchard, I., and Finlayson, L. H., 1976, The electrical activity of mechanoreceptive and neurosecretory neurons in the stick insect *Carausius morosus, J. Comp. Physiol.* **107**:327.

Orchard, I., and Finlayson, L. H., 1977, Electrically excitable neurosecretory cell bodies in the periphery of the stick insect, *Carausius morosus, Experientia* **33**:226.

Orchard, I., and Loughton, B. G., 1980, A hypolipaemic factor from the corpus cardiacum of locusts, *Nature (London)* **286**:494.

Orchard, I., Friedel, T., and Loughton, B. G., 1981, Release of a neurosecretory protein from the corpora cardiaca of *Locusta migratoria* induced by high potassium saline and compound action potentials, *J. Insect Physiol.* **27**:297.

Orr, C. W. M., 1964, The influence of nutritional and hormonal factors on the chemistry of the fat body, blood, and ovaries of the blowfly *Phormia regina* Meig., *J. Insect Physiol.* **10**:103.

Osanai, M., 1966, Ueber die Umfärbung des Raupen von *Hestina japonica* zu Beginn der Uberwinterung II, *Hoppe Seyler's Z. Physiol. Chem.* **347**:145.

Osanai, M., and Arai, Y., 1962, Ueber die Umfärbung der Raupen von *Hestina japonica* zu Beginn der Uberwinterung. I. Durchschnürungsversuche an der Umfärbung der *Hestina*-Raupe, *Gen. Comp. Endocrinol.* **2**:311.

Osborne, D. J., Carlisle, D. B., and Ellis, P. E., 1968, Protein synthesis in the fat body of the female desert locust, *Schistocerca gregaria* Forsk., in relation to maturation, *Gen. Comp. Endocrinol.* **11**:347.

Osborne, M. P., Finlayson, L. H., and Rice, M. J., 1971, Neurosecretory endings associated with striated muscles in three insects (*Schistocerca, Carausius,* and *Phormia*) and a frog (*Rana*), *Z. Zellforsch. Mikrosk. Anat.* **116**:391.

Oschman, J. L., and Wall, B. J., 1969, The structure of the rectal pads of *Periplaneta americana* L. with regard to fluid transport, *J. Morphol.* **127**:475.

O'Shea, M., and Adams, M. E., 1981, Pentapeptide (proctolin) associated with an identified neuron, *Science* **213**:567.

Ozbas, S., and Hodgson, E. S., 1958, Action of insect neurosecretion upon central nervous system *in vitro* and upon behavior, *Proc. Natl. Acad. Sci. USA* **44**:825.

Ozeki, K., 1962, Studies on the secretion of the juvenile hormone in the earwig, *Anisolabis maritima, Sci. Pap. Coll. Gen. Educ. Univ. Tokyo (Biol. Part)* **12**:65.

Ozeki, K., 1965, Studies on the function of the corpus allatum during the last nymphal stage in the earwig *Anisolabis maritima, Sci. Pap. Coll. Gen. Educ. Univ. Tokyo (Biol. Part)* **15**:149.

Ozeki, K., 1979, The brain controls of the corpus allatum function of the earwig, *Anisolabis maritima, Sci. Pap. Coll. Gen. Educ. Univ. Tokyo (Biol. Part)* **29**:55.

Padgham, D. E., 1976a, Control of melanization in first instar larvae of *Schistocerca gregaria, J. Insect Physiol.* **22**:1409.

Padgham, D. E., 1976b, Bursicon-mediated control of tanning in melanizing and non-melanizing first instar larvae of *Schistocerca gregaria, J. Insect Physiol.* **22**:1447.

Pammer, E., 1966, Auslösung und Steuerung des Spinnverhaltens und der Diapause bei *Philosamia cynthia* Dru. (Saturniidae, Lep.), *Z. Vgl. Physiol.* **53**:99.

Pan, M. L., 1977, Juvenile hormone and vitellogenin synthesis in the cecropia silkworm, *Biol. Bull. (Woods Hole, Mass.)* **153**:336.

Pan, M. L., and Wyatt, G. R., 1971, Juvenile hormone induces vitellogenin synthesis in the Monarch butterfly, *Science* **174**:503.

Pan, M. L., and Wyatt, G. R., 1976, Control of vitellogenin synthesis in the Monarch butterfly by juvenile hormone, *Dev. Biol.* **54**:127.

Panov, A. A., 1963, Distribution of neurosecretory cells in the abdominal section of the nerve cord of Orthoptera, *Dokl. Akad. Nauk SSSR Biol. Sci. Sect.* **145**:1409 [English translation].

Panov, A. A., 1964, Neurosecretory cells in the abdominal ganglia of Orthoptera (Insecta, Orthoptera), *Entomol. Rev. Entomol. Obozr.* **43**:403 [English translation].

Panov, A. A., 1975, On the similarity in composition of the brain neurosecretory cells in Lepidoptera-Trichoptera and Diptera, *Zool. Anz.* **194**:319.

Panov, A. A., 1976, The composition of the medial neurosecretory cells of the pars intercerebralis in *Calliphora* and *Lucilia* adults, *Zool. Anz.* **196**:23.

Panov, A. A., 1978, Insect pars intercerebralis: Diversity of PF-positive neurosecretory materials, in: *9th European Conference on Comparative Endocrinology, Giessen, Gen. Comp. Endocrinol.* **36**:104.

Panov, A. A., 1979, Brain neurosecretory cells and their axon pathways in the larva of *Chironomus plumosus* L. (Diptera: Chironomidae), *Int. J. Insect Morphol. Embryol.* **8**:203.

Panov, A. A., 1980, The cerebral neurosecretory system of Tipulidae (Diptera, Insecta), *Zool. Anz.* **205**:345.

Panov, A. A., and Bassurmanova, O. K., 1967, Corpora allata of some silkworms as depots of neurosecretory material from A'-cells, *Dokl. Akad. Nauk SSSR Biol. Sci. Sect.* **176**:577 [English translation].

Panov, A. A., and Davydova, E. D., 1976, Medial neurosecretory cells in the brain of Mecoptera and Neuropteroidea (Insecta), *Zool. Anz.* **197**:187.

Panov, A. A., and Kind, T. V., 1963, The neurosecretory cell system in the lepidopteran brain (Lepidoptera, Insecta), *Dokl. Akad. Nauk SSSR Biol. Sci. Sect.* **153**:1558 [English translation].

Pappas, C., and Fraenkel, G., 1978, Hormonal aspects of oögenesis in the flies *Phormia regina* and *Sarcophaga bullata*, *J. Insect Physiol.* **24**:75.

Pasteels, J. M., 1965, Description d'un système neuroendocrinien dans le ganglion infraoesophagien du Phasme *Carausius morosus* (Insecte, Orthoptère), *C. R. Acad. Sci.* **261**:3884.

Peacock, A. J., 1976, Effect of corpus cardiacum extracts on the ATPase activity of locust rectum, *J. Insect Physiol.* **22**:1631.

Peacock, A., Bowler, K., and Anstee, J. H., 1976, Properties of Na^+ K^+-dependent ATPase from the Malpighian tubules and hindgut of *Homorocoryphus nitidulus vicinus*, *Insect Biochem.* **6**:281.

Peled, Y., and Tietz, A., 1973, Fat transport in the locust, *Locusta migratoria:* The role of protein synthesis, *Biochim. Biophys. Acta* **296**:499.

Pener, M. P., 1965, On the influence of corpora allata on maturation and sexual behaviour of *Schistocerca gregaria*, *J. Zool.* **147**:119.

Pener, M. P., 1967, Effects of allatectomy and sectioning of the nerves of the corpora allata on oöcyte growth, male sexual behavior, and color change in adults of *Schistocerca gregaria*, *J. Insect Physiol.* **13**:665.

Pener, M. P., Girardie, A., and Joly, P., 1972, Neurosecretory and corpus allatum controlled effects on mating behaviour and color change in adult *Locusta migratoria migratorioides* males, *Gen. Comp. Endocrinol.* **19**:494.

Persaud, C. E., and Davey, K. G., 1971, The control of protease synthesis in the intestine of adults of *Rhodnius prolixus*, *J. Insect Physiol.* **17**:1429.

Pflugfelder, O., 1937, Bau, Entwicklung und Funktion der Corpora allata und cardiaca von *Dixippus morosus* Br., *Z. Wiss. Zool.* **149**:477.

Pflugfelder, O., 1938, Farbverränderungen und Gewebsentartungen nach Nervendurchschneidung und Exstirpation der Corpora allata von *Dixippus morosus* Br., *Verh. Dtsch. Zool. Ges.* **40**:127.

Pflugfelder, O., 1939, Wechselwirkungen von Drüsen innerer Sekretion bei *Dixippus morosus* Br., *Z. Wiss. Zool.* **152**:384.

Phillips, J. E., Mordue, W., Meredith, J., and Spring, J., 1980, Purification and characteristics of the chloride transport stimulating factor from locust corpora cardiaca: A new peptide, *Can. J. Zool.* **10**:1851.

Pilcher, D. E. M., 1970a, The influence of the diuretic hormone on the process of urine secretion by the Malpighian tubules of *Carausius morosus*, *J. Exp. Biol.* **53**:465.

Pilcher, D. E. M., 1970b, Hormonal control of the Malpighian tubules of the stick insect, *Carausius morosus*, *J. Exp. Biol.* **52**:653.

Pilcher, D. E. M., 1971, Stimulation of movements of Malpighian tubules of *Carausius* by pharmacologically active substances and tissue extracts, *J. Insect Physiol.* **17**:463.

Pipa, R. L., 1978, Locations and central projections of neurons associated with the retrocerebral neuroendocrine complex of the cockroach *Periplaneta americana* (L.), *Cell Tissue Res.* **193**:443.

Pipa, R. L., and Novák, F. J., 1979, Pathways and fine structure of neurons forming the nervi corporis allati II of the cockroach *Periplaneta americana* (L.), *Cell Tissue Res.* **201**:227.

Plotnikova, S. I., 1968, The structure of the sympathetic nervous system of insects, in: *Symposium on Neurobiology of Invertebrates* (J. Salanki, ed.), Plenum Press, New York, p. 59.

Plotnikova, S. I., and Govyrin, V. A., 1966, Distribution of catecholamine-containing nerve elements in some representatives of *Protostomia* and *Coelenterata, Arch. Anat. Hist. Embryol.* **50**:79.

Poras, M., 1975, Rupture de la diapause imaginale des femelles de *Tetrix undulata* (Sow.) (Orthoptère, Tetrigidae) par implantation d'un corps allate de femelle active de *Locusta migratoria, C. R. Acad. Sci. Ser. D* **281**:551.

Poras, M., 1977a, Activité des corps allates de femelles en diapause chez *Tetrix undulata* (Swrb.) (Orthoptère-Tetrigidae), *C. R. Acad. Sci. Ser. D* **284**:1301.

Poras, M., 1977b, Rupture de la diapause imaginale des femelles de *Tetrix undulata* (Swrb.) (Orthoptère Tetrigidae) par cautérisation de la région de la pars intercerebralis, *C. R. Acad. Sci. Ser. D* **284**:1441.

Possompès, B., 1953, Recherches expérimentales sur le déterminisme de la métamorphose de *Calliphora erythrocephala* Meig., *Arch. Zool. Exp. Gen.* **89**:203.

Possompès, B., Charbonnière, J., and Ralisoa, B. O., 1967, Evolution des cellules neurosécrétrices de la pars intercerebralis, croissance des ovocytes et ovoviviparité chez *Sarcophaga argyrostoma* (Dipt. Cyclorrhaphe), *Ann. Soc. Entomol. Fr.* **3**:1.

Post, L. C., 1972, Bursicon: Its effect on tyrosine permeation into insect haemocytes, *Biochim. Biophys. Acta* **290**:424.

Postlethwait, J. H., and Handler, A. M., 1979, The roles of juvenile hormone and 20-hydroxyecdysone during vitellogenesis in isolated abdomens of *Drosophila melanogaster, J. Insect Physiol.* **25**:455.

Prabhu, V. K. K., 1966, Interferometric studies on the neurosecretory cells of *Iphita limbata* Stal (Heteroptera: Insect), in: *Proceedings of the Seminar of International Cell Biology, Bombay* T. V. Chidambaran, University of Bombay, p. 428.

Pratt, G. E., and Davey, K. G., 1972, The corpus allatum and oögenesis in *Rhodnius prolixus* (Stål.). I. The effects of allatectomy, *J. Exp. Biol.* **56**:201.

Pratt, G. E., and Tobe, S. S., 1974, Juvenile hormones radiobiosynthesised by corpora allata of adult female locusts *in vitro, Life Sci.* **14**:575.

Prentø, P., 1972, Histochemistry of neurosecretion in the pars intercerebralis–corpus cardiacum system of the desert locust *Schistocerca gregaria, Gen. Comp. Endocrinol.* **18**:482.

Price, G. M., 1970, Pupation inhibiting factor in the larva of the blowfly *Calliphora erythrocephala, Nature (London)* **228**:876.

Prince, W. T., Berridge, M. J., and Rasmussen, H., 1972, Role of calcium and adenosine-3′:5′-cyclic monophosphate in controlling fly salivary gland secretion, *Proc. Natl. Acad. Sci. USA* **69**:553.

Proux, J., 1978a, Influence de saignées répétées sur le comportement pondéral et hydrique du criquet migrateur privé de cellules neurosécrétrices sous-ocellaires médianes diurétiques, *C. R. Acad. Sci. Ser. D* **286**:981.

Proux, J., 1978b, Section de la chaîne nerveuse ventrale et comportement hydrique après une saignée chez le criquet migrateur, *C. R. Acad. Sci. Ser. D* **287**:305.

Proux, J., 1978c, Ganglions de la chaîne nerveuse ventrale et équilibre hydrique chez le criquet migrateur. I. Ganglions abdominaux, *Arch. Biol.* **89**:297.

Proux, J., 1978d, Ganglions de la chaîne nerveuse ventrale et équilibre hydrique chez le criquet migrateur. II. Ganglions sous-oesophagien et thoraciques, *Arch. Biol.* **89**:313.

Proux, J., 1979, Influence de l'électrocoagulation de la pars intercerebralis et des cellules neurosécrétrices sous-ocellaires médianes sur le comportement pondéral, hydrique et alimentaire du criquet migrateur, *Bull. Soc. Zool. Fr.* **104**:89.

Proux, J., and Buscarlet, L. A., 1976, Etude à l'aide d'eau tritiée, du rôle des cellules neurosécrétrices sous-ocellaires médianes dans la régulation hydrique des mâles de *Locusta migratoria migratorioides* (R. et F.), *Acrida* **5**:311.

Proux, J., and Rougon-Rapuzzi, G., 1980, Evidence for vasopressin-like molecule in migratory locust. Radioimmunological measurements in different tissues: Correlation with various states of hydration, *Gen. Comp. Endocrinol.* **42**:378.

Proux, J., Rougon-Rapuzzi, G., and Rémy, C., 1980, Influence de la pars intercerebralis et des cellules neurosécrétrices sous-ocellaires médianes sur la teneur en substance apparentée à la vasopressine dans la chaîne nerveuse ventrale et l'hémolymphe du criquet migrateur. Etude radio-immunologique et immunohistologique, *J. Physiol. (Paris)* **76**:277.

Provansal, A., 1968, Mise en évidence d'organes neurohémaux métamériques associés à la chaîne nerveuse ventrale chez *Vespa crabro* L. et *Vespula germanica* Fabr. (Hymenoptère Vespidae), *C. R. Acad. Sci. Ser. D* **267**:864.

Provansal, A., 1971, Caractères particuliers des organes périsympathiques de la larve de *Diprion pini* L. (Hyménoptère, Symphyte, Diprionidae), *C. R. Acad. Sci. Ser. D* **272**:855.

Provansal, A., 1972, Les organes périsympathiques des Lépidoptères, *C. R. Acad. Sci. Ser. D* **274**:97.

Provansal, A., and Grillot, J.-P., 1972, Les organes périsympathiques des insectes holométaboles. I. Coléoptères, *Ann. Soc. Entomol. Fr.* **8**:863.

Provansal, A., Baudry, N., and Raabe, M., 1970, Recherches sur l'ultrastructure des organes neurohémaux périsympathiques des Vespidae (Hyménoptères). Les organes médians sphériques, *C. R. Acad. Sci. Ser. D* **271**:1115.

Provansal, A., Grillot, J.-P., and Delachambre, J., 1974, Dynamique de la sécrétion de la bursicon dans l'ensemble du système nerveux de *Tenebrio molitor* L. (Coléoptère), *C. R. Acad. Sci. Ser. D* **278**:2193.

Quennedey, A., 1969, Innervation de type neurosécréteur dans la glande sternale de *Kalotermes flavicollis* (Isoptera). Etude ultrastructurale, *J. Insect Physiol.* **15**:1807.

Raabe, M., 1961, Recherches sur le déterminisme des genèses de pigments chez le Phasme, *Carausius morosus*, *C. R. Acad. Sci.* **252**:3663.

Raabe, M., 1963a, Mise en évidence, chez des Insectes d'ordres variés, d'éléments neurosécréteurs tritocérébraux, *C. R. Acad. Sci.* **257**:1171.

Raabe, M., 1963b, Existence chez divers Insectes d'une innervation tritocérébrale des corpora cardiaca, *C. R. Acad. Sci.* **257**:1552.

Raabe, M., 1963c, Recherches expérimentales sur la localisation intra-cérébrale du facteur chromactif des Insectes, *C. R. Acad. Sci.* **257**:1804.

Raabe, M., 1964, Nouvelles recherches sur la neurosécrétion chez les Insectes, *Ann. Endocrinol.* **25**:107.

Raabe, M., 1965a, Recherches sur la neurosécrétion dans la chaîne nerveuse ventrale du Phasme, *Clitumnus extradentatus*: Les épaississements des nerfs transverses, organes de signification probablement neurohémale, *C. R. Acad. Sci.* **261**:4240.

Raabe, M., 1965b, Etude des phénomènes de neurosécrétion au niveau de la chaîne nerveuse ventrale des Phasmides, *Bull. Soc. Zool. Fr.* **90**:631.

Raabe, M., 1966, Recherches sur la neurosécrétion dans la chaîne nerveuse ventrale du Phasme, *Carausius morosus*: Liasion entre l'activité des cellules Bl et la pigmentation, *C. R. Acad. Sci. Ser. D* **263**:408.

Raabe, M., 1967, Recherches récentes sur la neurosécrétion dans la chaîne nerveuse ventrale des Insectes, *Bull. Soc. Zool. Fr.* **92**:67.

Raabe, M., 1972, Les organes périsympathiques des Dermaptères, *C. R. Acad. Sci. Ser. D* **275**:1925.

Raabe, M., and Demarti-Lachaise, F., 1973, Etude de l'action des organes périsympathiques sur la protéinémie du Phasme *Carausius morosus*, *C. R. Acad. Sci. Ser. D* **277**:2037.

Raabe, M., and Monjo, D., 1970, Recherches histologiques et histochimiques sur la neurosécrétion chez le Phasme, *Clitumnus extradentatus:* Les neurosécrétions de type C, *C. R. Acad. Sci. Ser. D* **270:**2021.

Raabe, M., and Provansal, A., 1972, Les organes périsympathiques des Paléoptères, *C. R. Acad. Sci. Ser. D* **275:**925.

Raabe, M., and Ramade, F., 1967, Observations sur l'ultrastructure des organes périsympathiques des Phasmides, *C. R. Acad. Sci. Ser. D* **264:**77.

Raabe, M., Cazal, M., Chalaye, D., and de Bessé, N., 1966, Action cardioaccélératrice des organes neurohémaux périsympathiques ventraux de quelques Insectes, *C. R. Acad. Sci. Ser. D* **263:**2002.

Raabe, M., Baudry, N., and Provansal, A., 1970, Recherches sur l'ultrastructure des organes neurohémaux périsympathiques des Vespidae (Hyménoptères). Les organes latéraux longitudinaux, *C. R. Acad. Sci. Ser. D* **271:**1210.

Raabe, M., Baudry, N., Grillot, J.-P., and Provansal, A., 1971, Les organes périsympathiques des Insectes Ptérygotes. Distribution. Caractères généraux, *C. R. Acad. Sci. Ser. D* **273:**2324.

Raabe, M., Panov, A. A., Davydova, E. D., and Chervin, D., 1979, Neurosecretory products diversity in the pars intercerebralis of Insects, *Experientia* **35:**404.

Rademakers, L. H. P. M., 1977a, Effects of isolation and transplantation of the corpus cardiacum on hormone release from its glandular cells after flight in *Locusta migratoria*. A quantitative electron microscopical study, *Cell Tissue Res.* **184:**213.

Rademakers, L. H. P. M., 1977b, Identification of a secretomotor centre in the brain of *Locusta migratoria,* controlling the secretory activity of the adipokinetic hormone producing cells of the corpus cardiacum, *Cell Tissue Res.* **184:**381.

Rademakers, L. H. P. M., and Beenakkers, A. M. Th., 1977, Changes in the secretory activity of the glandular lobe of the corpus cardiacum of *Locusta migratoria* induced by flight. A quantitative electron microscopic study. *Cell Tissue Res.* **180:**155.

Raina, A. K., and Bell, R. A., 1978, Morphology of the neuroendocrine system of the pink bollworm and histological changes in the neurosecretory cells of the brain during induction, maintenance and termination of diapause, *Ann. Entomol. Soc. Am.* **71:**375.

Ralph, C. L., 1962a, Action of extracts of cockroach nervous system on fat bodies *in vitro, Am. Zool.* **2:**362.

Ralph, C. L., 1962b, Heart accelerators and decelerators in the nervous system of *Periplaneta americana* L., *J. Insect Physiol.* **8:**431.

Ralph, C. L., and McCarthy, R., 1964, Effects of brain and corpus cardiacum extracts on haemolymph trehalose of the cockroach, *Periplaneta americana, Nature (London)* **203:**1195.

Ralph, C. L., and Matta, R. J., 1965, Evidence for hormonal effects on metabolism of cockroaches from studies of tissue homogenates, *J. Insect Physiol.* **11:**983.

Ramade, F., 1966, Sur l'ultrastructure de la pars intercerebralis chez *Musca domestica* L., *C. R. Acad. Sci. Ser. D* **263:**271.

Ramade, F., 1968, Sur la présence de cellules vacuolaires dans le cerveau de *Musca domestica* L., *C. R. Acad. Sci. Ser. D* **266:**2437.

Ramade, F., 1969a, Mise en évidence de cellules neurosécrétrices Gomori négatives dans la pars intercerebralis de *Musca domestica* L. par un étude comparative en microscopie ordinaire et électronique, *C. R. Acad. Sci. Ser. D* **268:**1945.

Ramade, F., 1969b, Données histologiques, histochimiques et ultrastructurales sur la pars intercerebralis de *Musca domestica* L., *Mem. Mus. Natl. Hist. Nat. Ser. A Zool.* **58:**113.

Ramade, F., and L'Hermite, P., 1971, Mise en évidence de neurones adrénergiques par la microscopie de fluorescence dans le système nerveux central de *Calliphora erythrocephala* Meig., et de *Musca domestica* L., *C. R. Acad. Sci. Ser. D* **272:**3314.

Ramade, F., and Rivière, J. L., 1970, Recherches histochimiques sur les protéines associées aux neurosécrétions protocérébrales de *Musca domestica* L., *C. R. Acad. Sci. Ser. D* **270:**1803.

Ramamurty, P. S., and Engels, W., 1977, Allatektomie-und Juvenilhormon-Wirkungen auf Synthese und Einlagerung von Vitellogenin bei der Bienenkönigin (Apis mellifica), Zool. Jahrb. Abt. Allg. Zool. Physiol. Tiere **81**:165.

Ramsay, J. A., 1954, Active transport of water by the Malpighian tubules of the stick insect, Dixippus morosus (Orthoptera, Phasmidae), J. Exp. Biol. **31**:104.

Rankin, M. A., and Riddiford, L. M., 1977, Hormonal control of migratory flight in Oncopeltus fasciatus: The effects of the corpus cardiacum, corpus allatum, and starvation on migration and reproduction, Gen. Comp. Endcrinol. **33**:309.

Rankin, M. A., and Riddiford, L. M., 1978, Significance of haemolymph juvenile hormone titer changes in timing of migration and reproduction in adult Oncopeltus fasciatus, J. Insect Physiol. **24**:31.

Rasenick, M. M., and Berry, S. J., 1981, Regulation of brain adenylate cyclase and the termination of silkmoth diapause, Insect Biochem. **11**:387.

Rasenick, M. M., Neuburg, M., and Berry, S. J., 1976, Brain cyclic levels and the initiation of adult development in the cecropia silkmoth, J. Insect Physiol. **22**:1453.

Rasenick, M. M., Neuburg, M., and Berry, S. J., 1978, Cyclic nucleotide activation of the silkmoth brain—cellular localization and further observations on the patterns of activation, J. Insect Physiol. **24**:137.

Ratnasari, N. P., and Fraenkel, G., 1974, The physiological basis of anterior inhibition of puparium formation in ligated fly larvae, J. Insect Physiol. **20**:105.

Ray, A., Kumar, D., and Ramamurty, P. S., 1980, Histological study on the retro-cerebral–endocrine complex with special reference to neurohaemal involvement of aorta and pericardial cells in Coccinella septempunctata (L.) (Coccinellidae–Coleoptera), Z. Mikrosk. Anat. Forsch. (Leipzig) **94**:1141.

Raziuddin, M., Khan, T. R., and Singh, S. B., 1979, Neuropilar neurosecretory fibres in the grasshopper, Poekilocerus pictus (Fabr.), and the cockroach, Periplaneta americana L., Zool. Anz. **202**:209.

Reddy, G., Hwang-Hsu, K., and Kumaran, A. K., 1979, Factors influencing juvenile hormone esterase activity in the wax moth, Galleria mellonella, J. Insect Physiol. **25**:65.

Rehm, M., 1955, Morphologische und histochemische Untersuchungen an neurosekretorischen Zellen von Schmetterlingen, Z. Zellforsch. Mikrosk. Anat. **42**:19.

Reinecke, J. P., and Adams, T. S., 1977, A novel muscle complex on the hindgut of lepidopteran larvae, Int. J. Insect Morphol. Embryol. **6**:239.

Reinecke, J. P., Cook, B. J., and Adams, T. S., 1973, Larval hindgut of Manduca sexta (L.) (Lepidoptera: Sphingidae), Int. J. Insect Morphol. Embryol. **2**:277.

Reinecke, J. P., Gerst, J., O'Gara, B., and Adams, T. S., 1978, Innervation of hindgut muscle of larval Manduca sexta (L.) (Lepidoptera: Sphingidae) by a peripheral multinucleate neurosecretory neuron, Int. J. Insect Morphol. Embryol. **7**:435.

Rembold, H., Eder, J., and Ulrich, G. M., 1980, Inhibition of allatotropic activity and ovary development in Locusta migratoria by anti-brain-antibodies, Z. Naturforsch. Sect. C Biosci. **35**:1117.

Rémy, C., and Dubois, M. P., 1981, Immunohistological evidence of methionine enkephalin-like material in the brain of the migratory locust, Cell Tissue Res. **218**:271.

Rémy, C., and Girardie, J., 1980, Anatomical organization of two vasopressin–neurophysin-like neurosecretory cells throughout the central nervous system of the migratory locust, Gen. Comp. Endocrinol. **40**:27.

Rémy, C., Girardie, J., and Dubois, M. P., 1977, Exploration immunocytologique des ganglions cérébroïdes et sous-oesophagien du phasme Clitumnus extradentatus: Existence d'une neurosérétion apparentée à la vasopressine–neurophysine, C. R. Acad. Sci. Ser. D **285**:1495.

Rémy, C., Girardie, J., and Dubois, M. P., 1978, Présence dans le ganglion sous-oesophagien de la Chenille processionnaire du pin (Thaumetopoea pityocampa Schiff) de cellules révélées en immunofluorescence par un corps anti-α-endorphine, C. R. Acad. Sci. Ser. D **286**:651.

Rémy, C., Girardie, J., and Dubois, M. P., 1979, Vertebrate neuropeptide-like substances in the suboesophageal ganglion of two insects: *Locusta migratoria* R. and F. (Orthoptera) and *Bombyx mori* L. (Lepidoptera). Immunocytological investigation, *Gen. Comp. Endocrinol.* **37**:93.

Rensing, L., 1966, Zur circadianen rhythmik des Hormonsystems von *Drosophila, Z. Zellforsch. Mikrosk. Anat.* **74**:539.

Ressin, W. J., 1980, The effect of juvenile hormone on pupal pigmentation of *Pieris brassicae* L., *J. Insect Physiol.* **26**:295.

Retnakaran, A., and Joly, P., 1976, Neurosecretory control of juvenile hormone inactivation in *Locusta migratoria* L., *Coll. Int. CNRS* **251**:317.

Reynolds, S. E., 1974, Pharmacological induction of plasticization in the abdominal cuticle of *Rhodnius, J. Exp. Biol.* **61**:705.

Reynolds, S. E., 1975, The mechanism of plasticization of the abdominal cuticle in *Rhodnius, J. Exp. Biol.* **62**:81.

Reynolds, S. E., 1976, Hormonal regulation of cuticle extensibility in newly emerged adult blowflies, *J. Insect Physiol.* **22**:529.

Reynolds, S. E., 1977, Control of cuticle extensibility in the wings of adult *Manduca* at the time of eclosion: Effects of eclosion hormone and bursicon, *J. Exp. Biol.* **70**:27.

Reynolds, S. E., Taghert, P. H., and Truman, J. W., 1979, Eclosion hormone titres and bursicon and the onset of hormonal responsiveness during the last day of adult development in *Manduca sexta* (L.), *J. Exp. Biol.* **78**:77.

Richard, N., and Charniaux-Cotton, H., 1970, Conséquences de la destruction expérimentale du pôle antérieur de l'embryon sur le développement des gonades du Doryphore *Leptinotarsa decemlineata* Say, *C. R. Acad. Sci.* **270**:1595.

Richter, K., 1973, Struktur und Funktion der Herzen wirbelloser Tiere, *Zool. Jahrb. Abt. Allg. Zool. Physiol. Tiere* **77**:477.

Richter, K., and Gersch, M., 1974, Untersuchungen über die Wirkung cholinerger, aminerger und peptiderger Substanzen am lateralen Herznerven und den funktionellen Zusammenhang dieser Wirkungen bei der Herzregulation von *Blaberus craniifer* Burm. (Insecta: Blattaria), *Zool. Jahrb. Abt. Allg. Zool. Physiol. Tiere* **78**:16.

Richter, K., and Stürzebecher, J., 1969, Elektrophysiologische Untersuchungen zum Wirkungsmechanismus von Neurohormon D am lateralen Herznerven von *Periplaneta americana* (L.), *Z. Wiss. Zool. Abt. A* **180**:148.

Riddiford, L. M., and Williams, C. M., 1971, Role of the corpora cardiaca in the behavior of saturniid moths. I. Release of sex pheromone, *Biol. Bull. (Woods Hole, Mass.)* **140**:1.

Robertson, H. A., and Steele, J. E., 1974, Octopamine in the insect central nervous system: Distribution, biosynthesis and possible physiological role, *J. Physiol. (London)* **237**:34.

Robinson, N. L., and Goldsworthy, G. J., 1974, The effects of locust adipokinetic hormone on flight muscle metabolism *in vivo* and *in vitro, J. Comp. Physiol.* **89**:369.

Robinson, N. L., and Goldsworthy, G. J., 1976, Adipokinetic hormone and flight metabolism in the locust, *J. Insect Physiol.* **22**:1559.

Rossignol, P. A., Feinsod, F. M., and Spielman, A., 1981, Inhibitory regulation of corpus allatum activity in mosquitoes, *J. Insect Physiol.* **27**:651.

Rounds, H. D., 1963, A functional division in the nervous system of the cockroach, *Blaberus giganteus* (L.), *Acta Zool. (Stockholm)* **44**:43.

Rounds, H. D., 1968, Diurnal variation in the effectiveness of extracts of cockroach midgut in the release of intestinal proteinase activity, *Comp. Biochem. Physiol.* **25**:1125.

Rounds, H. D., 1975, A lunar rhythm in the occurrence of bloodborne factors in cockroaches, mice and men, *Comp. Biochem. Physiol. C* **50**:193.

Rounds, H. D., and Gardner, F. E., 1968, A quantitative comparison of the activity of cardioac-celeratory extracts from various portions of cockroach nerve cord, *J. Insect Physiol.* **14**:495.

Roussel, J. P., 1963, Consommation d'oxygène après ablation des corpora allata chez des femelles adultes de *Locusta migratoria* L., *J. Insect Physiol.* **9**:721.

Roussel, J. P., 1966, Rôle des corpora allata sur la pigmentation de *Gryllus bimaculatus* Deg., *J. Insect Physiol.* **12**:1085.

Roussel, J. P., 1969a, Action de l'ablation et de l'implantation de corpora allata sur le rythme cardiaque de *Locusta migratoria* L., *C. R. Acad. Sci. Ser. D* **269**:371.

Roussel, J. P., 1969b, Action de l'ablation et de l'implantation de corpora cardiaca sur le rythme cardiaque de *Locusta migratoria* L., *C. R. Acad. Sci. Ser. D* **269**:1882.

Roussel, J. P., 1970, Action de l'électrocoagulation et de l'implantation de pars intercerebralis sur le rythme cardiaque de *Locusta migratoria* L., *C. R. Acad. Sci. Ser. D* **270**:3083.

Roussel, J. P., 1972, Innervation du coeur et régulation nerveuse du rythme cardiaque chez *Locusta migratoria* L., *Arch. Zool. Exp. Gen.* **113**:265.

Roussel, J. P., 1975a, Action cardiotrope de l'hormone juvénile en C17 (JH-II) chez *Locusta migratoria* L., *C. R. Acad. Sci. Ser. D* **281**:1741.

Roussel, J. P., 1975b, Action cardiotrope de l'hormone juvénile synthétique en C18 de *Hyalophora cecropia* chez *Locusta migratoria* L., *C. R. Acad. Sci. Ser. D* **280**:2579.

Roussel, J. P., and Cazal, M., 1969, Action *in vivo*, d'extraits de corpora cardiaca sur le rythme cardiaque de *Locusta migratoria* L., *C. R. Acad. Sci. Ser. D* **268**:581.

Rowell, C. H. F., 1967, Corpus allatum implantation and greenbrown polymorphism in three African grasshoppers, *J. Insect Physiol.* **13**:1401.

Ruegg, R. P., Kriger, F. L., Davey, K. G., and Steel, C. G. H., 1981, Ovarian ecdysone elicits release of a myotropic ovulation hormone in *Rhodnius* (Insecta: Hemiptera), *Int. J. Invert. Reprod.* **3**:357.

Ryerse, J. S., 1978, Ecdysterone switches off fluid secretion at pupation in insect Malpighian tubules, *Nature (London)* **271**:745.

Sabaratnam, M., 1974, Hormonal control of resilin deposition in the blowfly, *Calliphora erythrocephala, J. Insect Physiol.* **20**:935.

Safranek, L., and Williams, C. M., 1980, Studies of the prothoracicotropic hormone in the tobacco hornworm, *Manduca sexta, Biol. Bull. (Woods Hole, Mass.)* **158**:141.

Safranek, L., Cymborowski, B., and Williams, C. M., 1980, Effects of juvenile hormone on ecdysone-dependent development in the tobacco hornworm, *Manduca sexta, Biol. Bull. (Woods Hole, Mass.)* **158**:248.

Sägesser, H., 1960, Über die Wirkung der Corpora allata auf den Sauerstoffverbrauch bei der Schabe *Leucophaea maderae* (F.), *J. Insect Physiol.* **5**:264.

Sakurai, H., 1977, Endocrine control of oögenesis in the housefly, *Musca domestica vicina, J. Insect Physiol.* **23**:1295.

Samaranayaka, M., 1976, Possible involvement of monoamines in the release of adipokinetic hormone in the locust *Schistocerca gregaria, J. Exp. Biol.* **65**:415.

Samaranayaka, M., 1977, Role of acetylcholine in organophosphate-induced release of adipokinetic hormone in the locust *Schistocerca gregaria, Pestic. Biochem. Physiol.* **7**:283.

Sandifer, J. B., and Tombes, A. S., 1972, Ultrastructure of the lateral neurosecretory cells during reproductive development of *Sitophilus granarius* (L.) (Insecta: Coleoptera), *Tissue Cell* **4**:437.

Saunders, D. S., 1965, Larval diapause of maternal origin: Induction of diapause in *Nasonia vitripennis* (Walk.) (Hymenoptera: Pteromalidae), *J. Exp. Biol.* **42**:495.

Schaller, F., and Charlet, M., 1970, Evolution du système neurosécréteur de larves *d'Aeschna cyanea* Müll. (Insecte, Odonate) privées d'ecdysone, *C. R. Acad. Sci. Ser. D* **271**:2004.

Schaller, F., and Meunier, J., 1968, Etude du système neurosécréteur céphalique des insectes Odonates, *Bull. Soc. Zool. Fr.* **93**:233.

Scharrer, B., 1952, Neurosecretion. XI. The effects of nerve section on the inter-cerebralis–cardiacum–allatum system of the insect *Leucophaea maderae, Biol. Bull. (Woods Hole, Mass.)* **102**:261.

Scharrer, B., 1955, "Castration cells" in the central nervous system of an insect (*Leucophaea maderae* Blattaria), *Trans. N.Y. Acad. Sci.* **17**:520.

Scharrer, B., 1963, Neurosecretion. XIII. The ultrastructure of the corpus cardiacum of the insect *Leucophaea maderae, Z. Zellforsch. Mikrosk. Anat.* **60**:761.

Scharrer, B., 1964a, Histophysiological studies on the corpus allatum of *Leucophaea maderae.* IV. Ultrastructure during normal activity cycle, *Z. Zellforsch. Mikrosk. Anat.* **62**:125.

Scharrer, B., 1964b, The fine structure of Blattarian prothoracic glands, *Z. Zellforsch. Mikrosk. Anat.* **64**:301.

Scharrer, B., 1968, Neurosecretion. XIV. Ultrastructural study of sites of release of neurosecretory material in Blattarian insects, *Z. Zellforsch. Mikrosk. Anat.* **89**:1.

Scharrer, B., 1978, Histophysiological studies on the corpus allatum of *Leucophaea maderae, Cell Tissue Res.* **194**:533.

Scharrer, B., and Kater, S. B., 1969, Neurosecretion. XV. An electron microscopic study of the corpora cardiaca of *Periplaneta americana* after experimentally induced hormone release, *Z. Zellforsch. Mikrosk. Anat.* **95**:177.

Scharrer, B., and Von Harnack, M., 1958, Histophysiological studies on the corpus allatum of *Leucophaea maderae.* I. Normal life cycle in male and female adults, *Biol. Bull. (Woods Hole, Mass.)* **115**:508.

Scheurer, R., 1969a, Endocrine control of protein synthesis during oöcyte maturation in the cockroach *Leucophaea maderae, J. Insect Physiol.* **15**:1411.

Scheurer, R., 1969b, Haemolymph proteins and yolk formation in the cockroach, *Leucophaea maderae, J. Insect Physiol.* **15**:1673.

Scheurer, R., and Lüscher, M., 1968, Nachweis der Synthese eines Dotterproteins unter dem Einfluss der Corpora allata bei *Leucophaea maderae, Rev. Suisse Zool.* **75**:715.

Schlein, Y., 1972a, Postemergence growth in the fly *Sarcophaga falculata* initiated by neurosecretion from the ocellar nerve, *Nature (London)* **236**:217.

Schlein, Y., 1972b, Factors that influence the postemergence growth in *Sarcophaga falculata, J. Insect Physiol.* **18**:199.

Schleip, W., 1910, Der Farbenwechsel von *Dixippus morosus, Zool. Jahrb. Abt. Allg. Zool. Physiol. Tiere* **30**:45.

Schooneveld, H., 1969, Control of the activity of neurosecretory A-cells in the Colorado beetle, *Leptinotarsa decemlineata* Say, *Gen. Comp. Endocrinol.* **13**:530.

Schooneveld, H., 1970, Structural aspects of neurosecretory and corpus allatum activity in the adult Colorado beetle, *Leptinotarsa decemlineata* Say, as a function of daylength, *Neth. J. Zool.* **20**:151.

Schooneveld, H., 1972, Effects of juvenile hormone on the endocrine system during break of diapause in the Colorado beetle, *J. Endocrinol.* **57**:55.

Schooneveld, H., 1974a, Ultrastructure of the neurosecretory system of the Colorado potato beetle, *Leptinotarsa decemlineata* (Say). I. Characterization of the protocerebral neurosecretory cells, *Cell Tissue Res.* **154**:275.

Schooneveld, H., 1974b, Ultrastructure of the neurosecretory system of the Colorado potato beetle *Leptinotarsa decemlineata* (Say). II. Pathways of axonal secretion transport and innervation of neurosecretory cells, *Cell Tissue Res.* **154**:289.

Schooneveld, H., Otazo Sanchez, A., and de Wilde, J., 1977, Juvenile hormone-induced break and termination of diapause in the Colorado potato beetle, *J. Insect Physiol.* **23**:689.

Schooneveld, H., Kramer, S. J., Privee, H., and Van Huis, A., 1979, Evidence of controlled corpus allatum activity in the adult Colorado beetle, *J. Insect Physiol.* **25**:449.

Schreiner, B., 1966, Histochemistry of the A cell neurosecretory material in the milkweed bug, *Oncopeltus fasciatus* Dallas (Heteroptera, Lygaeidae), with a discussion of the neurosecretory material/carrier substance problem, *Gen. Comp. Endocrinol.* **6**:388.

Schreiner, B., 1977, The effect of the hormone(s) from the corpus allatum complex on the ovarian tissue of *Oncopeltus fasciatus*. A light and electron microscopic investigation, *J. Morphol.* **151**:81.

Schultz, R. L., 1960, Electron microscopic observations of the corpora allata and associated nerves in the moth *Celerio lineata*, *J. Ultrastruct. Res.* **3**:320.

Schwartz, L. M., and Reynolds, S. E., 1979, Fluid transport in *Calliphora* Malpighian tubules: A diuretic hormone from the thoracic ganglion and abdominal nerves, *J. Insect Physiol.* **25**:847.

Sedlak, B. J., 1981, An ultrastructural study on neurosecretory fibers within the corpora allata of *Manduca sexta, Gen. Comp. Endocrinol.* **44**:207.

Sehnal, F., and Granger, N. A., 1975, Control of corpora allata function in larvae of *Galleria mellonella* L., *Biol. Bull. (Woods Hole, Mass.)* **148**:106.

Sehnal, F., and Sláma, K., 1966, The effect of corpus allatum hormone on respiratory metabolism during larval development and metamorphosis of *Galleria mellonella* L., *J. Insect Physiol.* **12**:1333.

Sekeris, C., and Karlson, P., 1962, Zum Tyrosinstoffwechsel der Insekten. VII. Der Katabolische Abbau des Tyrosins und die Biogenese der Sklerotisierungssubstanz, *N*-Acetyldopamin, *Biochim. Biophys. Acta* **62**:103.

Seligman, I. M., and Doy, F. A., 1972, Studies on cyclic AMP mediation of hormonally induced cytolysis of the alary hypodermal cells and of hormonally controlled DOPA synthesis in *Lucilia cuprina, Isr. J. Entomol.* **7**:129.

Seligman, I. M., and Doy, F. A., 1973, Hormonal regulation of disaggregation of cellular fragments in the haemolymph of *Lucilia cuprina, J. Insect Physiol.* **19**:125.

Seligman, M., Friedman, S., and Fraenkel, G., 1969, Bursicon mediation of tyrosine hydroxylation during tanning of the adult cuticle of the fly, *Sarcophaga bullata, J. Insect Physiol.* **15**:553.

Seligman, M., Blechl, A., Blechl, J., Herman, P., and Fraenkel, G., 1977, Role of ecdysone, pupariation factors, and cyclic AMP in formation and tanning of puparium of the fleshfly *Sarcophaga bullata, Proc. Natl. Acad. Sci. USA* **74**:4697.

Seshan, K. R., 1968, Les phénomènes de neurosécrétion dans la chaîne nerveuse ventrale d'*Iphita limbata, C. R. Acad. Sci. Ser. D* **266**:619.

Shaaya, E., and Sekeris, C. E., 1970, The formation of protocatechuic acid-4-O, β-glucoside in *Periplaneta americana* and the possible role of the juvenile hormone, *J. Insect Physiol.* **16**:323.

Shimada, S., and Yamashita, O., 1979, Trehalose absorption related with trehalase in developing ovaries of the silkworm, *Bombyx mori, J. Comp. Physiol.* **131**:333.

Siakotos, A. N., 1960, The conjugated plasma proteins of the American cockroach. I. The normal state, *J. Gen. Physiol.* **43**:999.

Sieber, R., and Benz, G., 1977, Juvenile hormone in larval diapause of the codling moth, *Laspeyresia pomonella* L. (Lepidopterae, Tortricidae), *Experientia* **33**:1598.

Sieber, R., and Benz, G., 1980, The hormonal regulation of the larval diapause in the codling moth, *Laspeyresia pomonella* (Lep. Tortricidae), *J. Insect Physiol.* **26**:213.

Siew, Y. C., 1965, The endocrine control of adult reproductive diapause in the chrysomelid beetle *Galeruca tanaceti* (L.). II, *J. Insect Physiol.* **11**:463.

Siew, Y. C., and Gilbert, L. I., 1971, Effects of moulting hormone and juvenile hormone on insect endocrine gland activity, *J. Insect Physiol.* **17**:2095.

Singh, H. H., and Sehnal, F., 1979, Lack of specific neurons in the ventral nerve cord for the control of prothoracic glands, *Experientia* **35**:1117.

Singh, R. K., Singh, S. B., and Khan, T. R., 1978, Correlative studies on the neuroendocrine system during postembryonic development in the bug, *Chrysocoris stolli* Wolff (Heteroptera: Pentatomidae), *Folia Morphol. (Prague)* **26**:107.

Sivasubramanian, P., Friedman, S., and Fraenkel, G., 1974, Nature and role of proteinaceous hormonal factors acting during puparium formation in flies, *Biol. Bull. (Woods Hole, Mass.)* **147**:163.

Sláma, K., 1964, Hormonal control of respiratory metabolism during growth, reproduction, and diapause in female adults of *Pyrrhocoris apterus* L. (Hemiptera), *J. Insect Physiol.* **10**:283.

Sláma, K., 1965, Effects of hormones on the respiration of body fragments of adult *Pyrrhocoris apterus* L. (Hemiptera), *Nature (London)* **205**:416.

Sláma, K., and Zdarek, J., 1974, Effect of hormones on water metabolism in *Pyrrhocoris apterus* L., *Zool. Jahrb. Abt. Allg. Zool. Physiol. Tiere* **78**:1397.

Smalley, K. N., 1970, Median nerve neurosecretory cells in the abdominal ganglia of the cockroach, *Periplaneta americana, J. Insect Physiol.* **16**:241.

Smith, D. S., 1963, The organization and innervation of the luminescent organ in a firefly, *Photuris pennsylvanica* (Coleoptera), *J. Cell Biol.* **16**:323.

Smith, D. S., 1968, *Insect Cells: Their Structure and Function,* Oliver & Boyd, Edinburgh.

Smith, N. A., 1969, Observations on the neural rhythmicity in the cockroach cardiac ganglion, *Experientia Suppl.* **15**:200.

Smith, U., 1971, Uptake of ferritin into neurosecretion terminals, *Philos. Trans. R. Soc. London Ser. B* **261**:391.

Smith, U., and Smith, D. S., 1966, Observations on the secretory processes in the corpus cardiacum of the stick insect, *Carausius morosus, J. Cell Sci.* **1**:59.

Sokolove, P. G., and Loher, W., 1975, Role of eyes, optic lobes, and pars intercerebralis in locomotory and stridulatory circadian rhythms of *Teleogryllus commodus, J. Insect Physiol.* **21**:785.

Sonobe, H., and Keino, H., 1975, Diapause factor in the brains, subesophageal ganglia and prothoracic ganglia of the silkworm, *Naturwissenschaften* **62**:348.

Sonobe, H., and Ohnishi, E., 1971, Silkworm *Bombyx mori* L.; Nature of diapause factor, *Science* **174**:4835.

Spencer, I. M., and Candy, D. J., 1976, Hormonal control of diacyl glycerol mobilization from fat body of the desert locust, *Schistocerca gregaria, Insect Biochem.* **6**:289.

Spring, J. H., and Phillips, J. E., 1980, Studies on locust rectum. II. Identification of specific ion transport processes regulated by corpora cardiaca and cyclic-AMP, *J. Exp. Biol.* **86**:225.

Spring, J., Hanrahan, J., and Phillips, J., 1978, Hormonal control of chloride transport across locust rectum, *Can. J. Zool.* **56**:1879.

Srivastava, B. B. L., and Hopkins, T. L., 1975, Bursicon release and activity in haemolymph during metamorphosis of the cockroach, *Leucophaea maderae, J. Insect Physiol.* **21**:1985.

Srivastava, K. P., Singh, H. H., and Tiwari, R. K., 1975, Neuroendocrine organs of the lemon-butterfly, *Papilio demoleus* L. I. The corpora cardiaca–allata complex of the larva, *Z. Mikrosk. Anat. Forsch.* **89**:415.

Srivastava, K. P., Tiwari, R. K., and Kumar, P., 1977, Effect of sectioning the prothoracic gland nerves in the larva of the lemon-butterfly, *Papilio demoleus* L., *Experientia* **33**:98.

Srivastava, R. C., 1969, A note on the neurosecretory pathways in *Pyrilla perpusilla* Walker (Fulgoridae: Homoptera), *Experientia* **25**:1097.

Sroka, P., and Barth, R. H., 1975, Lipid metabolism during oöcyte growth in the ovoviviparous cockroaches *Eublablerus posticus* and *Nauphoeta cinerea, Insect Biochem.* **5**:637.

Sroka, P., and Gilbert, L. I., 1971, Studies on the endocrine control of post-emergence ovarian maturation in *Manduca sexta, J. Insect Physiol.* **17**:2409.

Staal, Ir. G. B., 1961, *Studies on the Physiology of Phase Induction in Locusta migratoria migratorioides R. et F.,* Publicatie 40 LEB Fonds, H. Veenman and N. V. Zonen, Wageningen.

Starratt, A. N., and Brown, B. E., 1975, Structure of the pentapeptide proctolin, a proposed neurotransmitter in insects, *Life Sci.* **17**:1253.

Starratt, A. N., and Brown, B. E., 1977, Synthesis of proctolin, a pharmacologically active pentapeptide in insects, *Can. J. Chem.* **55**:4238.

Stay, B., and Tobe, S. S., 1977, Control of juvenile hormone biosynthesis during the reproductive cycle of a viviparous cockroach. I. Activation and inhibition of corpora allata, *Gen. Comp. Endocrinol.* **33**:531.

Stay, B., and Tobe, S. S., 1978, Control of juvenile hormone biosynthesis during the reproductive cycle of a viviparous cockroach. II. Effects of unilateral allatectomy, implantation of supernumerary corpora allata and ovariectomy, *Gen. Comp. Endocrinol.* **34**:276.

Stay, B., Friedel, T., Tobe, S. S., and Mundall, E. C., 1980, Feedback control of juvenile hormone synthesis in cockroaches: Possible role for ecdysterone, *Science* **207**:898.

Steel, C. G. H., 1977, The neurosecretory system in the aphid *Megoura viciae*, with reference to unusual features associated with long distance transport of neurosecretion, *Gen. Comp. Endocrinol.* **31**:307.

Steel, C. G. H., 1978a, Nervous and hormonal regulation of neurosecretory cells in the insect brain, in: *Comparative Endocrinology* (P. J. Gaillard and H. H. Boer, eds.), Elsevier/North-Holland, Amsterdam, p. 327.

Steel, C. G. H., 1978b, Some functions of identified neurosecretory cells in the brain of the aphid, *Megoura viciae, Gen. Comp. Endocrinol.* **34**:219.

Steel, C. G. H., and Lees, A. D., 1977, The role of neurosecretion in the photoperiodic control of polymorphism in the aphid *Megoura viciae, J. Exp. Biol.* **67**:117.

Steele, J. E., 1961, Occurrence of a hyperglycaemic factor in the corpus cardiacum of an insect, *Nature (London)* **192**:680.

Steele, J. E., 1963, The site of action of insect hyperglycemic hormone, *Gen. Comp. Endocrinol.* **3**:46.

Steele, J. E., 1964, The activation of phosphorylase in an insect by adenosine 3′,5′-phosphate and other agents, *Am. Zool.* **4**:237.

Steele, J. E., and Tolman, J. H., 1980, Regulation of water transport in the cockroach rectum by the corpora cardiaca–corpora allata system: The requirement for Na, *J. Comp. Physiol.* **138B**:357.

Stone, J. V., Mordue, W., Batley, K. E., and Morris, H. R., 1976, Structure of locust adipokinetic hormone, a neurohormone that regulates lipid utilisation during flight, *Nature (London)* **263**:207.

Stone, J. V., Mordue, W., Broomfield, C. E., and Hardy, P. M., 1978, Structure–activity relationships for the lipid mobilising action of locust adipokinetic hormone: Synthesis and activity of a series of hormone analogues, *Eur. J. Biochem.* **89**:195.

Strambi, A., and Strambi, C., 1973, Etude histochimique et ultrastructurale des sécrétions élaborées par les péricaryones neurosécréteurs de la pars intercerebralis chez la guêpe *Polistes, Acta Histochem.* **46**:101.

Strambi, C., Strambi, A., Cupo, A., Rougon-Rapuzzi, G., and Martin, N., 1978, Etude des taux d'une substance apparentée à la vasopressine dans le système nerveux de grillons soumis à différentes conditions hygrométriques, *C. R. Acad. Sci. Ser. D* **287**:1227.

Strambi, C., Rougon-Rapuzzi, G., Cupo, A., Martin, N., and Strambi, A., 1979, Mise en évidence immunocytologique d'un composé apparenté à la vasopressine dans le système nerveux du grillon *Acheta domesticus, C. R. Acad. Sci. Ser. D* **288**:131.

Strejčková, A., Servít, Z., and Novák, V. J. A., 1964, Effect of neurohormones Cl and Dl on spontaneous electrical activity of the central nervous system of the cockroach, *J. Insect Physiol.* **10**:889.

Strong, L., 1965a, The relationships between the brain, corpora allata and oöcyte growth in the central American locust, *Schistocerca* sp.-I. The cerebral neurosecretory system, the corpora allata, and oöcyte growth, *J. Insect Physiol.* **11**:135.

Strong, L., 1965b, The relationships between the brain, corpora allata and oöcyte growth in the central American locust, *Schistocerca* sp.-II. The innervation of the corpora allata, the lateral neurosecretory complex, and oöcyte growth, *J. Insect Physiol.* **11**:271.

Strong, L., 1966, On the occurrence of neuroglandular axons within the sympathetic nervous system of a locust, *Locusta migratoria migratorioides*, *J. R. Microsc. Soc.* **86**:141.

Strong, L., 1968, The effect of enforced locomotor activity on lipid content in allatectomized males of *Locusta migratoria migratorioides*, *J. Exp. Biol.* **48**:625.

Suzuki, A., Isogai, A., Horii, T., Ishizaki, H., and Tamura, S., 1975, A simple procedure for partial purification of silkworm brain hormone, *Agric. Biol. Chem.* **39**:2157.

Suzuki, A., Matsumoto, S., Ogura, N., Isogai, A., and Tamura, S., 1976, Extraction and partial purification of the hormone inducing cuticular melanization in armyworm larvae, *Agric. Biol. Chem.* **40**:2307.

Tager, H. S., and Kramer, K. J., 1980, Insect glucagon-like peptides: Evidence for a high molecular weight form in midgut from *Manduca sexta* (L.), *Insect Biochem.* **10**:617.

Tager, H. S., Markese, J., Kramer, K. J., Speirs, R. D., and Childs, C. N., 1976, Glucagon-like and insulin-like hormones of the insect neurosecretory system, *Biochem. J.* **156**:515.

Taghert, P. H., Truman, J. W., and Reynolds, S. E., 1980, Physiology of pupal ecdysis in the tobacco hornworm, *Manduca sexta*. II. Chemistry, distribution, and release of eclosion hormone at pupal ecdysis, *J. Exp. Biol.* **88**:339.

Takeda, N., 1972, Activation of neurosecretory cells in *Monema flavescens* (Lepidoptera) during diapause break, *Gen. Comp. Endocrinol.* **18**:417.

Takeda, N., 1978, Hormonal control of prepupal diapause in *Monema flavescens* (Lepidoptera), *Gen. Comp. Endocrinol.* **34**:123.

Takeda, S., 1977, Induction of egg diapause in *Bombyx mori* by some cephalo-thoracic organs of the cockroach, *Periplaneta americana*, *J. Insect Physiol.* **23**:813.

Takeda, S., and Ogura, N., 1976, Induction of egg diapause by implantation of corpora cardiaca and corpora allata in *Bombyx mori*, *J. Insect Physiol.* **22**:941.

Teissier, G., 1947, Fonctionnement des chromatophores de la larve de Corèthre, *C. R. Acad. Sci.* **225**:204.

Tembhare, D. B., and Thakare, V. K., 1977, Neurosecretory system of the ventral ganglia in the dragonfly, *Orthetrum chrysis* (Selys) (Odonata: Libellulidae), *Cell Tissue Res.* **177**:269.

Thomas, A., 1964, Recherches expérimentales sur le contrôle endocrine de l'ovogenèse chez *Gryllus domesticus* (L.) (Orthoptère), *C. R. Acad. Sci.* **259**:1561.

Thomas, A., 1968, Recherches sur le contrôle neuro-endocrine de l'oviposition chez *Carausius morosus* Br. (Phasmides-Chéleutoptères), *C. R. Acad. Sci. Ser. D* **267**:518.

Thomas, A., 1969, Etude du déterminisme de l'oviposition chez *Carausius morosus* Br. (Phasmides, Chéleutoptères), *C. R. Acad. Sci. Ser. D* **269**:2424.

Thomas, A., 1979, Nervous control of egg progression into the common oviduct and genital chamber of the stick-insect *Carausius morosus*, *J. Insect Physiol.* **25**:811.

Thomas, A., and Mesnier, M., 1973, Le rôle du système nerveux central sur les mécanismes de l'oviposition chez *Carausius morosus* et *Clitumnus extradentatus*, *J. Insect Physiol.* **19**:383.

Thomas, A., and Raabe, M., 1974, Les organes périsympathiques des Orthoptères, *Bull. Soc. Zool. Fr.* **99**:187.

Thomas, K. K., 1972, Studies on the synthesis of lipoproteins during larval-pupal development of *Hyalophora cecropia*, *Insect Biochem.* **2**:107.

Thomas, K. K., and Gilbert, L. I., 1969, The hemolymph lipoproteins of the silkworm *Hyalophora gloveri*: Studies on lipid composition, origin and function, *Physiol. Chem. Phys.* **1**:293.

Thomas, K. K., and Nation, J. L., 1966a, Control of a sex-limited haemolymph protein by corpora allata during ovarian development in *Periplaneta americana* (L.), *Biol. Bull. (Woods Hole, Mass.)* **130**:254.

Thomas, K. K., and Nation, J. L., 1966b, R.N.A., protein and uric acid content of body tissues of *Periplaneta americana* (L.) as influenced by corpora allata during ovarian development, *Biol. Bull. (Woods Hole, Mass.)* **130**:442.

Thomsen, E., 1949, Influence of the corpus allatum on the oxygen consumption of adult *Calliphora erythrocephala* Meig., *J. Exp. Biol.* **26**:137.

Thomsen, E., 1952, Functional significance of the neurosecretory brain cells and the corpus cardiacum in the female blow-fly, *Calliphora erythrocephala* Meig., *J. Exp. Biol.* **29**:137.

Thomsen, E., and Lea, A. O., 1969, Control of the medial neurosecretory cells by the corpus allatum in *Calliphora erythrocephala, Gen. Comp. Endocrinol.* **12**:51.

Thomsen, E., and Møller, I., 1959, Neurosecretion and intestinal proteinase activity in an insect, *Calliphora, Nature (London)* **183**:1401.

Thomsen, E., and Møller, I., 1963, Influence of neurosecretory cells and of corpus allatum on intestinal protease activity in the adult *Calliphora erythrocephala* Meig., *J. Exp. Biol.* **40**:301.

Thomsen, E., and Thomsen, M., 1970, Fine structure of the corpus allatum of the female blow-fly *Calliphora erythrocephala, Z. Zellforsch. Mikrosk. Anat.* **110**:40.

Thomsen, E., Hansen, B. L., Hansen, G. N., and Jensen, P. V., 1980, Ultrastructural immunocytochemical localization of vitellogenin in the fat body of the blowfly, *Calliphora vicina* Rob.-Desv. (*erythrocephala* Meig.) by use of the unlabeled antibody–enzyme method, *Cell Tissue Res.* **208**:445.

Thomsen, M., 1943, Effect of corpus cardiacum and other insect organs on the colour-change of the shrimp, *Leander adspersus, Biol. Medd. K. Dan. Vidensk. Selsk.* **19**:1.

Thomsen, M., 1954, Neurosecretion in some Hymenoptera, *Dan. Biol. Skr.* **7**:1.

Thomsen, M., 1965, The neurosecretory system of the adult *Calliphora erythrocephala.* II. Histology of the neurosecretory cells of the brain and some related structures, *Z. Zellforsch. Mikrosk. Anat.* **67**:693.

Thomsen, M., 1969, The neurosecretory system of the adult *Calliphora erythrocephala.* IV. A histological study of the corpus cardiacum and its connections with the nervous system, *Z. Zellforsch. Mikrosk. Anat.* **94**:205.

Tobe, S. S., and Stay, B., 1979, Modulation of juvenile hormone synthesis by an analogue in the cockroach, *Nature (London)* **281**:481.

Tobe, S. S., and Stay, B., 1980, Control of juvenile hormone biosynthesis during the reproductive cycle of a viviparous cockroach. III. Effects of denervation and age on compensation with unilateral allatectomy and supernumerary corpora allata, *Gen. Comp. Endocrinol.* **40**:89.

Tobe, S. S., Chapman, C. S., and Pratt, G. E., 1977, Decay in juvenile hormone biosynthesis by insect corpus allatum after nerve transection, *Nature (London)* **268**:728.

Tolman, J. H., and Steele, J. E., 1980a, The control of glycogen metabolism in the cockroach hindgut: The effect of the corpora cardiaca–corpora allata system, *Comp. Biochem. Physiol. B* **66**:59.

Tolman, J. H., and Steele, J. E., 1980b, The effect of the corpora cardiaca–corpora allata system on oxygen consumption in the cockroach rectum: The role of Na^+ and K^+, *J. Comp. Physiol. B* **138**:347.

Tombes, A. S., 1976, Myoneural junctions and neurosecretory endings on spermathecal muscle fibers of two weevils, *Sitophilus granarius* and *Hypera postica, J. Insect Physiol.* **22**:1573.

Tombes, A. S., and Malone, T. A., 1977, Ultrastructure of the frontal ganglion of *Chaoborus* and observations on neurosecretory cells during the annual cycle, *J. Insect Physiol.* **23**:639.

Tombes, A. S., and Smith, D. S., 1970, Ultrastructural studies on the corpora cardiaca–allata complex of the adult alfalfa weevil, *Hypera postica, J. Morphol.* **132**:137.

Traina, M. E., Bellino, M., and Frontali, N., 1974, A comparison between neurohormone D and "peak 1" extracted from cockroach corpora cardiaca, *Zool. Jahrb. Abt. Allg. Zool. Physiol. Tiere* **78**:424.

Traina, M. E., Bellino, M., Serpietri, L., Massa, A., and Frontali, N., 1976, Heart-accelerating peptides from cockroach corpora cardiaca, *J. Insect Physiol.* **22**:323.

Treherne, J. E., and Willmer, P. G., 1975, Hormonal control of integumentary water-loss: Evidence for a novel neuroendocrine system in an insect (*Periplaneta americana*), *J. Exp. Biol.* **63**:143.

Truman, J. W., 1971, Physiology of insect ecdysis. I. The eclosion behaviour of saturniid moths and its hormonal release, *J. Exp. Biol.* **54**:805.

Truman, J. W., 1972, Physiology of insect rhythms. I. Circadian organization of the endocrine events underlying the moulting cycle of larval tobacco hornworms, *J. Exp. Biol.* **57**:805.

Truman, J. W., 1973a, How moths "turn on": A study of the action of hormones on the nervous system, *Am. Sci.* **61**:700.

Truman, J. W., 1973b, Physiology of insect ecdysis. II. The assay and occurrence of the eclosion hormone in the Chinese oak silkmoth, *Antheraea pernyi*, *Biol. Bull. (Woods Hole, Mass.)* **144**:200.

Truman, J. W., 1973c, Physiology of insect ecdysis. III. Relationship between the hormonal control of eclosion and of tanning in the tobacco hornworm, *Manduca sexta*, *J. Exp. Biol.* **58**:821.

Truman, J. W., 1978, Hormonal release of stereotyped motor programmes from the isolated nervous system of the *cecropia* silkmoth, *J. Exp. Biol.* **74**:151.

Truman, J. W., and Riddiford, L. M., 1970, Neuroendocrine control of ecdysis in silkmoths, *Science* **167**:1624.

Truman, J. W., and Riddiford, L. M., 1971, Role of the corpora cardiaca in the behavior of saturniid moth. II. Oviposition, *Biol. Bull. (Woods Hole, Mass.)*, **140**:8.

Truman, J. W., and Riddiford, L. M., 1974, Physiology of insect rhythms. III. The temporal organization of the endocrine events underlying pupation of the tobacco hornworm, *J. Exp. Biol.* **60**:371.

Truman, J. W., and Sokolove, P. G., 1972, Silk moth eclosion: Hormonal triggering of a centrally programmed pattern of behavior, *Science* **175**:1491.

Truman, J. W., Fallon, A. M., and Wyatt, G. R., 1976, Hormonal release of programmed behavior in silk moths: Probable mediation by cyclic AMP, *Science* **194**:1432.

Truman, J. W., Mumby, S. M., and Welch, S. K., 1979, Involvement of cyclic GMP in the release of stereotyped behaviour patterns in moths by a peptide hormone, *J. Exp. Biol.* **84**:201.

Truman, J. W., Taghert, P. H., and Reynolds, S. E., 1980, Physiology of pupal ecdysis in the tobacco hornworm, *Manduca sexta*. I. Evidence for control by eclosion hormone, *J. Exp. Biol.* **88**:327.

Truman, J. W., Taghert, P. H., Copenhaver, P. F., Tublitz, N. J., and Schwartz, L. M., 1981, Eclosion hormone may control all ecdyses in insects, *Nature (London)* **291**:70.

Unger, H., 1957, Untersuchungen zur neurohormonalen Steuerung der Herztätigkeit bei Schaben (*Periplaneta orientalis, P. americana, Phyllodromia germanica*), *Biol. Zentralbl.* **76**:204.

Unger, H., 1967, Der Einfluss der Neurohormone C und D auf Wasserhaushalt der Stab-heuschrecke (*Carausius morosus* Br.), *Biol. Rundsch.* **5**:31.

Unnithan, G. C., Bern, H. A., and Nayar, K. K., 1971, Ultrastructural analysis of the neuroen-docrine apparatus of *Oncopeltus fasciatus* (Heteroptera), *Acta Zool. (Stockholm)* **52**:117.

Vandenberg, R. D., and Mills, R. R., 1974, Hormonal control of tanning by the American cockroach: Cyclic AMP as a probable intermediate, *J. Insect Physiol.* **20**:623.

Vandenberg, R. D., and Mills, R. R., 1975, Adenyl cyclase in the haemocytes of the American cockroach, *J. Insect Physiol.* **21**:221.

Vanderberg, J. P., 1963, Synthesis and transfer of DNA, RNA, and protein during vitellogenesis in *Rhodnius prolixus* (Hemiptera), *Biol. Bull. (Woods Hole, Mass.)* **125**:556.

Van Der Horst, D. J., Van Doorn, J. M., and Beenakkers, A. M. Th., 1979, Effects of the adipokinetic hormone on the release and turnover of haemolymph diglycerides and on the formation of the diglyceride-transporting lipoprotein system during locust flight, *Insect Biochem.* **9**:627.

Van Handel, E., and Lea, A. O., 1965, Medial neurosecretory cells as regulators of glycogen and triglyceride synthesis, *Science* **149**:298.

Vedeckis, W. V., and Gilbert, L. I., 1973, Production of cyclic AMP and adenosine by the brain and prothoracic glands of *Manduca sexta, J. Insect Physiol.* **19**:2445.

Vedeckis, W. V., Bollenbacher, W. E., and Gilbert, L. I., 1974, Cyclic AMP as a possible mediator of prothoracic gland activation, *Zool. Jahrb. Abt. Allg. Zool. Physiol. Tiere* **78**:440.

Vedeckis, W. V., Bollenbacher, W. E., and Gilbert, L. I., 1976, Insect prothoracic glands: A role for cyclic AMP in the stimulation of α-ecdysone secretion, *Mol. Cell. Endocrinol.* **5**:81.

Vejbjerg, K., and Normann, T. C., 1974, Secretion of hyperglycaemic hormone from the corpus cardiacum of flying blowflies, *Calliphora erythrocephala, J. Insect Physiol.* **20**:1189.

Véron, J. E. N., 1973, Physiological control of the chromatophores of *Austrolestes annulosus* (Odonata), *J. Insect Physiol.* **19**:1689.

Vietinghoff, U., 1966a, Untersuchungen über die Funktion der Rektaldrüsen der Stabheuschrecke *Carausius morosus* Br., *Zool. Anz.* **29**:157.

Vietinghoff, U., 1966b, Einfluss der Neurohormone C1 und D1 auf die Absorptionsleistung der Rektaldrüsen der Stabheuschrecke (*Carausius morosus* Br.), *Naturwissenschaften* **53**:162.

Vietinghoff, U., 1967, Neurohormonal control of "renal function" in *Carausius morosus* Br., *Gen. Comp. Endocrinol.* **9**:503.

Vijverberg, A. J., 1970, The larval and pupal neurosecretory system of *Calliphora erythrocephala* Meigen. Histology and activity of neurosecretory cells in brain and sub-oesophageal ganglion, *Neth. J. Zool.* **20**:353.

Vincent, J. F. V., 1971, Effects of bursicon on cuticular properties in *Locusta migratoria migratorioides, J. Insect Physiol.* **17**:625.

Vincent, J. F. V., 1972, The dynamics of release and the possible identity of bursicon in *Locusta migratoria migratorioides, J. Insect Physiol.* **18**:757.

Vogt, M., 1949, Fettkörper und Oenocyten der *Drosophila* nach Exstirpation der adulten Ringdrüse, *Z. Zellforsch. Mikrosk. Anat.* **34**:160.

von Knorre, D., Gersch, M., and Kusch, T., 1972, Zur Frage der Beeinflussung des "tanning"-Phänomens durch zyklisches 3',5'-AMP, *Zool. Jahrb. Abt. Allg. Zool. Physiol. Tiere* **76**:434.

Von Weinbörmair, G., Pohlhammer, K., and Dürnberger, H., 1975, Das Axonsystem des ventromedianen neurosekretorischen Zellpaares im Suboesophagealganglion bei *Teleogryllus commodus* Walker. Darstellung im Totalpräparat mit Hilfe von Resorcinfuchsin, *Mikroskopie* **31**:147.

Vroman, H. E., Kaplanis, J. N., and Robbins, W. E., 1965, Effect of allatectomy on lipid biosynthesis and turnover in the female American cockroach, *Periplaneta americana* L., *J. Insect Physiol.* **11**:897.

Vuillaume, M., Seuge, S., and Bergerard, J., 1971, Photopériode et pigment tégumentaire vert des chenilles de *Pieris brassicae:* Conditionnement de la diapause, *C. R. Acad. Sci. Ser. D* **273**:1608.

Waku, Y., 1960, Studies on the hibernation and diapause in insects. 4. Histological observations of the endocrine organs in the diapause and non-diapause larvae of the Indian meal-moth, *Plodia interpunctella* Hübner, *Sci. Rep. Tohoku Univ. Ser. 4* **26**:327.

Waku, Y., and Gilbert, L. I., 1964, The corpora allata of the silkmoth *Hyalophora cecropia:* An ultrastructural study, *J. Morphol.* **115**:69.

Walker, G. P., and Denlinger, D. L., 1980, Juvenile hormone and moulting hormone titres in diapause- and non-diapause destined flesh flies, *J. Insect Physiol.* **26**:661.

Walker, P. R., and Bailey, E., 1970, Changes in enzymes associated with lipogenesis during development of the adult male desert locust, *J. Insect Physiol.* **16**:679.

Walker, P. R., and Bailey, E., 1971a, Effect of allatectomy on fat body lipid metabolism of the male desert locust during adult development, *J. Insect Physiol.* **17**:813.

Walker, P. R., and Bailey, E., 1971b, Effect of allatectomy on the growth of the male desert locust during adult development, *J. Insect Physiol.* **17**:1125.

Wall, B. J., 1965, Regulation of water metabolism by the Malpighian tubules and rectum in the cockroach *Periplaneta americana* L., *Zool. Jahrb. Abt. Allg. Zool. Physiol. Tiere* **71**:702.

Wall, B. J., 1966, Diuresis in the cockroach, *Am. Zool.* **6**:198.

Wall, B. J., and Oschman, J. L., 1973, Structure and function of rectal pads in *Blattella* and *Blaberus* with respect to the mechanism of water uptake, *J. Morphol.* **115**:69.

Wall, B. J., and Ralph, C. L., 1962, Factors from the nervous system of the cockroach that influence excretion rates of Malpighian tubules, *Am. Zool.* **2**:103.

Wall, B. J., and Ralph, C. L., 1964, Evidence for hormonal regulation of Malpighian tubule excretion in the insect, *Periplaneta americana* L., *Gen. Comp. Endocrinol.* **4**:452.

Weaver, R. J., 1981, Radiochemical assays of corpus allatum activity in adult female cockroaches following ovariectomy in the last nymphal instar, *Experientia* **37**:435.

Weber, V., and Emmerich, H., 1976, RNA biosynthesis in isolated prothoracic glands of *Tenebrio molitor in vitro, Experientia* **32**:1609.

Weber, W., and Gaude, H., 1971, Ultrastruktur des Neurohaemalorgans in Nervus corporis allati II von *Acheta domesticus, Z. Zellforsch. Mikrosk. Anat.* **121**:561.

Weed-Pfeiffer, I. W., 1945, Effect of the corpora allata on the metabolism of adult female grasshoppers, *J. Exp. Zool.* **99**:183.

Weeda, E., 1981, Hormonal regulation of proline synthesis and glucose release in the fat body of the Colorado potato beetle, *Leptinotarsa decemlineata, J. Insect Physiol.* **27**:411.

Wenig, K., and Joachim, J., 1936, Influence of insulin on silkworm, *Biochem. Z.* **285**:98.

Whitehead, D. L., 1969, New evidence for the control mechanism of sclerotization in insects, *Nature (London)* **224**:721.

Whitehead, A. T., 1971, The innervation of the salivary gland in the American cockroach: Light and electron microscopic observations, *J. Morphol.* **135**:483.

Whitehead, D. L., 1974, The retardation of puparium formation in Diptera: Could factors other than ecdysone control cuticle stabilisation in *Glossina* and *Sarcophaga* species?, *Bull. Entomol. Res.* **64**:223.

Whitmore, D., Jr., Whitmore, E., and Gilbert, L. I., 1972, Juvenile hormone induction of esterases: A mechanism for the regulation of juvenile hormone titer, *Proc. Natl. Acad. Sci. USA* **69**:1592.

Wiens, A. W., and Gilbert, L. I., 1965, Regulation of cockroach fat-body metabolism by the corpus cardiacum *in vitro, Science* **150**:614.

Wiens, A. W., and Gilbert, L. I., 1967, The phosphorylase system of the silkmoth, *Hyalophora cecropia, Comp. Biochem. Physiol.* **21**:145.

Wigglesworth, V. B., 1934, The physiology of ecdysis in *Rhodnius prolixus* (Hemiptera). II. Factors controlling moulting and metamorphosis, *Q. J. Microsc. Sci.* **77**:191.

Wigglesworth, V. B., 1936, The function of the corpus allatum in the growth and reproduction of *Rhodnius prolixus, Q. J. Microsc. Sci.* **79**:91.

Wigglesworth, V. B., 1940, The determination of characters at metamorphosis in *Rhodnius prolixus* (Hemiptera), *J. Exp. Biol.* **17**:201.

Wigglesworth, V. B., 1954, The breakdown of the thoracic gland in the adult insect, *Rhodnius prolixus, J. Exp. Biol.* **32**:485.

Wigglesworth, V. B., 1955, The role of the haemocytes in the growth and moulting of an insect, *Rhodnius prolixus* (Hemiptera), *J. Exp. Biol.* **32**:649.

Wilkens, J. L., 1968, The endocrine and nutritional control of egg maturation in the fleshfly *Sarcophaga bullata, J. Insect Physiol.* **14**:927.

Wilkens, J. L., 1969, The endocrine control of protein metabolism as related to reproduction in the fleshfly *Sarcophaga bullata, J. Insect Physiol.* **15**:1015.

Willey, R. B., 1961, The morphology of the stomodeal nervous system in *Periplaneta americana* (L.) and other Blattaria, *J. Morphol.* **108**:219.

Williams, C. M., 1946, Physiology of insect diapause: The role of the brain in the production and termination of pupal dormancy in the giant silkworm, Platysamia cecropia, Biol. Bull. (Woods Hole, Mass.) 90:234.

Williams, C. M., 1947, Physiology of insect diapause. II. Interaction between the pupal brain and prothoracic glands in the metamorphosis of the giant silkworm, Platysamia cecropia, Biol. Bull. (Woods Hole, Mass.) 93:89.

Williams, C. M., 1948, Extrinsic control of morphogenesis as illustrated in the metamorphosis of insects, Growth Symposium 126:61.

Williams, C. M., 1951, Morphogenesis and the metamorphosis of insects, Harvey Lect. 47:126.

Williams, C. M., 1952, Physiology on insect diapause. IV. The brain and prothoracic glands as an endocrine system in the cecropia silkworm, Biol. Bull. (Woods Hole, Mass.) 103:120.

Williams, C. M., 1959, The juvenile hormone. I. Endocrine activity of the corpora allata of the adult cecropia silkworm, Biol. Bull. (Woods Hole, Mass.) 116:323.

Williams, C. M., 1967, The present status of the brain hormone, in Insects and Physiology (J. W. L. Beament and J. E. Treherne, eds.), Oliver & Boyd, Edinburgh, p. 133.

Williams, C. M., and Adkisson, P. L., 1964, Physiology of insect diapause. XIV. An endocrine mechanism for the photoperiodic control of pupal diapause in the oak silkworm, Antheraea pernyi, Biol. Bull. (Woods Hole, Mass.) 127:511.

Wright, R. D., Sauer, J. R., and Mills, R. R., 1970, Midgut epithelium of the American cockroach: Possible sites of neurosecretion release, J. Insect Physiol. 16:1485.

Wyatt, G. R., 1961, Effects of experimental injury on carbohydrate metabolism in silkmoth pupae, Fed. Proc. 20:81.

Wyss-Huber, M., and Lüscher, M., 1966, Uber die hormonale Beeinflussbarkeit der Protein-synthese in vitro im Fettkörper von Leucophaea maderae (Insecta), Rev. Suisse Zool. 73:517.

Wyss-Huber, M., and Lüscher, M., 1972, In vitro synthesis and release of proteins by fat body and ovarian tissue of Leucophaea maderae during the sexual cycle, J. Insect Physiol. 18:689.

Yagi, S., 1980, Hormonal control of larval colouration associated with phase variation in the armyworm Spodoptera exempta, ICIPE Annual Report 7.

Yagi, S., and Akaike, N., 1976, Regulation of larval diapause by juvenile hormone in the European corn borer, Ostrinia nubilalis, J. Insect Physiol. 22:389.

Yagi, S., and Fukaya, M., 1974, Juvenile hormone as a key factor regulating larval diapause of the rice stem borer, Chilo suppressalis (Lepidoptera: Pyralidae), Appl. Entomol. Zool. 9:247.

Yamaoka, K., and Hirao, T., 1977, Stimulation of virginal oviposition by male factor and its effect on spontaneous nervous activity in Bombyx mori, J. Insect Physiol. 23:57.

Yamashita, O., and Hasegawa, K., 1964, Studies on the mode of action of diapause hormone in the silkworm, Bombyx mori. III. Effect of diapause hormone extract on 3-hydroxykynurenine content in ovaries of silkworm pupae, J. Seri. Sci. Jpn. 33:115.

Yamashita, O., and Hasegawa, K., 1966, The response of the pupal ovaries of the silkworm to the diapause hormone with special reference to their physiological age, J. Insect Physiol. 12:325.

Yamashita, O., and Hasegawa, K., 1967, The effect of the diapause hormone on the trehalase activity in pupal ovaries of the silkworm, (A preliminary note), Proc. Jpn. Acad. 43:547.

Yamashita, O., and Hasegawa, K., 1976, Diapause hormone action in silkworm ovaries incubated in vitro: ^{14}C-trehalose incorporation into glycogen, J. Insect Physiol. 22:409.

Yamazaki, M., and Kobayashi, M., 1969, Purification of the proteinic brain hormone of the silkworm, Bombyx mori, J. Insect Physiol. 15:1981.

Yin, C. M., and Chippendale, G. M., 1973, Juvenile hormone regulation of the larval diapause of the southwestern corn borer, Diatraea grandiosella, J. Insect Physiol. 19:2403.

Yin, C. M., and Chippendale, G. M., 1975, Insect frontal ganglion: Fine structure of its neurosecretory cells in diapause and non-diapause larvae of *Diatraea grandiosella*, *Can. J. Zool.* **53**:1093.

Yin, C. M., and Chippendale, G. M., 1976, Hormonal control of larval diapause and metamorphosis of the southwestern corn borer *Diatraea grandiosella*, *J. Exp. Biol.* **64**:303.

Yin, C. M., and Chippendale, G. M., 1979, Diapause of the southwestern corn borer, *Diatraea grandiosella:* Further evidence showing juvenile hormone to be regulator, *J. Insect Physiol.* **25**:513.

Zalokar, M., 1968, Effect of corpora allata on protein and RNA synthesis in colleterial glands of *Blattella germanica*, *J. Insect Physiol.* **14**:1177.

Zdarek, J., and Fraenkel, G., 1969, Correlated effects of ecdysone and neurosecretion in puparium formation (pupariation) of flies, *Proc. Natl. Acad. Sci. USA* **64**:565.

Zdarek, J., and Fraenkel, G., 1971, Neurosecretory control of ecdysone release during puparium formation of flies, *Gen. Comp. Endocrinol.* **17**:483.

Zdarek, J., Rohlf, R., Blechl, J., and Fraenkel, G., 1981, A hormone effecting immobilization in pupariating fly larvae, *J. Exp. Biol.* **93**:51.

Ziegler, R., 1979, Hyperglycaemic factor from the corpora cardiaca of *Manduca sexta* (L.) (Lepidoptera: Sphingidae), *Gen. Comp. Endocrinol.* **39**:350.

Species Index

Subject Index